IET CIRCUITS, DEVICES AND SYSTEMS SERIES 21

TCAD for Si, SiGe and GaAs Integrated Circuits

Other volumes in this series:

TCAD for Si, SiGe and GaAs Integrated Circuits

G.A. Armstrong
Queen's University Belfast, Northern Ireland

C.K. Maiti
Indian Institute of Technology, Kharagpur

The Institution of Engineering and Technology

Published by The Institution of Engineering and Technology, London, United Kingdom

© 2007 The Institution of Engineering and Technology

First published 2007

The Institution of Engineering and Technology
Michael Faraday House
Six Hills Way, Stevenage
Herts, SG1 2AY, United Kingdom

www.theiet.org

British Library Cataloguing in Publication Data

Maiti, C.K.
  TCAD for Si, SiGe and GaAs integrated circuits. - (Circuits, devices & systems; v. 21)
  1. Integrated circuits - Design and construction
  2. Computer-aided design
  I. Title II. Armstrong, G.A. III. Institution of Engineering and Technology
  621.3'815

ISBN 978-0-86341-743-6

Typeset in India by Newgen Imaging Systems (P) Ltd, Chennai
Printed in the UK by Athenaeum Press Ltd, Gateshead, Tyne & Wear

Contents

Preface

The 2005 International Technology Roadmap for Semiconductors (ITRS) predicts that the use of Technology computer-aided design (TCAD) will provide as much as a 40 per cent reduction in technology development costs by reducing the number of experimental lots and shortening development time for the semiconductor industry. In a highly competitive manufacturing environment, one needs to take the full advantage of the predictive power of TCAD to reduce cost and time for product development. Currently, process and device simulation has established itself as an indispensable tool for developing and optimising device and microelectronic process technologies in the R&D phase. With the device compact model parameter extraction for circuit analysis, TCAD can be extended into manufacturing for advanced process control and parametric yield analysis.

In addition to the rising costs of semiconductor process and product development, each technology generation is driving towards higher performance and increased complexity. The challenge of analysing advanced semiconductor processes and devices through numerical simulations have been our main motivation for this monograph. The purpose is to bring various aspects of TCAD for microelectronic applications into one resource, presenting a comprehensive perspective of the field. The aim is also to provide device and process engineers with examples of the TCAD simulation applied to a number of different technologies, illustrating a quantitative link between the basic technological parameters and electrical behaviour of microelectronic devices.

In Chapter 1, the key developments in TCAD from the early research to the present day, leading to a brief summary of the current status and role of TCAD tools for nanoelectronic devices, are presented. Chapter 2 describes key technology issues for microelectronic fabrication and introduces the reader to commercial TCAD simulation tools. This chapter also considers how process simulation can be linked to device simulation to determine key indicators of device performance, illustrated by simulation results of several device structures based on AlGaAs/InGaAs/GaAs and strained-Si heterostructures. Chapter 3 describes issues relating to oxidation and diffusion modelling for Si heterostructures (SiGe and SiGeC) and their implementation in Silvaco tool ATHENA.

Chapter 4 deals with the topical field of strain-engineered MOSFETs. Various strategies for enhancement of electron and hole mobility, including process-induced and substrate-induced strain are explained. The impact of both strain and stress in the performance enhancement of deep sub-micron heterojunction field effect transistors (hetero-FETs) is considered. Non-classical CMOS device structures (ultrathin single- and multi-gate devices, such as SOI and FinFET) will play a fundamental role in assuring a successful continuation of device scaling trends. The significance of silicon-on-insulator (SOI) MOSFETs for extension of CMOS scaling is covered in Chapter 5.

Chapter 6 gives the background theory of the heterojunction bipolar transistors (HBTs), illustrated by simulation of several device structures based on SiGe. A retrospective look at the many years of development and production of SiGe/SiGeC devices and circuits reveals the significant challenges that have been overcome to successfully integrate SiGe/SiGeC HBTs and BiCMOS technology into successful products. Key technology developments including the growth of SiGe/SiGeC material and overall integration issues with emphasis on high-frequency performance and reliability are discussed in Chapter 7.

An important role of the predictive modelling using TCAD is to utilize detailed process and device simulation to facilitate the interaction between the circuit designer and process technologist, until the product data is incorporated into the design kit. The objective of the remaining chapters of this monograph is to bring together the diversity of activities and the advancements made in compact bipolar, MOS and emerging device models, as well as models for the active and passive components for RF circuit applications.

Accurate MOSFET models that include the observed physical phenomena as well as models for both analog and digital circuits are crucial to design and optimise advanced very large-scale integration/ultra large-scale integration (VLSI/ULSI) circuits for nanoscale CMOS technology. The number of different devices for which compact models are required is enormous: ranging from bipolar to MOSFETs, HBTs, MESFET and high electron mobility transistors (HEMTs). In Chapter 8, we discuss the currently available MOSFET models and the impact of the new modelling paradigm on MOSFET circuit simulation with particular attention to RF issues. Chapter 9 provides an overview of different bipolar device compact models and the relevance of their associated internal circuit components. This chapter also provides an overview of d.c./RF parameter extraction techniques for SiGe HBTs for VBIC model and implementation in SPICE with the help of TCAD.

Chapter 10 presents a comprehensive comparison of three state-of-the-art heterojunction bipolar transistors; the AlGaAs/GaAs HBT, the Si/SiGe HBT and the InGaAs/InP HBT. The trade-offs between the devices are examined, focusing on their intrinsic properties and electrical parameters, and on how these properties and parameters relate to the reported and potential device performance.

Device simulation and simulators are reviewed and the electronic properties of various semiconductors, on which heterostructure devices are based, is discussed. Models for the lattice, thermal, band structure and transport properties, as well as models for important high-field and high-doping effects that occur in the devices, are presented.

Finally, Chapter 11 gives a broad overview of the impact of TCAD on passive components design. The design concept and modelling approach of both passive spiral inductors and active inductors are addressed. Various types of varactor, MIM capacitors and resistor structures currently in use are also discussed.

Each chapter ensures coverage of up-to-date TCAD research results for state-of-the-art devices and a comprehensive list of seminal references. In summary, this book fills a gap in the literature in a rapidly evolving field, as it blends together a wide-ranging description of TCAD activities in Si, SiGe and GaAs materials, technology, device and their applications. An extensive reference list provided in each chapter will help the reader identify the key stages in the development of TCAD from early research through to its integration in current manufacturing.

The monograph is aimed at research and development engineers and scientists who are actively involved in microelectronics technology and device design via technology CAD, and may also serve as a reference for postgraduate and research students in the field of electrical engineering and solid-state physics, for TCAD users and developers. We hope the monograph will help the reader to be in a strong position to address future challenges in semiconductor process, device and circuit development via TCAD.

G.A. Armstrong and C.K. Maiti

Chapter 1

Introduction

Since the invention of the transistor more than 50 years ago [1, 2], the successful downscaling of silicon devices and advances in processing technologies have now enabled us to achieve highly complex electronic systems. The microelectronics industry has relied mainly on the shrinking of transistor geometries for improvements in circuit performance and cost per function for over four decades. Technology computer-aided design (TCAD) is used in the computer-aided design and engineering semiconductor device design, fabrication process design, technology characterisation for circuit design, manufacturing yield optimisation and process centring, and computer integrated manufacturing [3].

The strong foundation of the Si semiconductor industry and our ability to make the transistors smaller and smaller (downscaling), have now led the devices to their fundamental limits and more transistors on a single chip. This trend has led us into the ultra large-scale integration (ULSI), and the giga scale integration (GSI) era is approaching. Continued transistor scaling will not be as straight forward in the future as it has been previously because fundamental materials and process limits are rapidly being approached in the 'classical' bulk-Si CMOS technology. To sustain the silicon scaling over the next 10–15 years, the industry is currently facing difficult challenges. The deep submicron technology calls for new design and fabrication methodologies to cope with the complexity and coupling among different stages of a design and fabrication.

A chip design starts with the product specification, followed by the front-end and back-end designs. The design and implementation of an integrated circuit, be it the whole system on a chip or a circuit, can be represented by five distinctively different levels of abstraction such as system level, logic level, circuit level, layout level and technology level. The front-end design usually starts from the system level with a top-down synthesis approach, which is technology independent (i.e., at this level, it is 'irrelevant' whether the design will be implemented in CMOS or bipolar technology). The design is then transformed into the circuit level, in which the logic functionality, timing delays, speed and power, and so on are the primary concerns. Although this level is technology dependent, it is relatively process-independent.

At the back-end, the final design needs to be translated into the physical layout representation that is used to be implemented in wafer fabrication. In the 'conventional' hierarchy, technology development is relatively independent of the design. For this phase, electronic computer-aided design (ECAD) tools have been developed, which are now so powerful that the logic design can be synthesised from a high-level description language, the circuit netlist can be extracted from the logic functional description and the layout can be extracted from the circuit and logic-level descriptions.

Starting from the product idea, several sequential steps are followed in an integrated circuit development and production flow. The chip design and wafer fabrication involve a complex and iterative process involving the design, simulation, manufacturing, characterisation and verification. The main stages of semiconductor fabrication, from design to the final product, include mainly the following:

Design: The integrated circuit is designed as a schematic taking into account the needs for the specific application. It is now a standard practice to use ECAD tools to simulate the behaviour of any schematic design via detailed circuit simulation models and design rules that are specific to a particular process.

Layout: The resulting integrated circuit is drawn as a layout on the specific layers, depending on the available semiconductor process technology. The combination of multiple layers, like implantation masks and etch masks, defines the shape and functionality of the electronic devices in the integrated circuit.

Mask: The layout data is processed to fabricate the physical masks that are generally written using laser- or e-beam lithography equipment.

Processing: With the starting wafer, process flow typical to a technology is followed in semiconductor fabrication. During this, wafers are subjected to numerous unit process steps like oxidation, ion implantation, deposition and etching of semiconductor, dielectric and metal, diffusion of dopants and lithography to create specific device structure with various deposited layers using different masks.

Test: After the fabrication cycle, wafers with the functional integrated circuits are tested electrically. First, on single device level with process control monitors (PCMs), and second, on integrated circuit level including inking the bad dies.

Assembly: After testing, the good dies are scribed into pieces, and packaging of the single circuits are done. After the packaging, a final electrical test is performed.

The use of TCAD in semiconductor manufacturing is two-fold: First, it models the complex flow of semiconductor fabrication steps and ends up with detailed information on geometric shape and doping profile distribution of a semiconductor device (see Figure 1.1). Second, the device simulation uses the information of the first step to calculate the characteristics of semiconductor devices. The resulting device characteristics are used for fitting circuit simulation models.

In TCAD, a broad range of modelling and analysis activities that consist of detailed simulation of IC lithographic and fabrication processes, single- or multiple-device

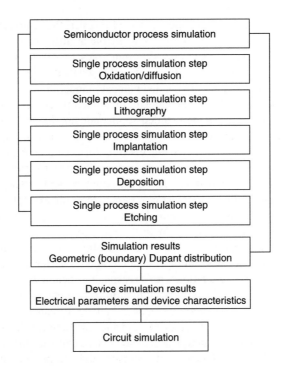

Figure 1.1   A typical process simulation flow

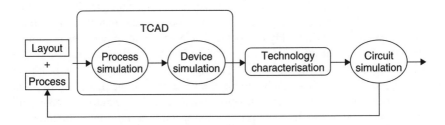

Figure 1.2   The conventional role of TCAD in IC processing

electrical performance and extraction of discrete parameters for equivalent circuit models are involved. TCAD represents our physical understanding of processes and devices in terms of computer models of semiconductor physics. The conventional role of TCAD in IC processing is shown in Figure 1.2. TCAD is used to design, analyse and optimise semiconductor technologies and devices with fundamental and accurate models.

TCAD is used in the semiconductor industry via two main approaches. The first approach is in predictive simulation, which is used in a conceptual device design phase for the prediction of trends in process options, device operation and circuit behaviour. The second approach is to use experimental data to calibrate the models in the simulator and then to use the calibrated tools in the optimisation phase of the development process. TCAD tools are available at different levels for different applications. The different tools play different roles in any technology development strategy. TCAD also comprises many other elements, including equipment simulation, pattern transfer simulation and compact modelling.

Process simulation models semiconductor manufacturing processes. Process simulation methodology requires the detailed semiconductor fabrication process flow including numerous fabrication details, like angle of incidence of ions implanted in ion implantation process steps or etch rate distribution as a function of the local angle of the etched layer surface. The simulation starts with the blank wafer and ends with device/circuits. Process simulation predicts the structural and doping profiles of the devices on the basis of the key processing parameters, for example, implant dose, energy and anneal time and temperature. However, so far the process simulation has been the weaker element of TCAD. With the maturity of TCAD tools, real wafer fabrication can now be emulated by process simulation, from which realistic device structures and doping profiles can be generated, and transistor performance can be predicted through device simulation with a reasonable accuracy.

One of the challenges of the deep-submicron technology is due to the fact that as the transistor dimensions become smaller, their performance gets more closely coupled with the fabrication parameters, and the interconnect delay is becoming dominant now. However, the interconnect delays can be extracted through technology characterisation with three-dimensional accuracy, and can provide information for design rule checker (DRC) and layout parasitic extraction (LPE) tools in the physical design. SPICE parameters can also be extracted from the 'virtual device'.

The 2005 International Technology Roadmap for Semiconductors (ITRS) predicts that the use of TCAD will provide as much as a 40 per cent reduction in technology development costs. In highly competitive semiconductor product environment, advantages of use of the predictive power of TCAD is now well established to reduce cost and time in product development. Through proper calibration with prototype wafers, TCAD can also accurately predict the behaviour in the manufacturing line for new technology nodes. This information, coupled with design tools, enables the manufacturing line to fine tune the process for specific designs and, ultimately, to achieve higher yields. From a technology and modelling point of view, a combination of process and device simulation is at the core of TCAD, as it bridges between a given technology and circuit design. TCAD is now playing an increasingly important role in virtual wafer fabrication (VWF) approach to design and technology development. Among various possible approaches for analysing microelectronic systems, the most common approach may be classified into three levels: device, circuit and systems. However, in multi-level modelling approach, electrical characterisation and analysis of analog/digital systems are classified at various levels as follows:

- **Technology Level**, is related to the detailed device layer structures and doping profiles and their dependence on process variations.
- **Device Level** is referred to as the analytical description of the device terminal characteristics.

- **Circuit Level** refers to the solution of linear/non-linear systems of equations by various matrix solution techniques, such as nodal or modified nodal analysis.
- **System Level** refers to digital or analog blocks that make up a given system.

Traditionally, the above four levels are relatively independent or loosely coupled. However, over the past decade, improved process and device simulation capability has become increasingly important in the design and development of complex integrated circuits. The increased importance has been primarily due to the growing complexity of semiconductor fabrication processes, ever shrinking device dimensions and the need to reduce to competitive levels the time and cost associated with very large-scale integration (VLSI) and ULSI design and development. However, with the rapid progress in the IC technology, maintaining an advanced and effective simulation capability requires a sustained commitment of resources. There are three main areas that need to be addressed for the development of state-of-the-art simulation: namely, physical model development, software development and experimental verification.

The manufacturing of an IC requires many complex processing steps, many of which are illustrated in Figure 1.3, with each of these steps containing numerous possible variations in process controls, and may take several weeks to complete. During the design stages of a device/circuit, a large number of experiments have to be undertaken to determine appropriate settings of certain process parameters, as well as of the layout topology and the geometrical dimensions best suited to meet the device specifications. Optimising these parameters by experimentation alone, would be quite impractical. After a process has been well characterised, simulation can be used to investigate the effect of parameter variation with the necessity of experimental design. Although the sophistication of process simulation is increasing, its accuracy remains a problem. There is generally a time lag between the introduction of a particular process and its accurate modelling and characterisation.

One-dimensional (1D) simulations of IC processes and devices were adequately sufficient for technologies of the 1970s with deeper junction depths and channel lengths of several microns. It was possible to extrapolate quasi-two-dimensional (2D)

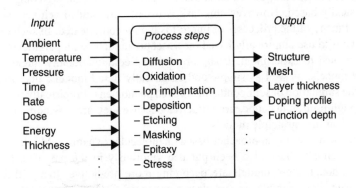

Figure 1.3   Process steps involved in the manufacturing of an IC. Several hundred steps are involved

profiles from 1D profiles. At the end of the 1970s, integrated circuit technology came to a point where the use of simple ad hoc models to predict the outcome of processing steps was no longer adequate.

Demand grew for more sophisticated computer-aided design of IC processes and devices. The 1980s witnessed the onset of aggressive MOSFET downscaling and the rise of CMOS for VLSI technology, especially as power consumption and reduced voltage supply issues became critical. With dimensions moving below 1 μm gate lengths (and commensurate junction and gate oxide scaling), TCAD assumed a critical role in MOSFET scaling and 2D modelling of effects such as local oxide isolation (LOCOS) and dependency of threshold voltage on channel implant. Numerous parasitic effects associated with the substrate such as latchup all required coordinated process and device modelling. Moreover, the physical effects such as enhanced dopant diffusion, due to point defects generated during oxidation and ion implantation, required new models and extensive characterisation. The challenges of relentless scaling mandated by Moore's law pushed lithography (and topography) issues to the forefront of TCAD challenges [4].

Equipment and topography modelling/simulation tools also became essential to reduce the technology development time and cost. There are several unit process areas that have been tackled: chemical mechanical polishing (CMP), chemical vapour deposition (CVD), physical vapour deposition (PVD) and etch and thermal processing including rapid thermal processing (RTP). The sophistication of the models used in the equipment simulators varies greatly from the simple experimentally derived response surface models (RSMs) for CMP [5] to the computationally intensive and more physically based etching/deposition models, such as in SPEEDIE [6].

Device simulation is used for obtaining the characteristics, for example, static, time-dependent, large and small-signal frequency-dependent behaviour of semiconductor devices and their electrical behaviour and also SPICE parameter extraction. SPICE parameters are subsequently used in circuit simulation. Device simulation models the electrical, optical, mechanical and magnetic behaviour of semiconductor devices using structures obtained from process simulation. Advanced multi-dimensional process and device simulation calls for models that are comprehensive and physically based. However, this task is usually difficult to achieve, since the physics of many of the processes of interest is highly complicated. In such cases, an alternative is to use empirically based macroscopic models, with parameters chosen that are related to experiment. Both complex and computationally intensive first-principle models, as well as simple and empirically based models, play important roles in the simulation of semiconductor-related phenomena. Empirically based models, although limited in scope, are attractive because they are less computationally intensive than first-principle models.

Atomistic modelling approaches based on molecular dynamics, Monte Carlo and quantum chemistry have also been pursued vigorously in the past decade [7]. In general, although these models are becoming more predictive, their validation is still an issue. For example, in the plasma-enhanced CVD processes, there are so many possible chemical reactions that obtaining the reaction rates for the dominant ones is an extremely difficult task [8]. The ultimate challenge in process technology development is presented by the system-on-chip (SoC) in which microprocessor and analog cores must be implemented in a single chip with several types of memory. The modular process development strategy has been proven effective in reducing process development time while satisfying the SoC process specifications [9].

Numerical simulation of semiconductor device/circuit fabrication and operation is also important to the design and manufacture of integrated circuits because it provides insights into complex phenomena that cannot be obtained through experimentation or simple analytic models. Many of the phenomena involved in semiconductor processes can be described by partial differential equations (PDEs): as a result, much of the TCAD software is based on the solution of PDEs. Since most of these equations are intractable except for the simplest geometry and boundary conditions, approximate methods are inevitably utilized. The most common techniques utilized are the finite element and finite difference methods, although boundary element and variants of Green's method are also used. Finite difference methods approximate the solution to the governing PDEs with difference operators while finite element methods utilize an integral form.

Typically, the grid approximation is an orthogonal grid for finite differences and a triangular or tetrahedral grid for finite elements. The net of the approximation approach is taken because finite difference grids are much simpler to generate automatically and result in a simpler matrix solution scheme; this approach (approximation approach) however, restricts the forms of structures analysed to those of orthogonal geometrics. Finite elements require much more complex software to create a mesh, and 2D and 3D mesh editors remain a significant research topic.

Advanced process/device simulation software development calls for programs that are numerically robust, modular in nature, optimised in their execution and easy to learn and use. Some features of device simulation tool are shown in Figure 1.4. Much attention has been given to creation of user-friendly pre- and post-processors, data storage and retrieval, and other components that are needed for a complete simulation system. A strong experimentation and characterisation program is vital to successful process and device model development and verification. The predictive capability of any model is a strong function of its foundation in reliable experimental data. The role of predictive modelling is to utilize detailed process and device simulation, or TCAD, in place of hardware to facilitate the feedback loop between circuit designer and process technologist.

In this chapter, the development of TCAD and its strategic use for semiconductor technology development and the technology scaling from the early 1960s until the late 1990s is first highlighted. A brief history on the progress that has been made in simulating semiconductor device fabrication processes is indicated. The role of TCAD in the overall IC design flow, focusing on the design-manufacturing interface is described. The current state of process modelling is reviewed, and current and future challenges are discussed. The hierarchy of TCAD simulation tools ranging from the atomistic level to the equipment and manufacturing simulation levels is then presented. A discussion on the challenges one has to meet in future transistor design with the modelling strategies needed to meet these challenges will follow. Then, one needs to turn to the problem of handling the large amount of data generated

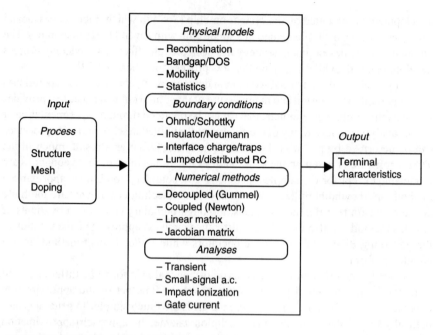

Figure 1.4   Steps involved in device simulation

by these programs, as well as the detailed data required as input to the programs. Focus is then turned to the emerging field of scientific visualisation, which is proving to be very helpful in analysing the generated data, as well as recent developments in the geometrical modelling of 3D structures, which is needed for characterising complicated device structures.

1.1   History of device simulation

The first truly numerical device simulator was reported by De Mari [10] for a one-dimensional problem and Scharfetter *et al.* [11] for the two-dimensional problem. A good early history of device simulation tools can be found in Reference 12. The advent of the 1970s saw the beginning of attempts to couple numerical process and device simulators, starting with the one-dimensional programs SUPREM and SEDAN at Stanford University [13]. Ouput from SUPREM (i.e., doping profile and geometry) formed a starting point for electrical simulation. Both programs formed an interrelated pair and were successfully used at a large number of research institutions. Although process simulation was a new field at that time, the fundamentals for (more physically based) device simulation were already established. In SEDAN, devices consisting of up to five metallurgical junctions could be analysed. SEDAN performed device analysis by simultaneously solving Poisson equation and transient current continuity equations to obtain the potential and carrier concentration in silicon. Updated SEDAN versions can handle SiGe, GaAs and AlGaAs devices.

During the 1980s a vast amount of literature became available on two-dimensional device simulation from various universities. Relatively important and widely used programs were MINIMOS [14], BAMBI [15], PISCES [16], BIPOLE [17] and HQUPETS [18]. All these programs solve the Poisson equation in two-dimensions along with the current continuity equations.

Device simulation programs rapidly became useful to process and device engineers. Some of the first detailed two-dimensional transistor device simulations were carried out by Slotboom [19] and Heimeier [20] at Phillips. Also, some three-dimensional process and/or device simulators were reported by researchers with large amounts of computer resources. Since those early days many other institutions have developed their software. These include IBM's FEDSS/FIELDAY and Agere's PROPHET and PADRE. MINIMOS, developed in the 1980s at the Technical University Vienna, and specifically targeted at a standard MOS transistor was the first two-dimensional numerical simulator for which the source code was made widely available. As its name implies, the original version of MINIMOS solved the fundamental semiconductor equations in a rectangular two-dimensional domain, where the solution domain was specifically a MOS transistor structure. The numerical schemes of MINIMOS have since been extended from its initial release in the early 1980s to today's version MINIMOS-NT, which is available for a number of hardware platforms and can handle planar and non-planar device structures in two- and three-dimensions. A Monte Carlo module can couple self-consistently to the Poisson equation and replace the drift-diffusion approximation in critical device areas. The current version of MINIMOS is capable of simulating Si, SiGe, GaAs and AlGaAs devices.

BIPOLE is a quasi-two-dimensional device simulator, developed at the University of Waterloo, Canada, which specifically focuses on rapid prediction of terminal electrical characteristics of bipolar transistors. Programme input comprises fabrication data such as mask dimensions, impurity profiles and physical parameters such as carrier lifetime. The calculation is based on the variable boundary regional approach using one-dimensional transport equations. Two-dimensional and quasi-cylindrical edge effects are handled by combining the vertical one-dimensional analysis with a coupled one-dimensional horizontal analysis of the transport equations in the base region. It is then used to solve the transport equations for diffusion and drift in the presence of arbitrary recombination in quasi-neutral regions. This scheme is extremely fast in terms of computing time. BIPOLE can also be used for accurate avalanche multiplication and breakdown studies in high-speed transistors with shallow collector regions and heterojunction structures, including graded base Ge fraction in SiGe devices, and emission across the thin interfacial oxide layer in a poly-Si emitter and offers a facility for automatic parameter extraction for SPICE models.

PISCES, developed at Stanford University was the first successful general-purpose two-dimensional device simulation program applicable to any arbitrary semiconductor device, whether its dominant mode of operation was MOS or bipolar [21]. It was subsequently adapted by different TCAD companies such as TMA (now Synopsys) or Silvaco to form the basis of their early software products released in the mid to late 1980s. PISCES uses a box integration scheme to solve the semiconductor equations at each node within the device. From the grid structure of the device and impurity doping profile, PISCES can provide two-dimensional contours and vectors yielding additional insight into how to enhance device performance.

Since the early 1980s commercial vendors such as SILVACO, ISE-TCAD, TMA and AVANTi developed their own graphical user interfaces (GUI) around existing frameworks, which facilitated the integration of process and device simulation tools by non-specialists within a wider engineering environment. The concept of general-purpose–process and device simulators that allows flexible simulation of different structures in different technologies became a reality.

As a specific example, ATLAS, a Silvaco International product, originally derived from PISCES, has been extensively enhanced providing general capabilities for numerical, physics-based, two- and three-dimensional simulation of semiconductor devices. It predicts the electrical behaviour of specified semiconductor structures and provides insight into the internal physical mechanisms associated with device operation.

ATLAS provides a comprehensive set of physical models such as drift-diffusion transport models, energy balance transport models and lattice heating. It is also capable of simulating graded or abrupt heterojunctions, hot carrier injection, band-to-band and Fowler–Nordheim tunnelling, non-local impact ionization, thermionic currents, optoelectronic interactions, and stimulated emission and radiation. ATLAS has a modular architecture that includes other simulation tools such as MIXED-MODE, which offers circuit simulation capabilities that couple compact analytical models to the device simulator. The increased role of statistically based simulation is reflected in the commercial interest. Many software platforms with statistical interfaces are now available from TCAD vendors, for example, Virtual Wafer Fabrication (Silvaco) and SENTAURUS (Synopsys).

Since the 1990s, there has been an increasing demand for simulators that provide the capability of modelling in a comprehensive manner for as many different types of semiconductor devices and processes as possible (see for example, an excellent review on 'Simulation in semiconductor technology development' [22]). This goal can either be achieved by incorporating new features in a simulator or by establishing links between simulators with different capabilities, such as coupling of electronic design (ECAD) with technology design (TCAD) (see Figure 1.5). Supporting the process and device engineers in an efficient and intuitive way in exploring the design space calls for a number of capabilities of the process simulation environment, such as free interaction and analysis of all aspects of a device, transparency with respect to real-life data and control to facilitate the management of large-scale and iterative experiments.

1.2   History of process simulation

Process simulation models manufacturing steps such as oxidation, implantation, diffusion, etching, deposition and lithography. Process simulation has evolved with increasing levels of sophistication from the earliest models for process steps. For example in ion implantation, the LSS theory came to be widely used for describing

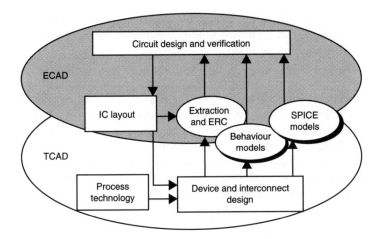

Figure 1.5   Coupling of electronic design (ECAD) with technology design (TCAD)
            [After R. W. Dutton et al., Proc. IEEE ISSCC, 1996(78–79)]

the ion range distributions in various materials. According to this theory, an ion implant profile may be described to first order by a Gaussian distribution. Range statistics for a variety of ion and target combinations were determined and made available. In the area of diffusion, some of the progress dealt with dopant diffusion under extrinsic conditions. Early examples of models reported for arsenic, boron and phosphorus are provided in Reference 23.

With the trend moving towards smaller devices and more complex processes, a strong need developed to account for what had originally been considered to be second-order effects within certain processes. Early models relied primarily on analytical equations, which became inadequate under more complex circumstances. Consequently, numerically based models were necessary for general applicability. The computer implementation of such models led to the evolution of SUPREM simulator SUPREM I in 1977, followed by enhanced versions up to the two-dimensional version SUPREM IV [24].

In all versions of SUPREM the various process steps are entered in the order as they appear in the actual process. Both SUPREM III and SUPREM IV model the moving boundary between oxide and silicon during oxidation and diffusion. SUPREM IV offers several models for oxidation to describe the 2D shape of local oxidation structures. Each program offers advanced models for second-order effects such as dopant-enhanced diffusion, oxidation-enhanced diffusion, dopant clustering and thin oxide growth. It has been shown that the process models are sufficiently general and accurate to be very useful. However, the model coefficients need to be adjusted to give good agreement for a particular fabrication facility. This is particularly true for the model parameters related to oxidation rates and implantation profiles. As the models for ion implantation, diffusion and oxidation were further developed, more realistic models for etching and deposition were added. An object-oriented continuum-based modelling of silicon integrated circuit processing has also been introduced for implantation, diffusion and material growth [25]. For simulation of lithography, etching, and deposition steps [4], and other specialist software such as SAMPLE has been developed [26].

Over the last decade, as the shrinking of device dimensions has continued, 3D process simulation has become a necessity. Development of such programs were first reported in the early 1980s [27–29]. The required information about the three-dimensional doping profiles is usually developed from either analytic models or from rotations and stretching of doping profiles generated by two-dimensional process simulations. It should, however, be mentioned that although the ability to simulate in three dimensions looks impressive, such simulations are generally necessary only for extremely small devices. Another important aspect of semiconductor device simulation is the availability of accurate doping profiles. Particularly for very small devices, process simulator (either 2D or 3D) should be capable of predicting the doping profiles well enough for accurate device simulation purposes. Therefore, measured profiles should be used in these cases.

1.3   Evolution of TCAD

TCAD modelling based on computer simulations spans the interrelated disciplines of circuit design, device engineering, process development and integration into manu-facturing [30]. Computer simulations help in quantifying the details of behavioural models for ICs at the transistor and circuit levels and show the physical limitations at the process and manufacturing levels. The current generations of TCAD tools need to meet (1) heterogeneous device requirements and (2) atomic-scale limits both in the devices and processing that shift the emphasis to consideration of the complete manu-facturing process. The introduction of new materials as well as the need to account for more effects as the technology is scaled (e.g., quantum mechanical modelling of gate oxide capacitance) require the use of TCAD in a larger scale. Major semiconductor manufacturers now routinely use TCAD for process and device development [31]. Various TCAD applications and challenges have also been discussed in Reference 32.

Key advantages of a simulation-based approach are rapid prototyping and the abil-ity to analyse the devices. The ability to investigate individual physical phenomena and interactions of different phenomena through the huge number of process parame-ters is especially useful. This leads frequently into more quantitative use of simulation in the determination of optimum processing conditions. The other use of TCAD is in exploration studies of new processes and new process sequences. A good example is the study of the use of isotropic deposition combined with anisotropic etching to planarise the wafer surface. Another important use of TCAD is the assessment of the impact of hypothetical and future technology advances. It is also possible to use TCAD simulation in diagnostic and discovery modes in which the physical models are modified to create new effects or establish plausible explanations of experimental observations [33].

The most famous of the Stanford TCAD software programs are SUPREM for process simulation and PISCES for device simulation. These are general-purpose simulators, designed to work with arbitrary semiconductor structures. The first commercial versions of these programs from Technology Modelling Associates (TMA) were entitled TSUPREM4 and MEDICI (now available from Synopsys). Silvaco later licensed the same software, and offered a commercial alternative ATHENA and ATLAS [34, 35]. A third major TCAD vendor Integrated Systems Engineering (ISE) has since offered its own equivalent products DIOS and DESSIS [36, 37] and is now merged with Synopsys.

Also the presence of passive elements poses major challenges to circuit and system designers. Now there is a transition from the gate-dominated delays towards resistance–capacitance (RC) interconnect-dominated delay limitations, and on-chip inductance will bring still greater complexity to modelling and verification. Materials and processing issues are becoming critical and the modelling is also intertwined with systems issues such as chip-scale layout. Earlier generation reliability issues for bipolar and CMOS have shifted now to the role of electrostatic discharge (ESD) and its interrelationship with I/O circuit scaling [38].

1.4 Process flow integration

In the 1980s, the advent of VLSI created the need to formalise the interface between the process and design engineers. To develop an IC product, there are number of engineers involved in the unit process development to the system-level specifications. Typically, the technology development may be split into two basic regimes: product design and fabrication process development. There are several aspects of technology and product development. The process development is being modularised, and process integration has now become a more structured process. Current technology scaling has put a bigger challenge to process simulation than to device simulation. To simplify the boundary between the product and process development, semiconductor companies standardised on a basic interface of SPICE model cards and the technology files. On the basis of the technology constraints and the target device performance, processes have been developed and the device performance has been abstracted for the product designers.

TCAD has been expanding from the software tools mostly used for technology development, to the software that is also used to build a bridge between the designers and manufacturing. The fact that the new interface needs to provide the information to the IC designer has been examined in detail by Dutton *et al.* [39]. Truly predictive process modelling has proven to be an elusive goal, because the controlling physics is complicated and difficult to investigate experimentally. The complexity of process variables, especially in coupled fabrication steps and for complex processes such as plasma etching, require exhaustive experimentation. Design of experiments (DOE) has now become an integral part of in-depth process/device simulation to avoid costly fabrication [40].

A key component of the design–manufacturing interface is the employment of statistical process/device simulation to predict the distributions of electrical properties and SPICE parameters based on the distributions of equipment controls [41].

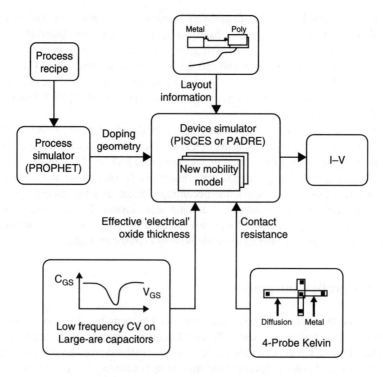

Figure 1.6   TCAD-based electrical characterisation [After R. W. Dutton et al., IEEE Trans. Computer-Aided Design of Integrated Circuits and Systems, Vol. 19, 2000(1544–1560). © 2000 IEEE]

The ability to use deterministic algorithms when extracting SPICE parameters is also a key part of the statistical characterisation process. It has been shown that it is possible to use the information available from TCAD process simulations (such as actual dopant profiles) to extract the SPICE parameters in ways that cannot be typically accomplished from I–V characterisation. Figure 1.6 gives a more detailed view of the several pieces of data and simulations required to achieve TCAD-based I–V characterisation. This methodology provides reliable data as well as support for relating physical effects to controlling parameters that relate directly to either process recipe or even processing equipment dependencies. This approach has been demonstrated with 0.25 μm technology with excellent agreement, and one parameter set fits all devices from 20 μm down to 0.25 μm channel lengths.

On the basis of the technology constraints and the target device performance, processes have been developed and the device performance has been abstracted for the product designers (see Figure 1.7). Although the full blown product/process co-design has not been accomplished yet, there have been several approaches to synthesise the process to meet the desired device performance objectives.

The design and fabrication of smaller and faster semiconductor devices relies on the proper numerical simulation of fabrication processes and electrical characteristics.

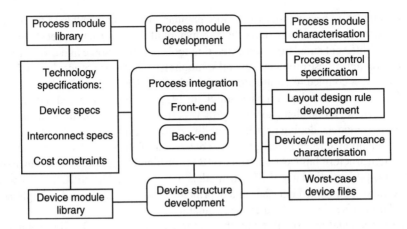

Figure 1.7 Process synthesis [After R. W. Dutton et al., IEEE Trans. Computer-Aided Design of Integrated Circuits and Systems, Vol. 19, 2000(1544-1560). © 2000 IEEE]

The integration of independently developed simulators is achieved by providing a binding function that makes the tool available to the TCAD framework. Process flows may either be defined by writing a text file using symbolic names for the simulators to invoke or by using a graphical process flow editor. Portability, communicability and modularisation of recipes pave the way for technology knowledge sharing and remote fabrication (virtual wafer) and have also been attempted. The VISTA framework and its modules form a programmable simulation environment for TCAD applications. Within VISTA tool [42], integration activities resulted in a simulation flow control module [43] that allows the specification of process sequences by means of simulator calls used in a plug-and-play fashion. A subset of the Profile Interchange Format (PIF) is used as the basis for an unambiguous wafer state description. DOE methods are frequently used for automatic generation of experiments. These methods are also available in other TCAD frameworks, namely, the Virtual Wafer Fabrication [44], the TMA Work Bench [45] and the NORMAN framework [46]. Although attempts are being made, no single solution exists that satisfactorily covers all the issues mentioned above.

1.5 TCAD and compact model

Computer-aided circuit simulation cannot provide reliable results without adequate specification of circuit elements and device models. The electrical behaviour of the devices can either be measured or simulated. When performing a device optimisation, fabricating and measuring each optimisation step would be very expensive as the existing device models use large sets of parameters, values of which must be determined very carefully to represent device characteristics accurately. Hence, device simulators became more and more popular, and the data obtained from these simulations can be used to extract the parameters of the compact model. However, the use of different simulators and extraction tools is cumbersome and error-prone.

To overcome these problems several solutions have been proposed where a device simulator was coupled to circuit simulator such as SPICE [47] or circuit simulation capabilities were added to a device simulator [48]. In the latter mixed-mode simulation, devices can be characterised by their performance in a circuit as a function of transport models, doping profiles, mobility models and so forth. This is of fundamental importance when investigating the behaviour of advanced submicron devices like heterostructure bipolar transistors (HBTs) or high electron mobility transistors (HEMTs) [49], particularly if SPICE models are not available. SPICE has a well-defined and documented interface and in principle, it is straightforward to implement a similar interface in the mixed-mode simulator.

Circuit simulation programs have the common feature that the electrical behaviour of the devices is modelled by means of a compact model, that is generally a set of analytical expressions for charge or current that are fitted to extensive measured data. Once a suitable compact model has been developed, its parameters need to be extracted. For example, in the widely used MOS compact model BSIM3 [50], more than 100 parameters need to be extracted for calibration purposes. The systematic identification of these parameters is obviously a time-consuming task, requiring the use of specialist software.

If the device design is known and further modification is not needed, these parameters need to be extracted only once and can be used for circuit design provided the accuracy of the models is sufficient. When there is a need to optimise a device using modified geometries and/or doping profiles the compact model parameters have to be extracted for each different layout. Although physics-based device modelling achieves satisfactory accuracy in a short development time, the associated computational cost prevents its application to circuits comprising more than a few devices [51]. Thus, the device modelling for circuit applications faces conflicting requirements of accuracy, efficiency and development time. Conventional compact models are based on simple geometry and process-based models as process and geometric factors control much of the device behaviour. This is particularly true in the predictive modelling mode, which for compact model generation demands simulation of the complete device structure, including all parasitic capacitances and resistances, and often represents a considerable expanse of silicon, over a wide range of bias conditions.

A fundamental difference between digital technology and analog/RF technology is the sensitivity of analog/RF circuits to many manufacturability/performance trade-offs that must be made. RF and microwave applications generally impose more serious design constraints than digital applications. It also puts continuously increasing demands on accuracy, smoothness and scaling of the compact models for RF/microwave devices. Moreover, the RF/microwave circuit design requires, in an integrated way, advanced compact modelling of active and passive devices including various parasitic effects, like substrate and thermal coupling as well as noise.

Early SPICE semiconductor models cannot model current devices accurately enough at high speed or high power levels. To meet the challenges of rapidly evolving semiconductor technologies yielding electronic circuits with higher performance new improved models are needed. The latest compact models are strongly based on device physics and can be exceedingly complex. An alternative is the use of macro-models, empirical models that can be constructed using elements already available in a circuit simulator. Physical modelling and macromodelling techniques yielding accurate predictive capability are now available for silicon power devices, gallium arsenide and silicon-germanium high-speed devices using various circuit simulation environments, such as ICCAP and APLAC [52].

One potential use of TCAD is in the development of the compact models themselves. For example, frequency-dependent compact models for MOS capacitors and p-n junction have been developed using TCAD to verify the models [53,54]. A SPICE sub-circuit was developed for ESD protection device with extensive use of TCAD tools [55]. Other approaches for compact model development based on the direct and automatic extraction of equivalent circuits from device simulation have also been reported [56, 57]. Equivalent circuits are automatically generated from device simulation, and the circuit topology and elements may be directly interpreted in terms of device physics. The accuracy that can be achieved with this technique largely depends on the validity of the assumptions made when reducing continuous functions, as obtained from the physical device simulation. Simulation-based compact models are advantageous for high-frequency effects where analytic formula may not be adequate [58].

Extension of compact models to process correlation is also necessary, which will be the key to studying process effects on system performance [59]. A robust circuit design requires knowledge about the variation in key device parameters and TCAD is finding applications in the generation of statistical compact models, which are useful for yield and performance improvement. Starting with a baseline process simulation, known process variations can be simulated in a TCAD tool to generate statistical distributions of device or even circuit parameters. A physics-based compact model, based on technology characterisation, which serves as the building block for a multi-level circuit simulator (based on subcircuit expansion approach) has been reported, in which process variation can be captured through device compact models [30]. There are certainly opportunities for further interactions as devices continue to scale down and new device structures emerge.

1.6   Parameter extraction

Parameter extraction is an integral part of compact modelling. The goal of parameter extraction is to determine the values of device model parameters that minimise the total differences between a set of measured characteristics and results obtained by evaluation of the device model. This minimisation process is often called fitting of model characteristics to the measurement data. Several different approaches to parameter extraction have been proposed in the literature. They include general or specialised and direct or iterative extraction methods. Specialised methods extract some subsets of model parameters; for example, model resistances, or capacitances or d.c. parameters only, while general methods determine all parameters of the model. Several programs for parameter extraction are available on a commercial basis. Hewlett-Packard offers an integrated circuit characterisation and analysis program (ICCAP) [60].

The most important test structures in an IC manufacturing process are the devices themselves. It is imperative that these devices are accurately characterised so that the most accurate model parameter set for the device under test can be extracted. Silvaco's Universal Transistor Modelling Software (UTMOST) [61] is a data acquisition and parameter extraction tool with applications in the areas of device characterisation and modelling. UTMOST can accept data from direct measurements, measurements stored in a measurement log file, process and device simulation or from other model parameter sets in order to create a suitable parameter set for a chosen model.

UTMOST can control a wide range of commercial measurement equipment and probers, so that a user has maximum flexibility in the configuration of a measurement system. UTMOST supports the simulation of d.c., transient, capacitance and s-parameter characteristics. UTMOST can incorporate commercial device models, user-defined device models and macro models. UTMOST also supports the conversion of model parameter sets from one model to another. This is invaluable in situations where the model in use has changed and new parameter sets are needed without the requirement of measuring new data. UTMOST supports eight different technology modules – MOS, Bipolar, Diode, JFET, GaAs, SOI, TFT and HBT. Using batch-mode, device characteristics resulting from process and device simulations can be automatically converted into UTMOST format, and detailed parameter extractions can be performed. In this way, it is possible to develop nominal and even worst-case models for a process which is under development.

1.7   Statistical process control and yield analysis

Statistical fluctuations inherent in any IC manufacturing process cause variations in device and circuit performance. Product yield and manufacturing problems necessitate costly redesign cycles. Inverse modelling of the statistical variations in the processes by using TCAD methods is becoming important. Presently, by inserting statistical variations in the compact model parameters from experimental test data and batch data, design of circuits is generally performed. The main drawback of this method is the lack of information for the correlation in the statistical data; as a consequence, too pessimistic designs rules are adopted. TCAD may be used for the inclusion of process variations into the design of circuits, as it provides realistic input for the statistical variations as well as their correlation and therefore improves the yield of the production process.

Silvaco's SPAYN [62] is a statistical analysis software package specifically designed for the semiconductor industry for analysing variances from model parameter extraction sequences, electrical test routines and circuit test measurements and their impact on device performance. It offers the state-of-the-art solutions for statistical/worst-case circuit design and process control and generates accurate and realistic worst-case or 'process corner' parameter sets and SPICE models, reflecting variations in process parameters. SPAYN helps to identify the relationship between device or circuit performance variations and the underlying specific process fluctuations. In SPAYN, model parameters and device characteristics can be stored. It offers unique and extremely fast relational database search, merge, append and split capabilities.

SPAYN has many built-in general-purpose statistical analysis features to aid in the examination of parametric data. Because not all parametric data is adequately represented by a Gaussian distribution, SPAYN also offers exponential, log-normal and Gamma distribution options. Scatter plots are used to analyse the relationships between parameters by using a least squares method to fit one or more of the following functions: linear, logarithmic, parabolic, reciprocal, hyperbolic, exponential, power, root, or third-order polynomial. Resulting error residuals, ANOVA information and correlation coefficients are generated. Resulting variations in circuit performance caused by process fluctuations can then be predicted early in the design cycle. SPAYN helps to identify an efficient process monitoring strategy by identifying the minimum set of dominant factors that must be monitored in production, in order to control the yield.

1.8 Virtual wafer fabrication

As discussed above, simulation tools have now been developed to follow the actual wafer fabrication process very closely. Each fabrication step can now be modelled, and they are incorporated into 'process simulators'. Hence it is possible to 'construct' new semiconductor structures and predict their performance using these simulation tools. This is often referred to as a 'Virtual Wafer Fabrication', which provides a physical insight and intuition to illustrate physical principles of what real integrated circuits structures will look like. These tools also provide the ability to 'look inside' structures or devices and to visualise physical phenomena that often cannot be easily observed in traditional experiments.

Currently strong efforts are being made to build simulation platforms for the predictive simulation of micro- and nanodevices and circuits, as this constitutes the cost-effective and time-saving alternative to the traditional experimental approach. Today, the rapid progress in microelectronics technology is strongly supported by process/device modelling methodologies and dedicated simulation tools constituting a 'virtual laboratory' on the computer, which enables the visualisation and detailed analysis of the operating behaviour of single device. A virtual IC factory has the following capabilities:

- integrated graphical environment for physical simulation;
- comprehensive data management;
- post-simulation graphical and statistical data analysis;
- design of experiment capabilities;

- design of experiments for efficient design of simulated experiments based on the number of input variables;
- RSM rapid approximation and responses in a design space that can be further explored for optimisation;
- split run design comparison and trade-off;
- easy-to-use GUI.

1.9   Optimisation

The knowledge gained through TCAD simulation can be used to optimise processes and to identify potential yield problems that need to be addressed in the design before volume production. In order to minimise the time taken for optimisation, the time consumed in a single device simulation should be relatively short, since hundreds of such simulations are often required. With the availability of tools like MATLAB, attempts are being made for the implementation of the optimisation via MATLAB, alongside device simulation using commercial packages. High-performance optimisation algorithms, both local and global, are used to achieve an efficient design in shortest possible time. Such a design tool works with an arbitrary number of design parameters, such as device dimension, doping level and profile. For device design tuning, Nelder-Mead (Simplex) optimisation algorithm [63] is preferable as it is known to be a very efficient local optimisation method.

For its high flexibility, the optimisation tool is often implemented in an object-oriented way, divided into three independent parts: device, material and optimisation criteria. With small modifications the optimisation package can be adapted for any device simulator with text-based user interface. One important feature is the ability to change the design parameters that are optimised, so that important parameters for different performance measures can be optimised individually. The tool can be used for optimisation of different kinds of semiconductor devices using various performance measurements. After the implementation of a new device, it can immediately be optimised on the basis of previously used materials and goals. It has been shown that optimisation of semiconductor devices by iterative simulations is very efficient in design and evaluation of new device structures.

1.10   TCAD for nanoelectronics

In the previous sections, the strategic role of TCAD in technology development, both past and present, and a variety of manufacturing challenges important for scaling below the 100 nm gate length regime have been discussed. The challenges are now in terms of new materials and their scaling to atomic dimensions move outside the domain of currently available technology or tools, and fundamentally new developments are needed. Atomic-scale limits to device design and manufacturing present a key challenge for future technology generations.

Transformation of the microelectronics into nanoelectronics is taking place very rapidly and a subsequent transformation should also affect the virtual wafer fabrication for the development of future nanodevices. Currently used simulation tools for microelectronic technology processes will soon be obsolete since the scaling down of semiconductor devices requires atomic scale design. The technology roadmap of the ITRS indicates that TCAD tools will possibly lead to a 30–40 per cent saving of the overall costs dedicated to the research and development of future devices in the next decade [64]. However, the currently used process/device simulators implemented in the different TCAD suites, based on continuum model, will soon become obsolete to the process development.

The impact of quantisation of charge in metallic or semiconducting structures in conventional electronic devices is usually apparent in simulation output. However, when the smallest feature size is less than 10–20 nm and the capacitance becomes very small and the charging energy is larger than the thermal energy, the change in free energy associated with the addition or subtraction of a single electron to and from an island, or a quantum dot, becomes significant. If an electron can enter or exit islands only via tunnelling, this is often referred to as single electron tunnelling. Single electron tunnelling phenomenon (Coulomb blockade of single electron tunnelling) in the transport of electrons in ultrasmall structures has long been a subject of condensed matter physics since the 1950s. However, only in the late 1980s was it realised by Likarev that such Coulomb blockade of single electron could be controlled by an external gate, and this artificial structure was named as a single electron transistor (SET) [65]. In SETs and single electron circuits, the change of single electron can determine the device characteristics, and as a result, the logic level that is now being considered as a strong candidate for elements of future low power high density chips.

Single-electron devices show very promising characteristics, such as ultimate low power consumption, downscaling to atomic dimensions and high switching speed. Simulation tools can play a crucial part in the development of single electron devices and circuits. In analogy to existing TCAD software for conventional devices, where process simulation produces geometry and material-related input data for device simulation, one needs to develop similar methods for SET simulation. Since single electron devices are extremely charge sensitive, making them strongly dependent on stray capacitances and trapped charges, accurate computation of capacitances in three-dimensional structures will be needed [66].

There has been a great deal of emphasis in the past few years on examining nanodevices in Si and Si-based heterostructures. The most extensively studied material system in this direction is that of strained-Si, SiGe and SiGeC, which allows the creation of Si electron quantum wells. In strained-Si/SiGe/Si, which leaves Si quantum wells under tensile strain, the electron and hole transport properties are modified and allow for investigation of many novel device structures.

Figure 1.8 shows broadly the time and space axes of fabrication processes: at the highest levels are the equipment models and at the lowest levels are ab initio and quantum modelling. In contrast to the well-known hierarchical chip design process, the crossing of boundaries between the hierarchy shown in Figure 1.8 is still very much a research challenge, let alone able to be automated. Considering a rather specific example of plasma processing, the equipment level might involve bulk plasma

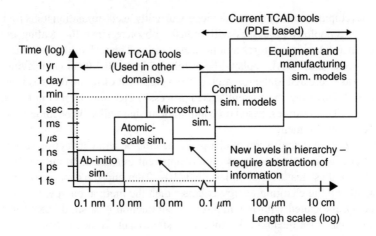

Figure 1.8   Simulation tool hierarchy [After R. W. Dutton et al., IEEE Trans. Computer-Aided Design of Integrated Circuits and Systems, Vol. 19, 2000(1544–1560). © 2000 IEEE]

simulations, gas flow and thermal modelling, all at the macro-scales (millimetre, centimetre and metre scales). At the same time, the wafer, die and structure levels of modelling move progressively in the other direction (down to the nanometer scale) and finally the surface chemistry. The above example illustrates a broad range of multi-scale physical effects that are critical to understanding and control of manufacturing at the atomic scale.

Two ambitious approaches to SoC technology development via TCAD to satisfy product requirements have been proposed by Kohyama [67] in which calibrated TCAD tools are applied to generate an optimised process recipe while Pinto [68] suggested the integration of several technologies in a mix-and-match manner. To cope with increasing speed and complexity in digital, analog and mixed-signal circuit simulation, a hierarchy of simulators and algorithms, together with generations of device models, and their integration in a TCAD tool has become essential.

For the realisation of atomic-scale devices, TCAD tools need to be developed on the basis of a reliable theoretical framework essential for accuracy and feasibility. However, use of an unique methodology is unlikely, since the complete description of atomic-level devices and physical phenomena occurring in them is beyond the capability of such a single approach and the choice of appropriate methodology will depend on the particular problem. In fact, such a task could be achieved by means of the integration of complementary theoretical investigations allowing the simulation of system kinetics on the basis of the microscopic material properties [69]. First-principle computations, based on the density functional theory within the local density approximation (DFT-LDA) and the tight-binding molecular dynamics (TBMD), are being employed for such simulations [70–72].

Organic semiconductors have witnessed a considerable development in recent years, mainly pushed by the realisation of light-emitting diodes (LEDs) and displays whose cost and performance are potentially lower with respect to more conventional solutions [73, 74]. Attention has been given also to organic thin film transistors (OTFTs). Such devices have been proposed at the beginning of the 1980s [75], but only since the mid-1990s have they attracted the interest of the scientific community [76]. For TFTs the motivations are due to the possibility of realising low-cost, large-area devices on flexible substrates. Up to now, however, the application of organic TFTs have been restricted to smart cards, electronic barcodes and low-cost memory devices. The main limitation to a larger exploitation of TFTs comes from their poor mobility.

Few attempts have been made to perform a physical simulation of the OTFT. Alam *et al.* [77] have analysed the operation of OTFT using a two-dimensional drift-diffusion simulator. They showed that under certain conditions, one can describe the electrical characteristics of organic transistor by using tools developed for inorganic semiconductors. Analytic models predict the overall behaviour of the device, even though they cannot describe the differences between the OTFTs and the inorganic ones. Such differences are mainly due to the different transport mechanism involved in the device operation.

1.11   Interconnect

As the critical dimensions in VLSI circuits continue to shrink, system performance of the integrated circuits are increasingly dominated by interconnect's performance. For the technology generations approaching 45 nm, innovative circuit designs and new interconnect materials and architecture will be required to meet the projected system performance. New interconnect materials solutions, such as copper and low-k dielectric offer only a limited improvement in system performance. Significant and scalable solutions to interconnect delay problems will require fundamental changes in system architecture, design and fabrication technologies.

The TCAD analysis of multi-layer interconnects is still in early stages of development. Figure 1.9 illustrates a major physical challenge for both modelling and manufacturing, driven largely by performance constraints of both the intrinsic device and interconnect layers of ICs. This figure shows both the history and prospects of candidate materials that are currently in use and may be used in future advanced IC processes. Problems such as signal integrity coming from inductive cross-talk give still another example where analysis complexity poses new challenges for TCAD in the future. Accompanied with these advancements is the evolution of advanced models and simulators at various levels of abstraction that are used to design and simulate these ULSIs.

1.12   Summary

Over the past decade, technology CAD has become essential for developing and optimising semiconductor process technologies ranging from nanoscale microprocessors

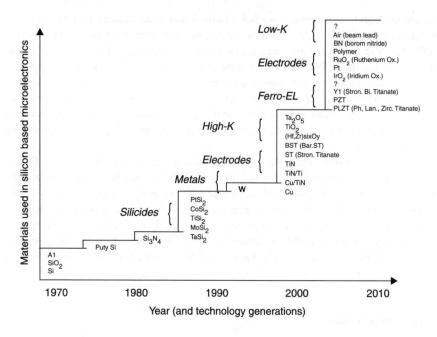

Figure 1.9   Materials used in integrated circuits [After R. W. Dutton et al., IEEE Trans. Computer-Aided Design of Integrated Circuits and Systems, Vol. 19, 2000(1544-1560). © 2000 IEEE]

to large-scale, high-voltage power devices. TCAD today plays an important role in the design and development of modern microelectronics technology. In order to meet the continuously growing demand of this industry, TCAD software tools are evolving as huge and complex computer programs for a wide range of individual simulation disciplines, including process simulation, device simulation, compact model parameter extraction, interconnection simulation and optimisation. In addition, new processes and equipment are inevitably utilized for new technologies and present modelling challenges for process simulation developers.

However, it is increasingly recognised that, with the movement of individual TCAD tools from predominantly research and development groups to more production-oriented environments, the usage and maintenance of coupled TCAD tools becomes cumbersome and requires a significant level of user experience. This fact has motivated a rapid development of TCAD systems in a uniform framework that supports the effective usage of multiple simulators with simple user interface, convenient and standardised data transfers and visualisation.

In this chapter, technology development, based on projections of new and emerging TCAD tools and methodologies, that offers exciting opportunities in realising new paradigms for the design-manufacturing interface is presented. It is important to have a better interface between process and product designers that will provide both sides with an understanding of how choices affect the overall chip performance, reliability and yield. As this new interface is created and explored, TCAD tools will find a new relevance in the IC design community. This will create new opportunities for innovation for TCAD research. The role of TCAD in compact model development was also presented.

Despite the intensive efforts to understand, characterise and model the process physics, industrial use still requires continual calibration of model parameters, and the predictive range of any process modelling capability can vary significantly between process modules. Much like process simulation, device simulation has reached an advanced level of physical sophistication. The approach used in the pioneering continuum-based device simulation efforts continues to be the foundation of most TCAD efforts. It is important to note that the TCAD of the near future will be challenged by a number of fundamental problems such as microscopic diffusion mechanisms, quantum mechanical transport, molecular dynamics, quantum chemistry and high-frequency interconnect simulation. Next-generation TCAD can and will play a key role in quantifying potential road blocks, indicating new solutions, and keeping technology development at par. Thus, the feedback between IC designers and technologists during technology development is crucial, which, however, is difficult to realise in practice.

References

1 W. Shockley. US Patent No. 2,569,347, 1951.
2 H. Kroemer, 'Der Drifttransistor', *Naturwissensch.*, vol. 40, p. 5789, 1953.
3 M. D. Giles, D. S. Boning, G. R. Chin, W. C. Dietrich, M. S. Karasick, M. E. Law, *et al.*, 'Semiconductor wafer representation for TCAD', *IEEE Trans. Computer-Aided Design of Integrated Circuits and Systems.*, vol. 13, pp. 82–95, 1994.
4 W. Oldham, A. Neureuther, C. Sung, J. Reynolds, and S. Nandgaonkar, 'A general simulator for VLSI lithography and etching processes: Part II – application to deposition and etching', *IEEE Trans. Electron Devices*, vol. ED-27, pp. 1455–1459, 1980.
5 B. Stine, D. Ouma, R. Divecha, D. Boning, J. Chung, D. L. Hetherington, *et al.*, 'Rapid characterization and modelling of pattern dependent variation in chemical mechanical polishing', *IEEE Trans. Semiconduct. Manufact.*, vol. 11, pp. 129–140, 1998.
6 Z. K. Hsiau, E. C. Kan, J. P. McVittie, and R. W. Dutton, 'Robust, stable and accurate boundary movement for physical etching and deposition simulation', *IEEE Trans. Electron Devices*, vol. 44, pp. 1375–1385, 1997.
7 R. J. Hoekstra, and M. J. Kushner, 'Comparison of two-dimensional and three-dimensional models for profile simulation of poly-Si etching of finite length trenches', *J. Vac. Sci. Technol. A*, vol. 16, pp. 3274–3280, 1998.
8 A. J. Strojwas, D. J. Collins, and D. White, Jr., 'A CFD model for the PECVD of silicon nitride', *IEEE Trans. Semiconduct. Manufact.*, vol. 7, pp. 176–183, 1994.
9 D. Chin, 'Executing system on a chip's requirements for a successful SOC implementation', in *IEEE IEDM Tech. Dig.*, pp. 3–8, 1998.

10 A. de Mari, 'An accurate numerical one-dimensional solution of the p-n junction under arbitrary transient conditions', *Solid State Electron.*, vol. 11, pp. 1021–1053, 1968.

11 D. L. Scharfetter, and H. K. Gummel, 'Large-signal analysis of a silicon read diode oscillator', *IEEE Trans. Electron Devices*, vol. ED-16, pp. 64–77, 1969.

12 J. P. Krusius, 'Process modelling for submicron complementary metal-oxide-semiconductor very large scale integrated circuits', *J. Vac. Sci. Technol. A*, vol. ED-24, pp. 905–911, 1986.

13 D. C. D'Avanzo, M. Vanzi, and R. W. Dutton, *One-Dimensional Semiconductor Device Analysis (SEDAN)*. Report G-201-5, Stanford University, 1979.

14 S. Selberherr, A. Schutz, and H. W. Potzl, 'MINIMOS – A Two-dimentional MOST transistor analyzer', *IEEE Trans. Electron Devices*, vol. ED-27, pp. 1540–1550, 1980.

15 A. F. Franz, and G. A. Franz, 'BAMBI – A design model for power MOSFETs', *IEEE Trans. Comp. Aided Design*, vol. CAD-4, pp. 177–189, 1985.

16 R. E. Jewell, W. B. Grabowski, and M. R. Kump, 'TCAD software from technology modeling associates', in *Software Tools for Process, Device and Circuit Modeling*, Dublin: Boole Press, pp. 121–129, 1989.

17 BIPOLE3, 'User's Manual', 1993.

18 G. A. Armstrong and T. C. Denton, 'HQUPETS – a two-dimensional simulator for heterojunction bipolar transistors', in *Proc. IMA Conf. on Semiconductor Modeling, Loughborough*, pp. 16–17, 1991.

19 J. W. Slotboom, 'Computer-aided two-dimensional analysis of bipolar transistors', *IEEE Trans. Electron Devices*, vol. ED-20, pp. 669–679, 1973.

20 H. H. Heimeier, 'A two-dimensional numerical analysis of a silicon N-P-N transistor', *IEEE Trans. Electron Devices*, vol. ED-20, pp. 708–714, 1973.

21 Stanford University, *PISCES-2ET 2-D Device Simulator*, 1994.

22 D. C. Cole, E. M. Buturla, S. S. Furkay, K. Varahramyan, J. Slinkman, J. A. Mandelman, *et al.*, 'The use of simulation in semiconductor technology development', *Solid State Electron.*, vol. 33, pp. 591–623, 1990.

23 R. B. Fair, and J. C. C. Tsai, 'A quantitative model for the diffusion of phosphorus in silicon and the emitter dip effect', *J. Electrochem. Soc.*, vol. 124, pp. 1107–1117, 1977.

24 C. P. Ho, J. D. Plummer, S. E. Hansen, and R. W. Dutton, 'VLSI process modelling – SUPREM III', *IEEE Trans. Electron Devices*, vol. ED-30, pp. 1438–1453, 1983.

25 M. E. Law, and S. M. Cea, 'Continuum based modelling of silicon integrated circuit processing: An object oriented approach', *Comp. Mat. Sci.*, vol. 12, pp. 289–308, 1998.

26 E. W. Scheckler, and A. R. Neureuther, 'Models and algorithms for 3-D topography simulation with SAMPLE-3D', *IEEE Trans. Computer-Aided Design*, vol. 13, pp. 219–230, 1994.

27 A. Yoshii, H. Kitazawa, M. Tomizawa, S. Horiguchi, and T. Sudo, 'A three-dimensional analysis of semiconductor devices', *IEEE Trans. Electron Devices*, vol. ED-29, pp. 184–189, 1982.

28 N. Shigyo, M. Konaka, and R. Dang, 'Three-dimensional simulation of inverse narrow-channel effect', *Electron. Lett.*, vol. 18, pp. 274–275, 1982.

29 E. Buturla, P. Cottrell, B. Grossman, and K. Salsburg, 'Finite-element analysis of semiconductor devices: The FIELDAY program', *IBM J. Res. and Devices*, vol. 25, pp. 218–231, 1981.

30 X. Zhou, 'The missing link to seamless simulation', *IEEE Circuits Devices Mag.*, vol. 19, pp. 9–17, 2003.

31 J. Mar, 'The application of TCAD in industry', in *Proc. SISPAD*, pp. 64–74, 1996.

32 M. Duane, 'TCAD needs and applications from a user's perspective', *Trans. IEICE*, vol. E82-C, pp. 976–982, 1999.

33 R. W. Dutton, and R. J. G. Goossens, 'Technology CAD at Stanford University: physics, algorithms, software and applications', *Microelectronics J.*, vol. 26, pp. 99–111, 1995.

34 Silvaco International, *Silvaco-ATHENA User's Manual*, 2005.

35 Silvaco International, *Silvaco-ATLAS/BLAZE*, 2005.

36 ISE Integrated Systems Engineering AG., Zurich, Switzerland, *Tech. Rep. DIOS-ISE, Release 6.0*, 1999.

37 DESSIS-ISE, Release 8.5, 'Integrated systems engineering AG, Zurich, Switzerland', 2003.

38 A. Amerasekera, V. McNeil, and M. Rodder, 'Correlating drain junction scaling, salicide thickness, and lateral NPN behaviour, with the ESD/EOS performance of a 0.25 μm CMOS process', in *IEEE IEDM Tech. Dig.*, pp. 893–896, 1996.

39 R. W. Dutton, and A. J. Strojwas, 'Perspectives on technology and technology-driven CAD', *IEEE Trans. Computer-Aided Des. of Integrated Circuits and Systems*, vol. 19, pp. 1544–1560, 2000.

40 A. R. Alvarez, B.L. Abdi, D. L. Young, H. D. Weed, J. Teplik, and E. R. Herald, 'Application of statistical design and response surface methods to computer-aided VLSI device design', *IEEE Trans. Computer-Aided Design*, vol. 7, pp. 272–288, 1988.

41 W. Maly, and A. J. Strojwas, 'Statistical simulation of the IC manufacturing process', *IEEE Trans. Computer Aided Design*, vol. CAD-1, pp. 120–131, 1982.

42 R. Plasun, M. Stockinger, and S. Selberherr, 'Integrated optimization capabilities in the VISTA technology CAD framework', *IEEE Trans. Computer-Aided Design Integrated Circuits and Systems*, vol. 17, pp. 1244–1251, 1988.

43 Ch. Pichler, and S. Selberherr, 'Process flow representation within the VISTA framework', in *Proc. SISDEP*, pp. 25–28, 1993.

44 Silvaco International, Santa Clara, CA, *VWF Interactive Tools User's Manual*, 2004.

45 Technology Modelling Associates, Inc., Sunnyvale, CA, *TMA WorkBench Version 2.0 User's Manual*, 1996.

46 R. Booth, L. Dupas, and R. Cartuyvels, *NORMAN User's Manual*, 1993.

47 K. Mayaram, and D. O. Pederson, 'Coupling algorithms for mixed-level circuit and device simulation', *IEEE Trans. Computer-Aided Design*, vol. 11, pp. 1003–1012, 1992.

48 J. R. F. McMacken, and S. G. Chamberlain, 'CHORD: a modular semiconductor device simulation development tool incorporating external network models', *IEEE Trans. Computer-Aided Design*, vol. 8, pp. 826–836, 1989.

49 H. Brech, T. Grave, T. Simlinger, and S. Selberherr, 'Optimization of pseudomorphic hemt's supported by numerical simulations', *IEEE Trans. Electron Devices*, vol. 44, pp. 1822–1828, 1997.

50 Y. Cheng, M.-C. Jeng, Z. Liu, J. Huang, M. Chan, K. Chen, *et al.*, 'A physical and scalable I–V model in BSIM 3v3 analog/figital circuit simulation', *IEEE Trans. Electron Devices*, vol. 44, pp. 277–287, 1997.

51 C. Fiegna, 'Physics-based analysis of RF performance of small geometry MOSFETs: Methodology and application to the evaluation of the effects of scaling', in *IEEE IEDM Tech. Dig.*, pp. 543–546, 1999.

52 M. Valtonen, P. Heikkila, A. Kankkunen, K. Mannersalo, R. Niutanen, P. Stenius, T. Veijola, and T. Virtanen, 'APLAC – A new approach to circuit simulation by object-orientation', in *Proc. 10th European Conf. on Circuit Theory and Design.*, pp. 351–360, 1991.

53 J. Victory, C. C. McAndrew, and K. Gullapalli, 'A time-dependent,surface potential based compact model for MOS capacitors', *IEEE Electron Device Lett.*, vol. 22, pp. 245–247, 2001.

54 S. E. Laux, and K. Hess, 'Revisiting the analytic theory of p-n junction impedance: Improvements guided by computer simulation leading to a new equivalent circuit', *IEEE Trans. Electron Devices*, vol. 46, pp. 396–412, 1999.

55 J. Rodriguez, M. C. Smayling, and W. L. Wilson, 'ESD circuit synthesis and analysis using TCAD and SPICE', in *IEEE IEDM Tech. Dig.*, pp. 97–100, 1998.

56 S. Luryi, 'Development of RF equivalent circuit models from physics-based device models', in *Future Trends in Microelectronics, (S. Luryi, J. Xu, and A. Zaslavsky, Eds.)*, pp. 463–466, Wiley, New York, 1999.

57 A. Pacelli, M. Mastrapasqua, and S. Luryi, 'Generation of equivalent circuits from physics-based device simulation', *IEEE Trans. Computer Aided Design of Integrated Circuits and Systems*, vol. 19, pp. 1241–1250, 2000.

58 P. V. Voorde, 'TCAD for bipolar process development: a user's perspective', in *IEEE BCTM Proc.*, pp. 101–109, 1991.

59 N. S. Rankin, C. Ng, L. S. Ee, F. Boyland, E. Quek, L. Y. Keung, *et al.*, 'Statistical SPICE analysis of a 0.18 μm CMOS digital/analog technology during process development', in *Proc. ICMTS*, pp. 19–23, 2001.

60 Hewlett-Packard Company, San Francisco, CA, *IC-CAP User's Manual – High Frequency Model Tutorials*, 1997.

61 Silvaco International, Santa Clara, CA, *UTMOST User's Manual*, 2004.

62 Silvaco International, Santa Clara, CA, *SPAYN User's Manual*, 2004.

63 J. A. Nelder and R. Mead, 'A simplex method for function minimization', *Comput. J.*, vol. 7, pp. 308–313, 1965.

64 Semiconductor Industry Association, 'International technology roadmap for semiconductors', San Jose, Calif., 2005.

65 K. K. Likharev, 'Correlated discrete transfer of single electrons in ultrasmall tunnel junctions', *IBM J. Res. and Devices*, vol. 32, pp. 144–158, 1988.

66 Ch. Wasshuber, H. Kosina, and S. Selberherr, 'SIMON - A simulator for single-electron tunnel devices and circuits', *IEEE Trans. Computer-Aided Design of Integrated Circuits and Systems*, vol. CAD-16, pp. 937–944, 1997.

67 S. Kohyama, 'SoC solutions and technologies for digital hypermedia platform', in *IEEE IEDM Tech. Dig.*, pp. 8–13, 1999.

68 M. Pinto, 'Atoms to applets: Building systems IC's in the 21st century', in *IEEE ISSCC*, pp. 26–30, 2000.

69 A. La Magna, P. Alippi, L. Colombo, and M. Strobel, 'Atomic scale computer aided design for novel semiconductor devices', *Comp. Mat. Sci.*, vol. 27, pp. 10–15, 2003.

70 P. Hohenberg, and W. Kohn, 'Inhomogeneous electron gas', *Phys. Rev. B*, vol. 136, pp. 864–871, 1964.

71 L. J. Sham, and M. Schluter, 'Density-Functional theory of the energy Gap', *Phys. Rev. Lett.*, vol. 51, pp. 1888–1891, 1983.

72 W. Kohn, and L. J. Sham, 'Self-Consistent equations including exchange and correlation effects', *Phys. Rev.*, vol. 140, pp. A1133–A1138, 1965.

73 P. E. Burrows, G. Gu, V. Bulovic, Z. Shen, S. R. Forrest, and M. E. Thompson, 'Achieving full-color organic light-emitting devices for lightweight, flat-panel displays', *IEEE Trans. Electron Devices*, vol. 44, pp. 1188–1203, 1997.

74 A. Dodabalapur, Z. Bao, A. Makhija, J. G. Laquindanum, V. R. Raju, Y. Feng, H. E. Katz, and J. Rogers, 'Organic smart pixels', *Appl. Phys. Lett.*, vol. 73, pp. 142–144, 1998.

75 F. Ebisawa, T. Kurokawa, and S. Nara, 'Electrical properties of polyacety-lene/polysiloxane interface', *J. Appl. Phys.*, vol. 54, pp. 3255–3259, 1983.

76 D. J. Gundlach, Y. Y. Lin, T. N. Jackson, S. F. Nelson, and D. G. Schlom, 'Pentacene organic thin-film transistors-molecular ordering and mobility', *IEEE Electron Device Lett.*, vol. 18, pp. 87–89, 1997.

77 M. A. Alam, A. Dodabalapur, and M. R. Pinto, 'A two-dimensional simulation of organic transistors', *IEEE Trans. Electron Devices*, vol. 44, pp. 1332–1337, 1997.

Chapter 2

IC technology and TCAD tools

The semiconductor fabrication process involves several steps such as diffusion, oxidation, ion implantation, etching, deposition and the lithographic steps of image transfer and photoresist exposure and development. Integrated circuits are manufactured by using these sequential fabrication process steps that have their scientific foundations in applied physics and chemistry. Each process is a complex physical phenomenon and, as a result, the exact physics may not always be accurately defined. A long sequence of fabrication steps are necessary to build up the layers on the silicon substrate and to etch them into appropriate shapes to form diodes, transistors and interconnects. Each step has to be precisely controlled to give the best circuit performance and the complete sequence needs to be highly optimised for the production of ICs.

Today much of the development of semiconductor devices and processes is done by TCAD as it offers unique possibilities in visualisation of processing steps, description of the physical changes and understanding of the interrelation of the process variables. Modelling of processes provides a way to interactively explore the fabrication process, studying the effects of process choices, leading to a 'virtual wafer fabrication' (VWF) design environment.

TCAD simulation tools provide a controlled and repeatable numerical experiment that can yield information that cannot be measured experimentally. The main aim of a TCAD simulator tool is to match the simulation methodology as close as possible to the fabrication technology and their integration into the actual fabrication. For TCAD tools to be useful in a practical environment, they must be physically accurate, computationally robust and usable by semiconductor process engineers. A TCAD simulation flow generally consists of the following:

- interface from design and mask layout to generate process simulation input;
- interface between the process flow description and the process simulation;
- interface between electrical test and device simulation;
- interface between device characterisation and modelling.

In the TCAD process integration scheme, the usual procedure is to measure the device characteristics on test structures on the semiconductor wafer and compare these characteristics with device simulation. This approach can relate worst-case predictions to process parameter changes. Statistical variations on specific semiconductor process step parameters can also be predicted.

After fabrication, it is necessary to generate d.c. and a.c. electrical parameters such as threshold voltage, effective gate length of MOS transistors or cut-off frequency in a bipolar transistor. Measurements are carried out using automated test systems (e.g., Agilent or Keithley) at wafer level. The measurement procedures may hierarchically be oriented for a particular technology node. The interface between device characterisation and modelling deals with the generation of reliable device models for circuit modelling.

Simulation of the mechanical effects of the final semiconductor device structure, including metallisation and interconnection, is also of interest. This interest arises from the increasing complex and smaller geometries that are becoming quite challenging due to high stress effects that manifest themselves in the form of cracks, extrusions of materials, electromigration and other effects. The back-end of the line (BEOL) modelling consists of mechanical, thermal and field analyses, and electromigration models.

In this chapter, the modelling and simulation of processes that are involved in the creation of a device will be discussed. For example, the modelling and simulation methodology for plasma processes in the context of how models have evolved and the metrics for their application will be addressed. Modelling and simulation can make an important contribution to process engineering of each particular plasma process step as design of experiments (DOE) become more costly. The concept of VWF, which is becoming increasingly popular in the semiconductor industry, is then introduced. Then, several commercially available process and device design tools, illustrating their use for specific design, will be discussed.

2.1   Process simulation

Semiconductor process simulation involves the numerical solution of equations describing the physics of dopant diffusion, oxidation, lithography, ion implantation, etching and deposition steps, resulting in the creation of data files that describe the geometry and doping profiles that define a device. Each process step requires solution of a specific set of equations that are often non-linear and require specification of physical constants, some of which may have to be calibrated for a particular process.

The realisation of better physical models depends on the accuracy of the knowledge of materials and process parameters. The limitations in direct measurement methods for the determination of some key parameters is one of the challenges that needs to be addressed. For example, measurement techniques for dopant profiling have led to a better understanding and development of models for diffusion and ion implantation. However, while simple techniques are capable of measuring 1D profiles, more complex procedures offer quantitative determination of

2D and 3D profiles [1]. Such techniques may greatly aid in the development of multi-dimensional models for ion implantation and diffusion, which then provide higher accuracy in predicting device structures and their material properties. The various process steps that have to be modelled are now described.

2.1.1 Oxidation

Oxidation is modelled by solving the diffusion equation for the transport of oxygen through oxide and silicon. Silicon at the interface is converted to silicon dioxide after accumulating a sufficient concentration of oxidant. In local oxidation of Si (LOCOS), a deposited silicon nitride layer acts as a diffusion barrier in regions where oxidation is not desired. Historically, modelling of thermal oxidation of silicon has been based upon the well-known linear-parabolic growth model due to Deal and Grove in 1965 [2]. This model is, however, inaccurate in the 'very thin' oxide regime (~ 20 nm). Since this regime has become a very important state-of-the-art MOS technology, many more complex models [3] have been proposed to account for the growth of thin oxides.

The problem of modelling oxidation is compounded considerably in 2D, where one can no longer ignore the mechanical stress of the growing oxide. Specifically, this is true when oxide is known to exhibit fluid flow at high process temperatures, but there is no consensus as to whether it is viscous, viscoelastic or even plastic flow. However, using a comprehensive process simulation program, one can now simulate a complex structure and analyse stresses that may arise during thermal oxidation.

In general, oxidation simulation tools should include the following:

- non-linear viscoelastic model;
- stress-dependent oxidation;
- initial film stress specification;
- thermal expansion;
- glass reflow.

2.1.2 Ion implantation

Ion implantation is the most important technique for introducing dopants into semi-conductors. Commonly, dopants are implanted into silicon through silicon dioxide. However, silicon nitride, photoresist, polySi or various silicides are also used as masking materials. For VLSI technologies, low 'thermal budgets' are required to minimise profile broadening by diffusion, and correct prediction of ion-implanted profiles has become crucial. As modern annealing methods such as rapid thermal annealing (RTA) and flash annealing do not alter the implanted profile very much, the determination of the initial implantation profile has become a very important task. Thus, the simulation of ion implantation has gained in significance tremendously.

For ion-implantation simulation, three main techniques are used: the analytical description of the doping profile [4], and the solution of the Boltzmann transport equation [5] or the Monte Carlo method [6,7]. Simulators contain both analytic and Monte Carlo ion-implantation models so that a wider range of the required physical situations can be realistically simulated. Simpler analytical models based on a Gaussian or other distribution functions may also be used to define an ion-implanted profile prior to subsequent processing.

Models based on Gaussian, joined half-Gaussian and Pearson IV distributions are available for simulation of any impurity element into silicon or into films such as oxide or nitride on silicon. A Gaussian distribution may be used to account for scattering effects. Specific models for As, P and B are available. For all cases, angled implants and irregular surface geometries, including undercuts, are allowed.

When greater accuracy is required in defining the implant profile, Monte Carlo programs have been developed for accurately simulating the implantation of ions into semiconductor materials. For example, the TRIM Monte Carlo ion-implantation program [8] was developed for 1D and 2D implantation studies in multi-layered structures. Such a capability can greatly aid in the understanding of ion implantation (especially lateral scattering in 2D) and can be used to generate new and improved parameters for analytic models, or to extend their range. Another powerful Monte Carlo program for ion implantation in solids is MARLOWE [9]. An important feature of MARLOWE is its capability of treating crystalline materials, rather than simply assuming the semiconductor material to be amorphous. Consequently, this program can simulate the channelling of dopant atoms in crystalline materials.

In general, an ion-implantation simulation tool should have the following capabilities:

- Monte Carlo implantation;
- dual Pearson, Pearson and Gaussian distribution functions;
- depth-dependent lateral straggling;
- dose accumulation;
- primary ion distribution;
- range matching and dose matching;
- calibrated internal tables for B, BF_2, As, P, Sb, In and Ge;
- automated saving and reloading of 1D and 2D implanted profiles.

2.1.3   Diffusion

Accurate modelling of dopant diffusion is one of the key elements in process simulation. Point defects are responsible for most non-equilibrium effects in diffusion, and a complete diffusion model must be able to account for defect-dopant interactions. Presently, most process simulators model diffusion by a numerical solution of the diffusion equation for dopants, and incorporate non-linear effects through use of a concentration-dependent dopant diffusivity relationship. Non-equilibrium effects, typically due to thin film/substrate interfacial reactions, are incorporated via boundary conditions.

For diffusion modelling, the most general form requires the simultaneous solution of a set of non-linear coupled differential equations for dopants and defects, where the point defect distributions are explicitly calculated. Diffusion simulators are also capable of modelling a variety of explicit point-defect-mediated diffusion effects, such as oxidation-enhanced diffusion and nitridation-retarded diffusion. The simulation of rapid thermal annealing (RTA) processes are also necessary. As short times and high temperatures are associated with RTA, a highly non-equilibrium state where impurity and point defect 'reactions' tend to dominate 'diffusion' arises.

Hu [10] introduced the dual diffusion mechanism in silicon, involving the lattice vacancy and silicon self-interstitial for mediating the dopant diffusion. Oxidation enhanced/retarded diffusion [11] and retarded diffusion during thermal nitridation [12] and silicidation [13] are compatible with the dual mechanism. Also, the pair-diffusion advanced in diffusion modelling by hypothesising that dopant diffusion proceeds by dopant point-defect pairs [14–16]. High concentration effects like clustering and precipitation are inherently connected to the processes of activation and deactivation of dopants [17].

Commercially available simulators, such as TMA/Avanti's (now Synopsys) TSUPREM4 and Silvaco's ATHENA, have several models and parameters to predict the dopant diffusion behaviour in two-dimensions. Careful consideration has to be made to fit the experimental results. In general, diffusion simulation tools should have the following capabilities:

- advanced point defect model with full charge states for dopants and defects, clusters, solubility;
- advanced calibration parameters;
- equilibrium model for fast simulations of large structures;
- polySi model for grain boundary diffusion;
- epitaxy.

Modelling realistic diffusion processes requires the incorporation of as much physics as possible to obtain sufficient accuracy. To design a new simulation model, one has to make a compromise between the number of parameters and the underlying physical relationships. As an example, a physically based model, in the form of a two-dimensional simulation is necessary for dopant diffusion study in polySi. PolySi layers are used in modern IC fabrication processes as diffusion sources, for instance, for out-diffusion processes in forming an n-polySi gate for MOSFETs, or for emitter and base formation in high-performance bipolar technology [18].

Any advanced polySi diffusion model must include various phenomena such as clustering due to the excessively high dopant concentrations and segregation kinetics, to handle the exchange of dopants in the grain/grain-boundary network. To determine the impurity profile in the complex lattice polySi-silicon structure, it is also necessary to include a generation/recombination mechanism and grain growth kinetics.

2.1.4 Lithography

Lithography deals with the basic processes of pattern definition and transfer, which ultimately decide the performance and packing density in a circuit. Topography deals with the change in the shape on the wafer surface. As IC fabrication technology pushes towards finer line-widths, the topographical features of devices are becoming critical. In IC processing, lithography step involves a multi-step combination of resist processing, plasma etching of oxide/nitride and sputter deposition of metals to form the contact to the substrate.

The important issues are the types of physical features that can be built with adequate yield and their corresponding electrical performance. If process models can be established, algorithms can often be used to explore process effects by simulating the time evolution of the line edge profile (topography). Besides the advantages of controllability and observability, simulation can be done much more rapidly and economically. In order to predict the profile evolution with enough accuracy to represent the very small structures such as trenches, contacts or vias, 3D approaches to topography modelling have been developed [19, 20].

The full spectrum of these topography models covers lithography, etching and deposition modules. In the lithography area, the quantitative resist modelling is still the bottleneck and the macroscopic models developed by Dill still dominate [21]. The modelling of image formation has received a lot of attention mostly because of the optical proximity correction needs. There are numerous software tools for aerial imaging that result in good engineering approximations [22–24].

However, even these tools are not sufficient to handle the current lithography systems with large numerical apertures, off-axis illumination and pupil filtering. The complex nature of material etching and deposition processes has limited the development of physically based models for such processes. Consequently, many simulators include user-defined models for material removal and addition as well as models for isotropic and anisotropic deposition/etching.

Since the wavelength of visible light is comparable to device dimensions for submicron device structures, then for conventional photoresist exposure methods, the image obtained from a lithographic mask will not match the intended masking design. Moreover, other complicating factors often arise due to topographical variations in the photoresist and in the structures underneath the photoresist. Developing the photoresist introduces another factor that typically offsets the final shape from the intended shape, as well as giving the edge of the photoresist a non-vertical profile.

A good example of topography design occurs in sputter deposition of metal into small contact holes. In shrinking VLSI layout for high density, the size of the contact holes is one of the limiting factors. Not only must the lithography be capable of resolving these two-dimensional features, but the deposition must also be capable of adequately coating them. The thickness in the bottom of the contact and along the sides is greatly influenced by the aspect ratio of the opening and edge profile shape. A tapered process results in much better coverage and reduced sensitivity to electromigration induced by high current densities.

A variety of etching and deposition models have also been developed. Basic etching concepts have evolved from crystal growth models and have been applied to plasma etching reactive-ion etching and ion milling [25–29]. Sputter deposition and evaporation have also been simulated [30–33]. The conditions necessary for metal 'lift-off' and step coverage have been explored [32].

Electron beam lithography simulation has been based on Monte Carlo simulation of the latent energy deposition profile [34, 35] and a solubility-versus-dose model for PMMA resist. Profile simulation has been useful in characterising effects of machine, wafer and resist parameters on lithographic performance [36–41]. X-ray and ion beam lithography have also been simulated using energy deposition models with the PMMA resist model [42–44].

Historically, it has often been sufficient to simulate only the two-dimensional cross section of an integrated device feature. Programs like SAMPLE [45, 46], SPEEDIE [47] or COMPOSITE [48] use the well-known string algorithm [49] for fast and accurate computation of two-dimensional topography processes. These programs are widely available and have been applied with great success. However, the real problems are in three-dimensional topography simulation.

2.1.5 Etching

An important process in semiconductor manufacturing is the etching of silicon and polySi for device fabrication. Chlorine-based chemistries are commonly used in industry owing to the capability of highly anisotropic feature etching allowing the necessary submicron feature production. Plasma etching using high-density plasma (HDP) reactors is predominant in semiconductor fabrication due to its capability to produce highly anisotropic features in line-widths below $0.2\,\mu$m. In general, etching/deposition simulation tools should have the following capabilities:

- Etching: Isotropic, vertical, anisotropic, crystallographic, and polygon.
- Deposition: Isotropic and anisotropic.

The etch and deposition simulator SPEEDIE [47] is intended to simulate two-dimensional profile evolution during etching and deposition in gaseous systems. SPEEDIE predicts time evolution of etch profiles using physical models and parameters extracted from special test structures. The models in SPEEDIE assume that: etch and deposition reactor pressure is below several Torr, such that gas-phase collisions within the topological features can be ignored because the mean free path is very large in comparison with the characteristic geometrical dimension of IC devices. The fluxes for three types of species are calculated: ions, chemical radicals and deposition precursors. Multiple transport mechanisms that are modelled include direct gas-phase fluxes, neutral adsorption/re-emission, ion-induced redeposition, surface diffusion and ion reflection.

Several other tools [50–52] have been developed for the study of the hybrid plasma equipment model (HPEM). These include the plasma chemistry Monte Carlo model (PCMCM) and the Monte Carlo feature profile model (MC-FPM), which focus on the effect of the plasma on the wafer surface. Using the output from the main plasma simulation, the PCMCM self-consistently determines the energy and angular distributions of all plasma species at the wafer. This distribution information can then be used by the MC-FPM to determine the time evolution of etch features on the wafer based on an energy- and angular-dependent surface chemistry.

2.1.6 Metallisation

Semiconductor devices require conducting materials to provide gate contacts, metal-to-silicon contacts, interconnects between devices and bonding pads for final assembly. In the microelectronics industry, an increase in device density has generally been facilitated by the miniaturisation of the device and interconnect features. As a result of the decrease in feature sizes, however, the fabrication of integrated circuits has become dependent upon the fundamental properties of the thin film material and film–substrate interactions.

Specially, the contact metallisation has been greatly affected by miniaturisation of the features as the semiconductor technologies has moved into ultra large-scale integration. Processing has evolved from rather simple sputtering into elaborate techniques involving source-geometry adjustments, collimation, feature-shape adjustments, temperature-controlled processing and RF sputtering. Metallisation materials have also evolved from simple metal such as aluminium to aluminium alloys used in conjunction with silicides and refractory metal diffusion barrier layers. Alternative metals such as copper are now being used.

As component miniaturisation continues, new technologies must be developed and research tools must likewise evolve with the technology. Extensive processing and materials research is required to develop the reliable fabrication processes of sub-0.1 μm metallisation. A new feature scale simulator needs to be used, as a process design and development tool requires both modelling and process prediction capability.

GROFILMS [53], a two-dimensional thin film process simulator, simulates the growth and processing of thin films, while considering the microscopic fundamental physical phenomena that occur. On the basis of a line segment/nodal description of thin film microstructure, GROFILMS is capable of modelling diffusion, grain growth competition, anisotropic energetic and ballistic shadowing. Phenomena such as substrate wetting, grain boundary grooving and surface faceting are also accurately represented. From a process development stand point, GROFILMS employs system scale parameters to simulate thin film processing (deposition and anneal) over topographical features such as trenches and vias.

As feature sizes decrease, the Al/Si metallisation shows several limitations. For example, owing to the high solubility of Si in Al, interdiffusion at Al-Si contact interfaces can occur during processing at moderately high temperatures. The interdiffusion can cause the Si to dissolve into the Al creating a void. Al atoms move into the void forming a spike into the device region. Reliability failures due to this phenomenon can be avoided by the use of barrier layers at the Al/Si junction. Al alloys were first employed to further extend the viability of Al-based metallisation. The addition of 1 per cent Si saturates the Al, eliminates the driving force for dissolution and suppresses junction spiking. Eventually, new barrier layer materials are required as metal silicide layers do not always form stable contacts. Examples of common barrier layers include titanium (Ti), titanium nitride (TiN) and stuffed inter-metallic alloys such as titanium-tungsten with or without nitride formation at the grain boundaries.

The second major drawback of Al/Si technology arising from the increased current densities and the low melting point of Al is the susceptibility to electromigration failures. Electromigration is the mass transport of the conductor material under the influence of high current density. Electromigration results as a momentum transfer occurs during interactions between conductor atoms and conductor electrons.

The mass transport leads to a divergence in atom flux causing voids and open circuits and interconnect failures. Relationships exist between electromigration susceptibility, the metal film texture and the number and type of grain boundaries.

Electromigration becomes a serious problem as miniaturisation and high-density integration continue to decrease the metallisation line-widths and increase current densities. Copper (Cu) is added to Al to form Al-Cu alloys to decrease susceptibility to electromigration failure. The Cu segregates to the grain boundaries and lowers the effective rate of grain boundary diffusion within the alloy. There is thus an increase in resistance to electromigration.

The use of Al-Cu alloys as the interconnect material results in trade-offs with processing techniques and metallisation properties. As Cu does not easily form a volatile species, difficulties with dry etch processes exist. Increased corrosion rates due to the presence of Cu requires encapsulation of the interconnect lines. The line resistivity of the alloys also increases over that of pure Al. These problems restrict the amount of Cu in the alloy to less than 2 per cent, however, Al with 0.5 per cent Cu is generally used to increase the electromigration reliability of integrated circuit interconnections.

New and innovative circuit layout designs are necessary to reduce delays due to capacitive coupling of conduction lines and simultaneously increase the packing density of the devices. To this end, multi-level metallisation (MLM) design schemes are employed. The MLM design minimises interconnect distances and reduces signal transmission delays. IC operating speed is thus enhanced by stacking metallisation layers one on top of the other. MLM structures also utilize the vertical dimension, permitting increased device density.

The use of MLM, however, tends to leave topographical features with high aspect ratios. These features are to be filled with the metallisation material. Consequently, feature miniaturisation and design geometry of MLM schemes significantly affect the development of the deposition processes required for successful metallisation. For example, uniformity of the metal fill is a critical issue as feature dimensions decrease. The sputter deposition of Al is often inadequate for filling vias and forming reliable contacts. Tungsten (W) chemical vapour deposition is being used as an alternative contact and via fill metallisation. Tungsten CVD can be selectively deposited. Good coverage is achieved within high aspect ratio vias. However, increased RC time delays that result from the higher resistance of W (5.65 $\mu\Omega$-cm) can significantly reduce the device performance.

Challenges in interconnect technology include circuit layouts, substrate geometry and fabrication. Circuit designs consider the decreasing interconnect separations and high-frequency operation of the circuits. Increased RC time delays impose limitations on the circuit layouts. Dielectric deposition and planarisation and reliable metal deposition become more difficult.

The properties of the contact metals are crucial to the development of the interconnect technology and fabrication processes. As feature sizes continue to decrease, the requirements of the interconnect metal become increasingly rigorous. The metal must be reliably deposited into increasingly complex topographies, remaining electrically and structurally stable. It must handle the higher current densities resulting from

*Table 2.1   Requirements for contact metallisation for integrated
circuit interconnect technology*

Metallisation property	Issue
Low electrical resistivity	Performance
High electromigration resistance	Reliability
Resistance to stress-induced voiding	Reliability
Oxidisation/corrosion resistance	Processing/reliability
Thermal stability	Processing/reliability
Silicon process compatibility	Processing
Good adhesion to barrier layer or dielectric	Processing
Ease of deposition and patterning	Processing
Low production cost/high yield	Manufacturing

cross-sectional areas that exhibit decreasing conductivity. A list of the requirements for reliable metallisation as summarised in Table 2.1 and may be a starting point for the evaluation of a contact material.

2.2   Plasma processing

In this section, an overview of plasma chemistry, models and model inputs is presented as a methodology for the development and application of plasma process simulation tools in microelectronics. Plasmas have been used in microelectronic processing since the 1960s. Early applications included ashing and isotropic etching of silicon or silicon dioxide with halogen plasmas. In the mid-1970s the phenomenon of directional (anisotropic) etching of surfaces was recognised, and it is this development that has allowed the manufacture of integrated circuit chips with very small feature sizes. The semiconductor industry has, in the past, relied primarily on conventional empirical methods for the design of processes and the equipment used to facilitate them. Empirical design has worked very well so far as the semiconductor industry has not generally faced any significant technological barriers that would have necessitated a thorough theoretical understanding of manufacturing processes and tools. Simulation tools for low temperature plasma modelling and simulation have now reached a maturity level, allowing them to be used to address real process development on state-of-the-art technologies.

The plasma sources include electron cyclotron resonance (ECR), helicon wave, helical resonator and inductively coupled plasma (ICP) designs. Lieberman and Gottscho [54] have reviewed the design of various plasma sources. The sources have interesting characteristics that make them attractive from a processing viewpoint. It is possible to operate these tools at relatively high plasma density (10^{11}–10^{12} cm^{-3}) and at low neutral pressures (typical values for etching are 1–50 mTorr, compared to 100 mTorr to several Torr for capacitively coupled discharges). Low-pressure operation at relatively high plasma density results in nearly collisionless sheaths, and it is thought that this feature promotes good etch anisotropy. High plasma density also promotes high ion fluxes to surfaces, and thus high process rates. Non-capacitive power deposition, through the use of microwave power, helicon waves or inductive heating leads to low plasma potential (10–30 V) and allows, through separate radio frequency (RF) biasing of substrates, control of ion energies and minimisation of device damage.

The ICP has the additional advantage of simplicity of design. It requires no external magnetic fields for efficient power coupling, eliminating the need for design and optimisation of electromagnet or permanent magnet configurations. The ICP consists, essentially, of a multi-turn RF antenna coil coupled across a dielectric window to the plasma, which acts as a single-turn lossy conductor. In addition to inductive coupling, there may be substantial capacitive coupling between coil and plasma. ICPs are now becoming one of the most important discharge configurations for applications in semiconductor processing [54–58]. Applications include etching, plasma-enhanced chemical vapour deposition and physical vapour deposition.

Plasmas are complex physico-chemical systems, involving the collective behaviour of electrons and ions in a reactive neutral background. The past 20 years have witnessed significant advances in plasma equipment and process modelling. However, the analytical solution of plasma processes is possible only for simple systems, with many simplifying assumptions and limited kinetics. During the 1980s the plasma process and equipment modelling work involved one-dimensional simulation of capacitively coupled discharges [59]. The emphasis was mainly on the understanding of the basic plasma characteristics and refinement of modelling techniques, rather than the development of comprehensive plasma equipment models. As the computational resources evolved, the models were extended to two-dimensions, and more complex plasma chemistries were explored.

The emergence of inductively coupled plasmas in the early 1990s prompted the development of the first truly comprehensive plasma models that simultaneously captured most aspects of the plasma reactor behaviour. Plasma models have also been extended to three-dimensions, and comprehensive models have been developed for a variety of new plasma devices. Although most of the above plasma models are either fluid-based or hybrid (in which electrons are treated kinetically), fully kinetic particle-in-cell (PIC) simulations have also been developed. With the basic models in place, other aspects of the reactor physics such as non-local behaviour of electrons and plasma–circuit interaction have also been investigated.

Most plasmas in the process equipment are excited by RF or microwave sources, and accurate modelling of plasma process is critical for developing predictive models. Not only does the circuit help accurately describe power-coupling, but an accurate representation of harmonics and the like also plays a crucial role in describing how the species interact with surfaces (energy flux and angular distribution functions). Since external RF measurements are relatively easy to make, even on commercial reactors, RF measurements can be used to fine tune plasma chemistries and reactor models. Although the characteristics of individual sheaths have been investigated in considerable detail, it is important to understand the interaction of plasmas and circuits in multiple electrode systems or on assessing the impact of external RF system components on plasma characteristics.

Gas-phase deposition is very useful for material synthesis, particularly devoted to the deposition of thin solid films. In these processes, the film is obtained from a gaseous phase by a chemical reaction involving inorganic or metal-organic precursors. The versatility of these processes allows the deposition of stable, metastable or kinetic solid products. A wide variety of gases are used in the semiconductor industry for plasma processes. Rare gases, such as Ar, He, Ne, Xe and O_2 and N_2 are commonly used for both etching and deposition chemistries. Etching often utilizes halogenated gases such as Cl_2, BCl_3, CF_4, C_2F_6, C_4F_8, SF_6, HBr and NF_3.

Deposition can be chemically more complex, and the gases used include mixtures of CH_4, SiH_4, NH_3 or N_2O. The semiconductor industry uses low-k materials for interconnects, and complicated deposition gases (e.g., TEOS and SiH_x-$(CH_3)_{4-x}$) are used. It is important to identify what species are generated in these plasmas and how they interact with each other. Some examples of deposition reactions involved in CVD processes are summarised in Table 2.2. From the above examples, it is clear that different precursors or their combinations can be conveniently used for the deposition of the desired material.

The most important interactions in the plasma are often electron impact processes such as neutral dissociation, ionisation, dissociative ionisation, attachment and dissociative attachment. Because the electron energy distribution function (EEDF) is usually non-Maxwellian, it is best to know the cross-sections for these electron impact processes. Once the cross-sections are known, the reaction rates can be calculated using EEDF. In addition to the parent gases, the cross-sections are also needed for the radicals (e.g., CF, CF_2, CF_3, SF_2, SF_3, CHF_2, BCl_2) and secondary molecules (e.g., C_2F_4 in C_4F_8) that are often a major constituent of the plasma. Gas-phase plasma chemical mechanisms are complex. This is especially true for complex etching or deposition chemistries where it is difficult to identify all the relevant species in the plasma.

Table 2.2   Examples of CVD processes

System	Type	Chemical reaction	Temperature (K)
Si-epitaxy	Thermal	$SiH_{4-x}Cl_x + H_2 \rightarrow Si + HCl$	1050–1350
LPCVD	Thermal	$SiH_4 \rightarrow Si + 2H_2$	850–950
		$SiH_4 + N_2O \rightarrow SiO_2 + N_2$	900–1000
		$SiH_2Cl_2 + 4NH_3 \rightarrow Si_3N_4 + HCl + H_2$	1000–1100
		$WF_6 + 3H_2 \rightarrow W + HF$	500–700
VPE	Thermal	$GaCl + AsH_3 + As_2 + As_4 \rightarrow GaAs + HCl$	800–1100
MOCVD	Thermal	$Ga(CH_3)_3 + AsH_3 \rightarrow GaAs + CH_4$	800–900
		$In(CH_3)_3 + PH_3 \rightarrow InP + CH_4$	600–900
PECVD	Plasma	$SiH_4 \rightarrow a\text{-}Si{:}H + H_2$	300–600

Chemical mechanisms that capture the relevant physical/chemical phenomena are often more relevant than individual atomic and molecular data. In addition to fundamental atomic/molecular physics data, information on chemical mechanisms are derived from swarm measurements and empirically inferred from experiments. It is important to develop and systematically improve chemical mechanisms for plasma processing relevant gases, and research has resulted in the development of mechanisms for several gases relevant to etching (e.g., CHF_3, C_2F_6 and C_4F_8).

In addition to electron collision cross-sections, ion–ion, ion–neutral and neutral–neutral reaction rates are needed as well. Complete sets of relevant cross-sections are not known for many of the relevant gases. Heavy metals (e.g., Ta or W) are important for many semiconductor contact applications and their complicated atomic structure makes modelling of these plasmas a challenge. Using gaseous lasers, thermochemistry, atmospheric chemistry and improvement of atomic and molecular measurement techniques, substantial amounts of atomic and molecular physics data were generated before 1980 (e.g., Ar, O_2 and Cl_2) [60]. Using the data, several mechanisms have been proposed for some of the simpler processing relevant gases.

The interaction of the plasma with surfaces plays an important role in the discharge dynamics. Sputtering or reactive ion etching introduces new species in the discharge, and these species (e.g., Cu in IPVD or SiF_x during fluorocarbon etching of SiO_2) can greatly alter the discharge behaviour. Surface processes at the plasma–surface interface are not well characterised compared to gas-phase processes within the plasma. Although it is difficult to investigate surface processes in situ (in the presence of the plasma), beam studies have proven useful for investigating fundamental etching and deposition processes at surfaces. Molecular dynamics simulations have also been found useful to understand surface processes such as ion sputtering of Cu and Si. A very small subset of relevant surface processes has been characterised in this manner and, for most applications, it is fair to state that the pertinent surface data does not exist.

2.3 Chemical mechanical polishing

Chemical mechanical polishing (CMP) is used for planarisation in ultra large-scale integrated circuit fabrication. It is used for planarisation of inter-level dielectrics (ILD), shallow trench isolation (STI) and copper damascene metallisation. In spite of extensive use of CMP, many aspects of CMP are not well understood. The simulation of CMP is particularly important within integrated topography process simulation environments to allow for studying the interplay between etching, deposition and CMP process steps.

The coupling of the different topography simulation modules allows for parameter studies covering a number of subsequent process steps and is therefore highly useful for reducing the amount of experimental work needed for integrated topography process characterisation and optimisation. The integration of chemical mechanical polishing module and application to damascene processes have been recently demonstrated [61].

The most important issues for modelling of CMP include the removal rate (RR), which consists of a mechanical as well as a chemical contribution, the bending and the stress of the pad, which strongly influence the quality of the planarisation, and the transport and flow of the slurry. Whether the primary removal mechanism is mechanical or chemical depends on the layer being removed. There are generally four approaches to model wear mechanisms and material removal rates in CMP processes. The phenomenological approach is useful in practice for predicting dishing and erosion, but does not provide much insight in the actual CMP process.

The approach based on fluid hydrodynamics assumes slurry erosion to be the main wear mechanism and neglects mechanical abrasion by particles embedded in the surface of the polishing pad. Although global planarisation can be achieved by hydrodynamic effects if a continuous flow of slurry is present, but the removal rate predicted by fluid modelling is too low in this regime. The third approach is based on a contact-mechanics analysis of the CMP process. According to these models, elastic deformation of the polishing pad determines the pressure distribution between wafer and pad and thus plays an important role in determining the wafer surface profile after polishing. The fourth approach is based on the theories of contact mechanics and fluid hydrodynamics. Other models take into account the roughness of the polishing pad and describe CMP in terms of asperity contacts between wafer and pad. There exist only a few models to predict surface evolution on feature scale.

The model takes into account both the roughness and elastic deformation of the polishing pad for computing the pressure distribution. The physical feature-scale CMP model based on contact mechanics has the advantage that the parameters needed for the simulation input can be experimentally measured. Several CMP models have been published in the literature [62].

The feature-scale physical model based on contact-mechanics has been implemented in 2D simulator ISE-TCAD [63]. The simulator allows for two-step polishing of multi-line structures with different pads, slurries and polishing parameters. When used within an ISE-TCAD software environment, topography simulation can be coupled to various tools, for example, other process simulators such as etching and deposition simulation modules, and modules for electrical characterisation. The integration with the 3D simulation of barrier and copper deposition has been demonstrated for damascene processes [61]. The feature-scale model takes into account both the roughness and elastic deformation of the polishing pad for computing the pressure distribution.

2.4   Synopsys TCAD Tools

Recently, Synopsys has announced their TCAD Sentaurus tool suite [64]. Sentaurus has been created by combining the features from Synopsys and former ISE-TCAD products, together with a wide range of new features and capabilities, along with other software tools, such as, TSUPREM4, MEDICI, DAVINCI and Raphael. Sentaurus addresses most of the simulation challenges arising from semiconductor technologies. It includes a comprehensive suite of TCAD products for multi-dimensional process, device, and system simulations, embedded into a powerful user interface. Sentaurus TCAD tools provide the users with the capability of automatically adapting the mesh to resolve dopant profiles during implantation and annealing. Sentaurus can handle the complete range of semiconductor technologies, from deep-submicron logic, memory and mixed-signal to smart power, sensors, compound semiconductors optoelectronics and RF.

Sentaurus Process is an advanced multi-dimensional process simulator for developing and optimising silicon and compound semiconductor process technologies. Current trends in the manufacturing of nanoscale devices demand a more fundamental approach towards modelling. It is important to understand new physical phenomena in order to optimise process and enhance device performance. For example a model for anisotropic elasticity allows accurate modelling of the variation of the elastic behaviour of silicon along the crystal directions. The model accounts for wafer orientation and slice angle to compute the crystal directions correctly. Recently semiconductor manufacturers have used strained-Si to enhance transistor performance by adding stress in the channel to enhance mobility and therefore drive current.

Three-dimensional simulation is needed to comprehend fully the impact of new physical phenomena on device performance. Advances in 3D meshing through the MGOALS library within Sentaurus Process, along with advances in Sentaurus Structure Editor, have made 3D simulation much easier to perform. Many innovative features include kinetic Monte Carlo [65], adaptive meshing, laser and flash annealing, Monte Carlo implantation of silicon carbide and enhancements in 3D oxidation.

Sentaurus Device is a multi-dimensional device simulator capable of simulating the complete range of semiconductor devices including nanoscale CMOS, FinFETs, CMOS image sensors, flash memory, SiGe HBTs, power devices, compound semiconductors, analog/RF, light-emitting diodes and semiconductor lasers. In addition, it can be used to analyse electrostatic discharge (ESD) single event upset (SEU) and soft errors.

A clear understanding of stress effects on the transport properties in semiconductor devices is also crucial for developing increasingly smaller and faster transistors. Sentaurus provides users with two tools to help understand stress effects on device performance: the continuum simulator Sentaurus Device (SDevice) and the ensemble Monte Carlo device simulator MOCA. It features both a bulk mode and a 2D mode. The bulk mode assumes a homogeneous geometry and a constant electric field, and it is used to study and to calibrate the stress dependencies of kinetic material parameters. The continuum simulator uses semi-empirical models to calculate the impact of stress on carrier mobility. These models are fast and robust, but they must be calibrated to experimental data. Sentaurus Device incorporates an extensive ensemble of physical models and material parameters to address a broad range of device applications. It offers a comprehensive set of options, allowing flexibility in configuring application-specific solutions and interfaces directly with the Synopsys tools Sentaurus Process, Sentaurus Workbench, Sentaurus Structure Editor and TSUPREM4 to provide a complete state-of-the-art TCAD simulation environment.

Strain engineering in MOSFETs is very important for 65 nm and 45 nm technology node development and beyond. At every step in the process development, the need to relate carrier mobility to stress in the material has increased significantly. Current versions of MEDICI and DAVINCI from Synopsys include a stress-induced hole mobility model reported by Intel [66] for strain-engineered MOSFETs, which will be introduced in Chapter 4. Although the model is specific to stress-induced hole mobility, it can be adapted to describe stress-induced electron mobility (the piezoresistance model or the valley occupancy model) [67]. The latest release of TSUPREM4 provides users with new models for treating anisotropic elasticity in silicon, stress-dependent point defect diffusivity, stress-dependent reaction rates and the segregation of dopant point defect pairs at interfaces.

Interchangability is also an important feature for Synopsys tools. For example, recent file Input/Output enhancements in MEDICI and DAVINCI support automatic TIF file compression and decompression. This feature allows the programs to save and load simulation results in compressed TIF files that can be up to ten times smaller than the decompressed versions. This feature allows MEDICI and DAVINCI to read compressed TIF files created by TSUPREM4 directly. Also, DAVINCI can save device structures in TDR file format that can be read by Sentaurus Device. This allows devices that were originally simulated with DAVINCI to be read directly into Sentaurus Device and can now be simulated using the advanced features of Sentaurus Device.

With semiconductor devices becoming increasingly complex, the simulation of the topographic steps calls for models that capture the influence of process condition. Sentaurus topography, an integral part of the Sentaurus TCAD tool suite, allows users to evaluate and optimise critical topography processing steps in arbitrarily shaped, multi-layer and multi-material semiconductor structures.

2.5   SILVACO: TCAD simulation suite

In this section, a brief overview on the progress that has been made in simulating semi-conductor device/process simulation and fabrication processes using TCAD software from another vendor will be presented. Silvaco TCAD products such as ATHENA for process simulation and ATLAS for device simulation are widely used to develop and optimise the semiconductor processes. IC designers may use the SPICE-like simulator SmartSpice to design and simulate analog circuits. As manufacturing and design depend on extracted device models and accurate circuit simulation, a well-established tool UTMOST for data acquisition and parameter extraction is also available.

Physically based simulation is different from empirical modelling. The goal of empirical modelling is to obtain analytic formulae that approximate existing data with good accuracy and minimum complexity. Physically based simulation is very important for two reasons. First, it is almost always much quicker and cheaper than performing experiments. Second, it provides information that is difficult to measure. Empirical models provide efficient approximation and interpolation. However, they do not provide insight and have predictive capabilities. Both ATHENA and ATLAS are physically based simulators offering predictive capability and insight into the viability of a device or process. Silvaco also provides integrated simulation software for specific technologies, for example, silicon on insulator (SOI). Such software includes technology simulation, SPICE model extraction, interconnect parasitic analysis, SPICE circuit simulation and traditional CAD.

2.5.1   ATHENA

ATHENA is an advanced 2D process simulation tool for silicon and advanced materials technologies. Arbitrary 2D structures composed of different semiconductor materials with profiles of different impurities, diffused, implanted or incorporated into deposited layers can be incorporated. ATHENA combines high-temperature process modelling such as impurity diffusion and oxidation, topography simulation and lithography simulation in a single easy-to-use framework. ATHENA consists of four primary tools and several secondary tools. The primary tools in ATHENA are

- SSUPREM4: for simulating ion implantation, diffusion, oxidation and silicidation processes for silicon;
- Flash: for simulating implantation, activation and diffusion for advanced materials technologies;
- Elite: for 2D topography simulation;
- Optolith: for lithography simulation.

ATHENA provides additional optional modules for the modelling of silicides, Monte Carlo modelling of ion implantation, etching and deposition. Hence it offers a capability for fast and accurate simulation of all critical fabrication steps used in CMOS, bipolar, SiGe, SOI, III-V, optoelectronic and power device technologies.

2.5.2   ATLAS

ATLAS is a physically based device simulator, providing general capabilities for 2D and 3D simulation of semiconductor devices. ATLAS is often used in conjunction with ATHENA using the physical structures and doping profiles generated by ATHENA as input to predict the electrical characteristics associated with specified bias conditions. The combination of ATHENA and ATLAS makes it possible to determine the impact of process parameters on d.c., a.c. and transient device characteristics.

ATLAS can also be used in conjunction with the Virtual Wafer Framework (VWF) Interactive Tools. These include Deckbuild, Tonyplot, DevEdit, MaskView and Optimiser. Input to ATLAS specifies the device simulation problems by defining the physical structure to be simulated, the physical models to be used and the bias conditions for which electrical characteristics are to be simulated.

ATLAS performs self-consistent solution of Poisson equation, current continuity equations, with an option to incorporate energy balance equations and lattice heat equations, handling multiple materials such as Si, Ge, SiGe, GaAs, AlGaAs, SiC, GaN, InP as well as arbitrary user-defined materials. Electrical characteristics of devices for steady-state, transient and small-signal a.c. analysis are performed. Arbitrary doping from analytic functions, tables or process simulation can be incorporated. Automatic regridding, I–V curve tracing and parameter extraction can be used for device performance optimisation and model calibration.

Physical models available in ATLAS include recombination and generation, bandgap narrowing, tunnelling and temperature-dependent impact ionisation. A range of different mobility models with adjustable parameters are dependent on lattice temperature, impurity concentration, carrier concentration, carrier energy, parallel and perpendicular electric fields. Transport models incorporate either Fermi–Dirac or Boltzmann statistics.

All ATLAS simulations require two inputs: (1) a text file that contains commands for ATLAS to execute and (2) a structure file that defines the structure that will be simulated. The structure file may be generated internally in ATLAS, coupled through output from ATHENA or generated using DevEdit, an interactive mesh editor. The ATLAS input file contains data relevant to mesh generation, physical model selection, type of contact and the analysis to be carried out. ATLAS produces three types of output files: a run-time diagnostic, a log file that stores all terminal voltages and currents from the device analysis and a solution file, which stores 2D and 3D data relating to the values of solution variables within the device at a given bias point. Supplementary programs Deckbuild and Tonyplot, respectively, provide an interactive run time environment and offer scientific visualisation of either terminal characteristics or profiles of solution variables or other derived quantities such as electric field, generation rate.

A subsidiary optional module QUANTUM solves the coupled Schrodinger–Poisson equations for the simulation of various effects of quantum confinement of carriers in thin layers. The Schrodinger–Poisson solver allows calculation of bound state energies and associated carrier wave function self-consistently with electrostatic potential. A quantum moment transport model allows simulation of confinement effects on carrier transport once the bound state energies are known for each well.

The electrical characteristics predicted by ATLAS can be used as input by the UTMOST device characterisation and SPICE modelling software. Compact models based on simulated device characteristics can then be supplied to circuit designers for circuit design. Combining ATHENA, ATLAS, UTMOST and SmartSpice makes it possible to predict the impact of process parameters on circuit characteristics. ATLAS can be used as one of the simulators within the VWF automation tools to perform highly automated simulation-based experimentation linking simulation very closely to technology development.

2.6   Process-to-device simulation using SILVACO

TCAD tools present unique opportunities in term of visualisation of processing steps, description of the physical changes and understanding of the interrelation of the process variables. Modelling of processes provides a way to interactively explore the fabrication process and study the effects of process choices.

A full TCAD simulation of a process involves the following steps:

- virtual fabrication of the device using a process simulator;
- device simulation that solves the equations describing the device behaviour;
- post processing that is, generation of figures and plots.

In the following, the use of Silvaco TCAD tools for the simulation of (1) AlGaAs/InGaAs/GaAs HEMT and InGaAs/InP HBT device structures, and (2) advanced strained-Si MOSFET is demonstrated.

2.6.1   Device generation

A typical process usually involves deposition, etching, planarisation and implantation of different species. For example, a pseudomorphic HEMT with an AlGaAs/InGaAs/GaAs layer structure generated using the graphical structure editor DevEdit is shown in Figure 2.1. A recessed gate has been included in the design, as well as several buffer layers and delta doped regions. After a process simulation the final device is defined in terms of a grid, the different materials and their associated properties (AlGaAs, InGaAS, GaAs) and doping profile associated with each material. In this case, doping distributions are defined using analytical functions.

For creating a non-planar HBT structures, DevEdit performs automatic meshing (see Figure 2.2) for the illustrated InGaAs/InP HBT structure. Extra refinement of the grid is needed particularly near interfaces between different materials, at steep doping gradients or where abrupt discontinuities in either conduction or valence band is evident. Mesh editing of a device grid can be obtained manually or automatically, depending on complexity of the structure.

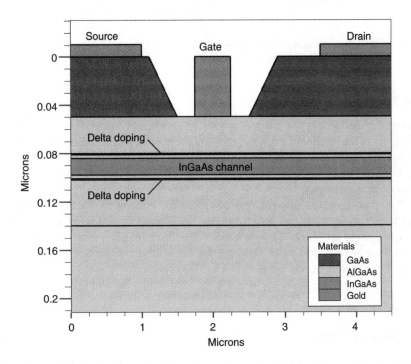

Figure 2.1   A pseudomorphic AlGaAs/InGaAs/GaAs HEMT device structure generated with DevEdit [Source: Silvaco International]

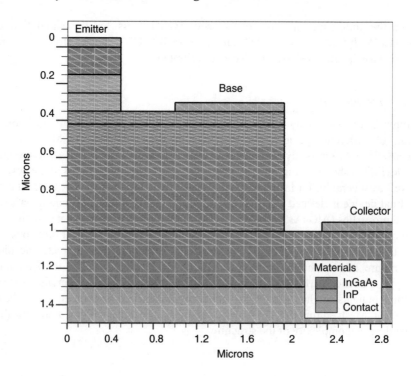

Figure 2.2   Automatic mesh generation with DevEdit for device simulation [Source: Silvaco International]

2.6.2 Device simulation

Device simulation has two main purposes: to understand and to depict the physical processes in the interior of a device, and to make reliable predictions of the behaviour of the next-generation devices. The quality of physical models is vital for understanding of the physical processes in semiconductor devices and for reliable prediction of device characteristics.

The fundamental semiconductor equations are Poisson equation and continuity equations for electrons and holes that may involve additional terms and complexity when using an energy balance model. Basic semiconductor equations must be discretised at each grid point. The current densities are expressed in terms of the quasi-Fermi levels ϕ_n and ϕ_p as

$$\vec{J_n} = -q\mu_n n\nabla\phi_n \tag{2.1}$$

$$\vec{J_p} = q\mu_p p\nabla\phi_p \tag{2.2}$$

where μ_n and μ_p are the electron and hole mobilities. The quasi-Fermi levels are then linked to the carrier concentrations and the potential through the two Boltzmann approximations:

$$n = n_{ie} \exp\left[\frac{q(\psi - \phi_n)}{kT_L}\right] \tag{2.3}$$

$$p = n_{ie} \exp\left[\frac{-q(\psi - \phi_p)}{kT_L}\right] \tag{2.4}$$

where n_{ie} is the effective intrinsic concentration and T_L is the lattice temperature. These two Equations (2.3) and (2.4) may then be rewritten to define the quasi-Fermi potentials:

$$\phi_n = \psi - \frac{kT_L}{q} \ln \frac{n}{n_{ie}} \tag{2.5}$$

$$\phi_p = \psi + \frac{kT_L}{q} \ln \frac{p}{n_{ie}} \tag{2.6}$$

By substituting Equations (2.5) and (2.6) into the current density expressions, the following current relationships are obtained:

$$\overrightarrow{J_n} = qD_n\nabla_n - qn\mu_n\nabla\psi - \mu_n n[kT_L\nabla(\ln n_{ie})] \tag{2.7}$$

$$\overrightarrow{J_p} = -qD_p\nabla_p - qp\mu_p\nabla\psi + \mu_p p[kT_L\nabla(\ln n_{ie})] \tag{2.8}$$

The final term accounts for the gradient in the effective intrinsic carrier concentration, which takes account of bandgap narrowing effects. Effective electric fields are normally defined, whereby

$$\overrightarrow{E_n} = -\nabla\left(\psi + \frac{kT_L}{q} \ln n_{ie}\right) \tag{2.9}$$

$$\overrightarrow{E_p} = -\nabla\left(\psi - \frac{kT_L}{q} \ln n_{ie}\right) \tag{2.10}$$

which then allows the more conventional formulation of drift-diffusion equations to be written [see Equations (2.7) and (2.8)].

$$\overrightarrow{J_n} = qn\mu_n \overrightarrow{E_n} + qD_n\nabla_n \tag{2.11}$$

$$\overrightarrow{J_p} = qp\mu_p \overrightarrow{E_p} - qD_p\nabla_p \tag{2.12}$$

It should be noted that this derivation of the drift-diffusion model has assumed that the Einstein relationship and in the case of Boltzmann statistics this corresponds to

$$D_n = \frac{kT_L}{q}\mu_n \tag{2.13}$$

$$D_p = \frac{kT_L}{q}\mu_p \tag{2.14}$$

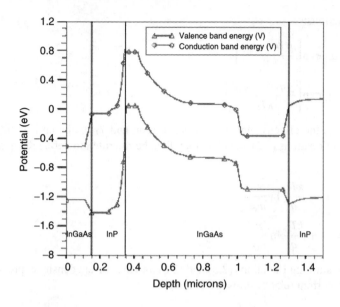

Figure 2.3   Band diagram for the InGaAs/InP HBT structure [Source: Silvaco International]

HBT devices constructed from different semiconductor layers can be solved self-consistently, if the above equations are applied. For the InGaAs/InP HBT structure, the band diagram is shown through the intrinsic region in Figure 2.3. An a.c. analysis of the HBT provides gain vs. frequency plots, s-parameter extraction and can predict the gain roll-off with frequency (see Figure 2.4). Impact ionisation models allow simulation of breakdown voltages and is shown in Figure 2.5. A Gummel plot is shown in Figure 2.6.

The selection of physical models is very important in order to get realistic results from the device simulator. This includes mobility models, carrier transport mechanism, velocity saturation, carrier generation and recombination, as well as quantum mechanical correction. The simplest device simulation is a d.c. simulation to predict I-V characteristics in a steady-state condition. But if a small sinusoidal signal is superimposed on the d.c. solution, the device can be treated as a two-port network where the complex small-signal admittance matrix can be calculated. By decomposing this matrix into a real part, conductance matrix and an imaginary part, capacitance matrix, the inter-electrode conductances and capacitances can be found. If the y-matrix is known, the other a.c. parameters such as s- and z-parameters can be calculated [68]. ATLAS can generate s-parameter data.

To study the non-quasistatic phenomena based on a steady-state solution a voltage or current increment can be applied to any of the device terminals. The simulator then calculates the transient response at any of the other terminals. This approach is relevant for simulation of device when switching state. Other types of simulations are also

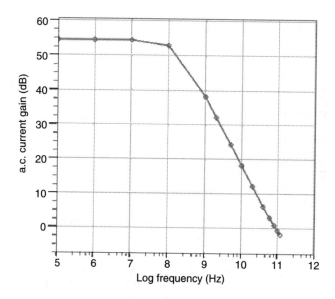

Figure 2.4   Alternating current gain roll-off with frequency for InGaAs/InP HBT structure [Source: Silvaco International]

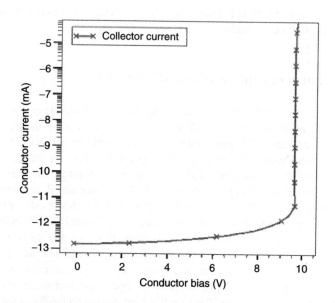

Figure 2.5   Breakdown voltage simulation in InGaAs/InP HBT structures [Source: Silvaco International]

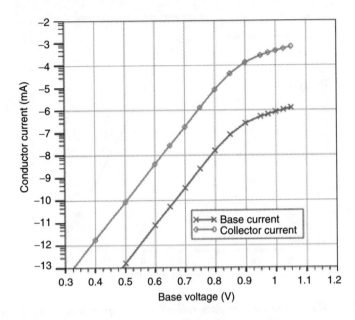

Figure 2.6   Gummel plots for InGaAs/InP HBTs [Source: Silvaco International]

possible for example in mixed-mode simulations, several devices can be connected in a circuit together with elements that have a compact model description. This approach is very useful, where a SPICE model of a novel device is not available or when a series resistance has to be included in an a.c. simulation.

2.7   Simulation example: strained-Si MOSFET

As another example of process-to-device simulation, in the following, the simulation for strained-Si n-MOSFETs using Silvaco simulation tools is discussed. Advanced MOSFETs have additional device features to enhance performance like HALO doping to suppress short channel effects, threshold implants to adjust the threshold voltage, anti-punch through implantations that locally increase doping concentration under the channel, source and drain extensions, silicided source and drain, and so forth. ATHENA was used for the process simulation of strained-Si n-MOSFET with a standard polysilicon gate down to 45 nm gate length. Typical Si CMOS fabrication process steps have been used in simulation. Implantation with As^+ and BF_2^+ created deep and shallow S/D regions and the p-well, respectively. The device structure obtained from the ATHENA process simulation with these features is illustrated in Figure 2.7, where both lightly doped and heavily doped regions of source and drain are shown. The channel region (top 10 nm) is the most critical region for heterostructure MOSFETs and needs to have a fine grid. Figure 2.8 shows the simulated doping profile.

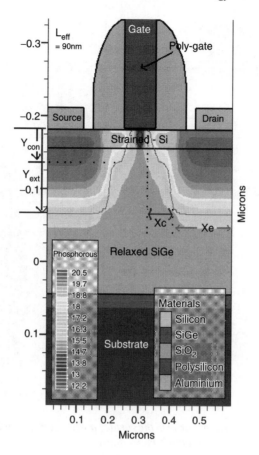

Figure 2.7   Simulated device structure obtained from process simulation for a strained-Si n-MOSFET with L_{eff} of 90 nm and oxide thickness 1.8 nm. This device was created using ATHENA process simulator [After A. R. Saha and C. K. Maiti, unpublished data]

To obtain realistic device simulation, an improved transport model is required. For relatively long channel devices, it is sufficient to use the basic drift-diffusion equations, but if the gate length is below 60 nm, an energy balance model is often recommended, since the standard drift-diffusion approach cannot reproduce velocity overshoot and often overestimates the impact ionisation rates. The energy balance model is a very good compromise and gives improved prediction of I-V characteristics, as shown in Figure 2.9. In the preceding device simulation the following physical effects were modelled:

- For the bandgap narrowing, the Slotboom model was used. Impact ionisation and the effects of polySi depletion due to the gate voltage were considered.

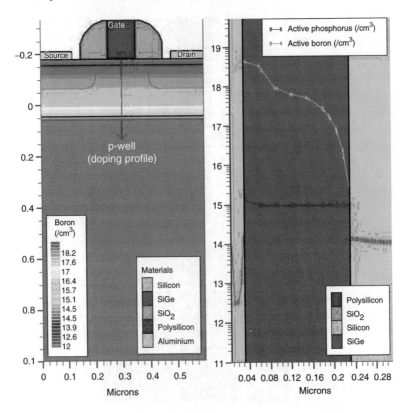

Figure 2.8   Depth profile of boron in the surface and buried channel [After A. R. Saha and C. K. Maiti, unpublished data]

- Doping-dependent mobility model was used with the following extensions: mobility degradation at interfaces, high field saturation (velocity saturation), that is, carrier drift velocity is no longer proportional to the electric field.
- Shockley–Read–Hall (SRH) model for recombination through deep levels in the band and model for trap-assisted tunnelling were used.

2.8   Summary

In this chapter, the modelling and simulation of unit processes such as ion implantation, diffusion, oxidation, lithography, etching, plasma processing, chemical mechanical polishing and contact metallisation have been addressed. With regard to plasma processing, only a small subset of commercial plasma reactors and processes can be modelled with reasonable accuracy. Very little fundamental plasma chemistry data is available for the new etching and deposition chemistries that are being explored in the semiconductor industry. It is important to have improved predictive

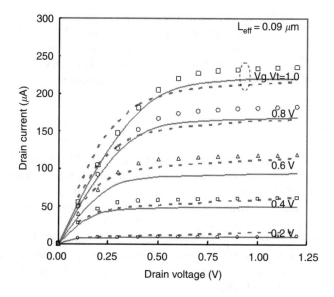

Figure 2.9 Calibrated I_d–V_d characteristics showing the usefulness of NEB model. Symbols – DD Simulation; solid lines – NEB Simulation; dashed lines – experimental (RimVLSI01) [After A. R. Saha and C. K. Maiti, unpublished data]

simulation capability for plasma processes and for establishing the larger need for simulation and modelling tools that can address process integration.

Some of the capabilities of state-of-the-art simulators used in characterising and predicting semiconductor process and device-related phenomena have been presented. An attempt will be made to outline the degree of sophistication of the physics incorporated in process/device simulation programs. The setup of the TCAD simulator is highly complicated and time-consuming. The main concept that should be considered is to match the simulation methodology as close as possible to the fabrication technology. A brief outline of some of the advances that have been made and some concerns that remain in simulator development is presented.

While using a simulation tool, the following points need to be remembered: (1) the simulator needs to replicate the integral physics, (2) default model parameters are generally used, unless more accurate values have been identified by the user, (3) validity of models requires experimental confirmation, (4) generation of appropriate mesh is important for consistency of output and (5) there is always a trade-off between accuracy and speed.

Use of TCAD tools to understand various design parameters on device performance has been described. Two different commercial tools from Synopsis and Silvaco have been introduced. As application examples, AlGaAs/InGaAs/GaAs HEMT, InGaAs/InP HBT device structures and strained-Si n-MOSFETs with 90-nm gate length have been simulated using Silvaco tools. It is important to note that the accurate TCAD simulations and modelling of physical devices depend significantly on calibrated physical models and appropriate input data. The interface between device characterisation and device simulation for the generation of reliable device models for circuit simulation is still not fully developed. This can limit the predictive power of the combined process and device simulation approach.

References

1 S. H. Goowin-Johansson, R. Subrahmanyan, C. E. Floyd, and H. Z. Massoud, 'Two-dimensional impurity profiling with emission computed tomography techniques', *IEEE Trans. Computer-Aided Design*, vol. 8, pp. 332–335, 1989.

2 B. E. Deal, and A. S. Grove, 'General relationship for the thermal oxidation of silicon', *J. Appl. Phys.*, vol. 36, pp. 3770–3778, 1965.

3 M. L. Green, E. P. Gusev, R. Degraeve, and E. L. Garfunkel, 'Ultrathin (<4 nm) SiO_2 and Si-O-N gate dielectric layers for silicon microelectronics: Understanding the processing, structure, and physical and electrical limits', *J. Appl. Phys.*, vol. 90, pp. 2057–2121, 2001.

4 J. F. Gibbons, 'Ion implantation in semiconductors – Part I: Range distribution theory and experiments', *Proc. IEEE*, vol. 56, pp. 295–313, 1968.

5 L. A. Christel, J. F. Gibbons, and S. Mylroie, 'An application of the Boltzmann transport equation to ion range and damage distribution in multilayered targets', *J. Appl. Phys.*, vol. 51, pp. 6176–6182, 1980.

6 M. T. Robinson, and O. S. Oen, 'Computer studies of the slowing down of energetic atoms in crystals', *Phys. Rev.*, vol. 132, pp. 2385–2398, 1963.

7 G. Hobler, and S. Selberherr, 'Monte Carlo simulation of ion implantation into two- and three-dimensional structures', *IEEE Trans. Computer-Aided Design of Integrated Circuits and Systems*, vol. 8, pp. 450–459, 1989.

8 J. F. Ziegler, J. P. Biersack, and U. Littmark, *The Stopping and Range of Ions in Solids*, vol. 1. Pergamon Press, New York, 1984.

9 B. Obradovic, G. Wang, Y. Chen, D. Li, C. Snell, and A. F. Tasch, 'UT-MARLOWE 5.0 with TOMCAT', The University of Texas at Austin, Austin, Texas, 1999.

10 S. M. Hu, 'Formation of stacking faults and enhanced diffusion in the oxidation of silicon', *J. Appl. Phys.*, vol. 45, pp. 1567–1573, 1974.

11 D. A. Antoniadis, and I. Moskowitz, 'Diffusion of substitutional impurities in silicon at short oxidation times: An insight into point defect kinetics', *J. Appl. Phys.*, vol. 53, pp. 6788–6796, 1982.

12 S. Mizuo, and H. Higuchi, 'Effect of back-side oxidation on B and P diffusion in Si directly masked with S_3N_4 films', *J. Electrochem. Soc.: Solid-State Sci. and Technol.*, vol. 129, pp. 2292–2295, 1982.

13 K. Maex, and L. Van Den Hove, 'The effect of silicides on the induction and removal of defects in silicon', *Mat. Sci. Engg. B*, vol. 4, pp. 321–329, 1989.

14 F. F. Morehead, and R. F. Lever, 'Enhanced tail diffusion of phosphorus and boron in silicon', *Appl. Phys. Lett.*, vol. 48, pp. 151–153, 1986.

15 R. B. Fair, 'Shallow junctions – Modeling the dominance of point defect charge states during transient diffusion', in *IEEE IEDM Tech. Dig.*, pp. 691–694, 1984.

16 S. M. Hu, 'Vacancies and self-interstitials in silicon: Generation and interaction in diffusion', *J. Electrochem. Soc.: Solid-State Sci. and Technol.*, vol. 139, pp. 2066–2075, 1992.

17 E. Guerrero, H. Potzl, R. Tielert, M. Grasserbauer, and G. Stingeder, 'Generalized model for the clustering of arsenic dopants in silicon', *J. Electrochem. Soc.: Solid-State Sci. and Technol.*, vol. 129, pp. 1826–1831, 1982.

18 Y. Tamaki, T. Shiba, T. Kure, K. Ohyu and T. Nakamura, 'Advanced process device technology for 0.3 μm high-performance bipolar LSIs', *IEEE Trans. Electron Devices*, vol. 39, pp. 1387–1391, 1992.

19 E. W. Scheckler and A. R. Neureuther, 'Models and algorithms for 3-D topography simulation with SAMPLE-3D', *IEEE Trans. Computer-Aided Design*, vol. 13, pp. 219–230, 1994.

20 M. Fujinaga, and N. Kotani, '3-D topography simulator (3-D MULLS) based on a physical description of material topography', *IEEE Trans. Electron Devices*, vol. 44, pp. 226–238, 1997.

21 F. H. Dill, A. R. Neureuther, T. A. Tuttle, and E. J. Walker, 'Modeling projection printing of positive photoresist', *IEEE Trans. Electron Devices*, vol. ED-22, pp. 456–464, 1975.

22 V. K. Malhotra, and F.-C. Chang, 'Verifying the correctness of your optical proximity correction designs', in *Proc. SPIE Conf.*, vol. 3679, pp. 130–137, 1999.

23 W. Maurer, C. Dolainsky, J. Thiele, C. Friedrich, and P. Karakatsanis, 'Process proximity correction using an automated software tool', in *Proc. SPIE Conf.*, vol. 3679, pp. 245–253, 1999.

24 K. D. Lucas, M. McCallum, and B. Falch, 'Design, reticle and wafer OPC manufacturability for the 0.18 μm lithography generation', in *Proc. SPIE Conf.*, vol. 3679, pp. 118–125, 1999.

25 H. W. Lehmann, L. Krausbauer, and R. Widmer, 'Redeposition – A serious problem in RF sputter etching of structures with micrometer dimensions', *J. Vac. Sci. Technol.*, vol. 14, pp. 281–284, 1977.

26 N. S. Viswanathan, 'Simulation of plasma etched lithographic structures', *J. Vac. Sci. Technol.*, vol. 16, pp. 388–390, 1979.

27 J. L. Reynolds and A. R. Neureuther, 'Simulation of dry etched line edge profiles', *J. Vac. Sci. Technol.*, vol. 16, pp. 1772–1775, 1979.

28 L. Mei, S. Chen, and R. W. Dutton, 'A surface kinetics model for plasma etching', in *IEEE IEDM Tech. Dig.*, pp. 831–832, 1980.

29 A. R. Neureuther, 'Simulating VLSI wafer topography', in *IEEE IEDM Tech. Dig.*, pp. 214–218, 1980.

30 I. A. Blech, 'Evaporated film profiles over steps in substrates', *Thin Solid Films*, vol. 6, pp. 113–118, 1970.

31 I. A. Blech, D. B. Fraser, and S. EoHasyko, 'Optimization of Al step coverage through computer simulation and scanning electron microscopy', *J. Vac. Sci. Technol.*, vol. 15, pp. 13–19, 1978.

32 A. R. Neureuther, C. H. Ting, and C. Y. Liu, 'Application of line edge pro-
    file simulation to thin-film deposition process', *IEEE Trans. Electron Devices*,
    vol. ED-27, pp. 1449–1455, 1980.
33 Y. Homma, A. Yajima, and S. Harada, 'Feature size limit analysis of lift-off
    metalization technology', in *IEEE IEDM Tech. Dig.*, pp. 570–573, 1981.
34 R. J. Hawryluk, 'Exposure and development models used in electron beam
    lithography', *J. Vac. Sci. Technol.*, vol. 19, pp. 1–17, 1981.
35 K. Murata, D. F. Kyser, and C. H. Ting, 'Monte Carlo simulation of fast sec-
    ondary electron production in electron beam resists', *J. Appl. Phys.*, vol. 52,
    pp. 4396–4405, 1981.
36 J. S. Greeneich, 'Impact of electron scattering on linewidth control in electron-
    beam lithography', *J. Vac. Sci. Technol.*, vol. 16, pp. 1749–1753, 1979.
37 K. Murata, E. Nomura, and K. Nogami, 'Experimental and theoretical study of
    cross-sectional profiles of resist patterns in electron-beam lithography', *J. Vac.
    Sci. Technol.*, vol. 16, pp. 1734–1736, 1979.
38 A. R. Neureuther, D. F. Kyser, and C. H. Ting, 'Electron beam resist edge
    profile simulation', *IEEE Trans. Electron Devices*, vol. ED-26, pp. 686–692,
    1979.
39 D. F. Kyser, and R. Pyle, 'Computer simulation of elctron beam resist profiles',
    IBM J. Res. and Develop., vol. 24, pp. 426–437, 1980.
40 M. G. Rosenfield, A. R. Neureuther, and C. H. Ting, 'The use of bias in electron-
    beam lithography for improved linewidth control', *J. Vac. Sci. Technol.*, vol. 19,
    pp. 1242–1247, 1981.
41 F. Jones, and J. Paraszczak, 'RD3D (Computer simulation of resist development
    in three-dimensions)', *IEEE Trans. Electron Devices*, vol. ED-28, pp. 1544–1552,
    1981.
42 A. R. Neureuther, 'Simulation of X-ray resist line edge profiles', *J. Vac. Sci.
    Technol.*, vol. 15, pp. 1004–1008, 1978.
43 K. Heinrich, H. Betz, A. Heuberger, and S. Pongraz, 'Computer simulations of
    resist profiles in x-ray lithography', *J. Vac. Sci. Technol.*, vol. 19, pp. 1254–1258,
    1981.
44 L. Karapiperis, I. Adesida, C. A. Lee, and E. D. Wolf, 'Ion beam exposure profiles
    in PMMA-Computer simulation', *J. Vac. Sci. Technol.*, vol. 19, pp. 1259–1263,
    1981.
45 W. G. Oldham, S. N. Nandgaonkar, A. R. Neureuther, and M. M. O'Toole, 'A gen-
    eral simulator for VLSI lithography and etching processes: Part I - Application to
    projection lithography', *IEEE Trans. Electron Devices*, vol. ED-26, pp. 717–722,
    1979.
46 W. Oldham, A. Neureuther, C. Sung, J. Reynolds, and S. Nandgaonkar, 'A general
    simulator for VLSI lithography and etching processes: Part II – Applica-
    tion to deposition and etching', *IEEE Trans. Electron Devices*, vol. ED-27,
    pp. 1455–1459, 1980.
47 J. McVittie, J. Rey, L.Y. Cheng, A. Bariya, S. Ravi and K. Saraswat, 'SPEEDIE:
    A profile simulator for etching and deposition', in *Proc. TECHNOCON*,
    pp. 16–19, 1990.

48 J. Lorenz, J. Pelka, H. Kyssel, A. Sachs, A. Seidel, and M. Svoboda, 'COM-POSITE – A complete modeling program of silicon technology', *IEEE Trans. Computer-Aided Design*, vol. 4, pp. 421–430, 1985.

49 R. E. Jewett, P. I. Hagouel, A. R. Neureuther, and T. Van Duzer, 'Line-profile resist development simulation techniques', *Polymer Eng. and Sci.*, vol. 17, pp. 381–384, 1977.

50 P. L. G. Ventzek, T. J. Sommerer, R. J. Hoekstra, and M. J. Kushner, 'Two-dimensional hybrid model of inductively coupled plasma sources for etching', *Appl. Phys. Lett.*, vol. 63, pp. 605–607, 1993.

51 R. J. Hoekstra, and M. J. Kushner, 'Predictions of ion energy distributions and radical fluxes in radio frequency biased inductively coupled plasma etching reactors', *J. Appl. Phys.*, vol. 79, pp. 2275–2286, 1996.

52 R. J. Hoekstra, and M. J. Kushner, 'Integrated plasma equipment model for polysilicon etch profiles in an inductively coupled plasma reactor with subwafer and superwafer topography', *J. Vac. Sci. Technol. A*, vol. 15, pp. 1913–1921, 1997.

53 L. J. Friedrich, S. K. Dew, M. Brett, and T. Smy, 'Thin film microstructure modelling through line-segment simulation', *Thin Solid Films*, vol. 266, pp. 83–88, 1995.

54 M. A. Leiberman, and A. J. Lichtenberg, *Principles of Plasma Discharges and Materials Processing*. John Wiley and Sons, New York, 1994.

55 H. F. Winters, and J. W. Cohurn, 'Surfacc science aspects of etching reactions', *Surf. Sci. Rep.*, vol. 14, pp. 161–269, 1992.

56 J. H. Keller, J. C. Forster, and M. S. Barnes, 'Novel radio-frequency induction plasma processing technique', *J. Vac. Sci. Technol. A*, vol. 11, pp. 2487–2491, 1993.

57 J. B. Carter, J. P. Holland, E. Peltzer, B. Richardson, E. Bogle, H. T. Nguyen, *et al.*, 'Transformer coupled plasma etch technology for the fabrication of subhalf micron structures', *J. Vac. Sci. Technol. A*, vol. 11, pp. 1301–1306, 1993.

58 D. B. Graves, 'Plasma processing in microelectronics manufacturing', *AIChE J.*, vol. 35, pp. 1–29, 1989.

59 D. B. Graves, and K. F. Jensen, 'A continuum model of d.c. and RF discharges', *IEEE Trans. Plasma Sci.*, vol. PS-14, pp. 78–91, 1986.

60 E. Aydil, and D. Economou, 'Theoretical and experimental investigations of chlorine RF glow discharges. I. Theoretical', *J. Electrochem. Soc.*, vol. 139, pp. 1396–1406, 1992.

61 P.-H. Nguyena, E. Bar, J. Lorenz, and H. Ryssel, 'Modeling of chemical-mechanical polishing on patterned wafers as part of integrated topography process simulation', *Microelectr. Eng.*, vol. 76, pp. 89–94, 2004.

62 E. Bar, J. Lorenz, and H. Ryssel, 'Simulation of the influence of via sidewall tapering on step coverage of sputter-deposited barrier layers', *Microelectr. Eng.*, vol. 64, pp. 321–328, 2002.

63 ISE Integrated Systems Engineering AG., Zurich, Switzerland, *ISE-TCAD Manual*, 2004.

64 Synopsys, 'TCAD News', October 2005.

65 N. Strecker, V. Moroz, and M. Jaraiz, 'Introducing Monte Carlo diffusion Simulation into TCAD tools', in *Proc. Intl. Conf. Modeling and Simulation of Microsystems*, pp. 462–465, 2002.

66 B. Obradovic, P. Matagne, L. Shifren, X. Wang, M. Stettler, J. He, and M. D. Giles, 'A physically-based analytic model for stress-induced hole mobility enhancement', in *IWCE-10 Tech. Dig.*, pp. 26–27, 2004.

67 Synopsys, 'TCAD News', June 2006.

68 S. E. Laux, 'Techniques for small-signal analysis of semiconductor devices', *IEEE Trans. Electron Devices*, vol. 32, pp. 2028–2037, 1985.

Chapter 3
Diffusion and oxidation of SiGe/SiGeC films

The scaling of silicon-based semiconductor devices continues to drive increased performance and reduced cost for many applications including wireless communications. Strained-SiGe/Si heterostructures and superlattices are essential for many advanced Si-based devices. In many ways, the ability to fabricate smaller devices controls the scaling process. While process modelling of silicon devices is well advanced, modelling of SiGe technology is still in its formative stage and many issues, specifically associated with this new material that affect passivation, oxidation, dopant diffusion and thermal stability, still need to be resolved. In particular, the ability to predict exact amounts of dopant diffusion is critical.

Diffusion of boron in SiGe is of great interest in view of obtaining boron-doped thin strained-SiGe layers resulting in a very good base layer in SiGe HBTs. One critical area is that of dopant profiles – in controlling the effect basewidth of HBTs that is, the ability to place a specific amount of impurity in one region of a device, and keep it there during the entire fabrication process.

In order to understand how to model dopant diffusion in any semiconductor, it is first important to understand the physical mechanisms governing dopant motion, which are dependent upon both the properties of the dopant and of the semiconductor. Dopants are known to diffuse through single crystal semiconductors via mediation with point defects. For a substitutional dopant to become mobile and diffuse through an undamaged silicon crystal, it must interact with a defect, where both defect and dopant may be charged. The activation energy for Frenkel pair generation in silicon is quite high; consequently, this is not believed to be a major source of point defects in silicon. However, for other semiconductors, such as GaAs, Frenkel pair generation can be a major source of point defects.

3.1 Boron out-diffusion

Low Ge-fraction silicon germanium (SiGe) has shown promise for providing faster, high gain transistors through bandgap engineering in a Si-based material system.

In a bipolar junction transistor, a small amount of boron diffusion will have only a very small effect on the transistor performance. However, for a p-type SiGe base in a heterojunction bipolar transistor, if the boron diffuses even slightly out of the SiGe base, the HBT performance is severely degraded [1]. The change of energy band influences the device current gain and frequency character greatly. When the base dopant out-diffusion length was 5 nm, current gain and f_T of a SiGe HBT dropped by 78 per cent and 58 per cent, respectively. At the same time, the basewidth and the height of parasitic conduction band barriers increased, which resulted in a decrease in the cut-off frequency. Extension of base dopant beyond the SiGe region occurs during thermal cycling, or improper control of the as-deposited profile [2]. Even small amounts of boron out-diffusion from a heavily-doped SiGe base into the Si emitter and collector cause parasitic barriers in the conduction band, which can drastically reduce the collector current enhancement [3–5]. This diffusion problem is resolved by doping only part of the SiGe layer and leaving the rest undoped, then assuming that the boron will diffuse during processing.

The roles of the out-diffusion of the boron from the $Si_{1-x}Ge_x$ base and the lifetime in the base have been investigated by Shafi *et al.* [6] using a two-dimensional drift-diffusion device simulator. It is shown that the increased base currents in the HBTs are caused by recombination in the neutral base, and that the lifetime in the $Si_{1-x}Ge_x$ is an important parameter in determining the base current. The device simulations were carried out using the HQUPETS simulator, which carries out a full drift diffusion analysis in two dimensions for both silicon bipolar transistors and Si/SiGe HBTs [7].

Figure 3.1 shows the effect of boron out-diffusion from the $Si_{1-x}Ge_x$ base on the collector current of the HBT. Results are shown for $Si_{1-x}Ge_x$ HBTs with two different base doping concentrations of 7×10^{19} cm^{-3} and 7×10^{18} cm^{-3}. Both devices show a decrease in the collector current with out-diffusion but the curve for the heavily doped base decreases much more rapidly than that for the lightly doped base. This result indicates that control of the out-diffusion of boron from the base is much more critical in devices with heavily doped bases.

The boron out-diffusion from the base can also affect the base current, and in particular how the base current varies with collector/base voltage. This effect is at maximum when the amount of boron out-diffusion is small, since the potential barrier at the collector/base junction can be most easily modulated by the collector/base bias in these circumstances. This is illustrated in Figure 3.2(a), which shows band diagrams for devices with 12 Å of out-diffusion and different values of collector/base voltage. The potential barrier at the collector/base junction decreases significantly as the collector/base reverse bias increases from 0 to 10 V. This leads to a significant decrease in the excess minority carrier concentration in the base, as illustrated in Figure 3.2(b). Since the recombination rate is dependent on the excess minority carrier concentration in the base, reducing the electron concentration also reduces the recombination, and hence the base current due to neutral base recombination. Since secondary ion mass spectrometry (SIMS) is unable to resolve very small amounts of out-diffusion, the best way of obtaining the out-diffusion length is to simulate the collector current. The value of out-diffusion can then be used, along with the

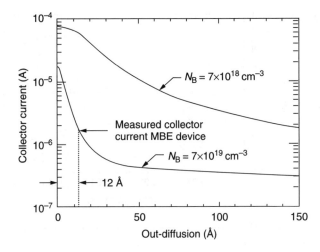

Figure 3.1   *Simulated collector current vs. boron out-diffusion for Si/Si$_{0.85}$Ge$_{0.15}$ HBTs with base doping concentrations of $N_A = 7 \times 10^{19}$ cm^{-3} and $N_B = 7 \times 10^{18}$ cm^{-3}. The bias conditions used were $V_{be} = 0.55\,V$ and $v_{cb} = 0\,V$ [After Z. A. Shafi et al., J. Appl. Phys., Vol. 78, 1995(2823–2829)]*

measured SIMS emitter profile, to model both the magnitude of the base current and its variation with collector/base bias.

For SiGe device simulation, knowledge of doping and Ge profile in the device is essential. The incorporation of high boron levels in the base region along with the Ge-induced effects makes prediction of boron diffusion through the SiGe layer difficult. Owing to the need to align doping and alloy (Ge) profiles in SiGe structures, it is important to understand dopant diffusion in SiGe so that the effect of processing can be properly simulated and final dopant profiles in the device are as desired. Published experimental data has shown that boron diffusion in SiGe differs from that in bulk silicon. Specifically, experimental evidence indicates that the boron diffusivity in SiGe is dependent on the concentration of germanium with boron diffusivity decreasing as the germanium content increases. If the boron diffuses less than expected and a significant intrinsic region exists in the base, the frequency performance of the device will be affected. Thus it is a critical need to simulate the amount of dopant diffusion that will occur during processing, so that an optimised device can be easily fabricated.

3.2   Diffusion: B in strained-SiGe

Simulation of diffusion of boron in SiGe is of great interest in view of obtaining boron-doped thin strained-SiGe layers resulting in a very good base layer. It has been found that B diffusivity in SiGe alloy can be reduced by an order of magnitude lower than that in Si [8]. Boron diffuses via an interstitial mediated mechanism. Moriya *et al.* [9] have shown that the degree of B diffusion reduction increases with Ge fraction in the

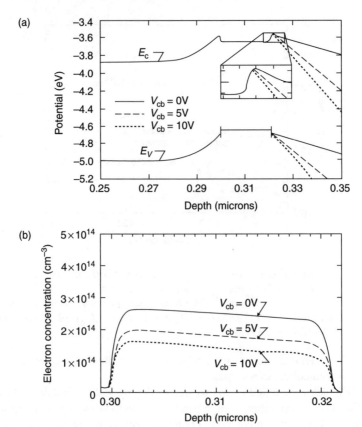

Figure 3.2   Effect of the collector/base bias on the band structure and carrier concentration in a Si/Si$_{1-x}$Ge$_x$ HBT. The out-diffusion was 12 Å, the emitter/base voltage, V_{be} = 0.65 V, and the lifetime in the base was τ_{nb} = 0.11 ns. (a) Simulated energy band diagram, (b) Simulated electron concentration in the base [After Z. A. Shafi et al., J. Appl. Phys., Vol. 78, 1995(2823–2829)]

strained-SiGe layers and explained that this was due to the change in the charged point defect concentration caused by bandgap narrowing. This explanation is based on the assumption that B diffusion is mediated by the positively charged point defects.

Another plausible explanation is the strain-dictated distribution of point defect population [10]. This explanation gets support from total energy calculation in which the authors computed the energies of formation for self-interstitials and vacancies in Si as a function of hydrostatic pressure. A linear increase in interstitial formation energy and decrease in vacancy formation energy were found with increasing pressure, corresponding to an outward relaxation of the lattice around the interstitial, and an inward relaxation of the vacancy [11]. Strain-induced changes in point defect formation energy has been experimentally investigated [10].

However, most Boron diffusion studies have been performed with strained-SiGe layers having uniform Ge content [12–14]. But the grading of the Ge across the quasi-neutral base induces a built-in electric field in the base, which accelerates electrons injected from the emitter to the collector, thereby reducing the base transit time. Rajendran and Schoenmaker [15] have studied the boron diffusion in strained-SiGe layers in which Ge is linearly graded from the e-b (emitter-base) to the c-b (collector-base) junction. However, the high doping concentration and the grading of the Ge composition in the base layer makes it difficult to predict the out-diffusion of B from the $Si_{1-x}Ge_x$ layer. Several workers [8, 10, 16–19] have investigated the reasons for the decrease in the macroscopic boron diffusivity in SiGe by examining changes in the various parameters that might affect it. It has been shown that boron in SiGe diffuses primarily via an interstitial mediated mechanism. Consequently, only parameters directly related to interstitial and mobile boron diffusion and the formation of mobile boron from substitutional boron are necessary to accurately simulate boron diffusion in SiGe.

On the basis of the diffusion studies of boron in compressively strained-$Si_{1-x}Ge_x$ epitaxial layers with graded Ge profiles grown by rapid pressure chemical vapour deposition during furnace and rapid thermal annealing, Rajendran *et al.* [17], have presented a simple and accurate empirical diffusion model that fits for various Ge concentrations, and thermal budget has been developed and implemented in a TAURUS-PMEI (physical model and equation interface). The diffusion model simulates the measured profiles over the range of Ge concentrations, and box and graded Ge profiles were considered.

The diffusion model used is essentially a modified model of Lever *et al.* [20], with the inclusion of the strain dependence of the activation energy of the diffusion coefficient [10]. Also the model for intrinsic carrier concentration of SiGe, drift field due to band-gap narrowing, modified strain model that includes the strain due to the fractions of Ge concentration and also the boron concentration are included. The boron flux, J_B, is given by

$$J_B = -D_B \zeta \left(\frac{\partial C_B}{\partial x} + \frac{C_B}{Q} \frac{\partial N_T}{\partial x} + \frac{2C_B n}{Q} \frac{\partial \ln n_i}{\partial x} \right) \tag{3.1}$$

where $Q = \sqrt{(N_T^2 + 4n_i^2)} = n + p$, $\zeta = (1 + \beta(p/n_i))/(1 + \beta)$, n and p are electron and hole concentrations, N_T, is the net doping, and n_i is the intrinsic carrier concentration. C_B is the boron concentration, ζ models the concentration dependence of the effective diffusion coefficient and D_B is the Boron diffusivity. The first term in the brackets is the conventional diffusion term (Fick's law) for boron diffusion in silicon; the second term describes the effect of the electric field caused by the dopant distribution. The third term is the contribution from the electric field created by the $Si_{1-x}Ge_x$ bandgap variation with the Ge fraction.

In their study, the SiGe layers have been grown in an ASM Epsilon 2000 RPCVD single wafer reactor. The Si-cap/$Si_{1-x}Ge_x$-layer/Si-spacer stack was grown epitaxially on p-Si (100) substrates for five different Ge fractions as well as graded Ge

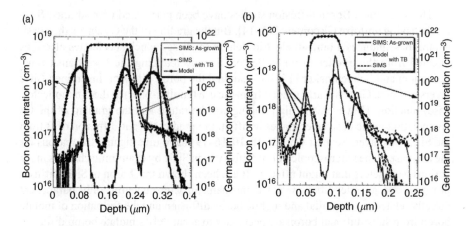

Figure 3.3   *Comparison of simulations to the SIMS profile of samples (a) with 10 per cent Ge, 7×10^{18} cm^{-3} for $850°C$ at 20 min and (b) with 15 per cent Ge, 2×10^{19} cm^{-3} for $900°C$ at 4 min [After K. Rajendran et al., IEEE Trans Electron Dev., Vol. 48, 2001(2022–2031). © 2001 IEEE]*

profiles. The samples were grown with target box B profiles with doping concentrations of 5×10^{18} cm^{-3}. Samples with boron box profiles in the Si-capping layer, $Si_{1-x}Ge_x$ and Si spacer have also been studied.

Figure 3.3 shows the results of simulation performed in TAURUS [21] through physical model and equation interface (PMEI), and measured profiles for the box-type Ge with fractions and thermal budgets of (1) 10 per cent, 850 °C, 20 min and (2) 15 per cent, 900 °C, 4 min. It appears that the calculated and SIMS diffused profiles fit very well and the diffusivity of B decreases with the increase of Ge content. This shows that the reduction of the B concentration in the underlying Si substrate of $Si_{1-x}Ge_x$/Si is caused by the reduction of the effective B diffusivity in the $Si_{1-x}Ge_x$/Si interface region. Figure 3.3 also shows a comparative analysis of the present simulation results with SIMS for various thermal budgets and Ge content. The SIMS of the B initial profiles are also given in Figure 3.3 and shows the different B peaks studied.

Figure 3.4 shows the B diffusivity as a function of reciprocal temperature for different Ge contents and also shows a B diffusivity in a depth scale. Graded Ge profiles show a lower value compared to Box-type Ge profiles. Figure 3.4 shows the simulated and measured profiles of graded-Ge (15 per cent) for (a) rapid thermal annealing (RTA) at 1050 °C for 10 s and (b) furnace annealing (FA) and RTA at 850 °C for 20 min and 1000 °C for 10 s. It is interesting to note that the simulated and SIMS diffused profiles fit very well and the diffusivity of B increases with the increase of annealing temperature. In the case of lowering annealing time, the B concentration in the $Si_{1-x}Ge_x$ layer becomes higher than the B concentration in Si. This phenomena is due to the decrease of B atoms that have redistributed from $Si_{1-x}Ge_x$ layer to the Si substrate. For some of the thermal budgets the diffusion of B in strained-$Si_{1-x}Ge_x$

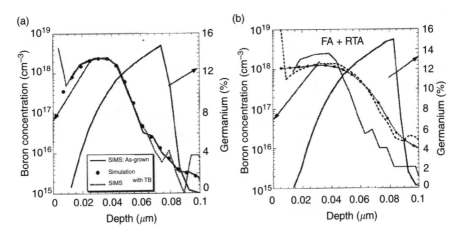

Figure 3.4 Comparison of simulation with SIMS diffusion profile with Ge: 15 per cent (Ge-graded profile), 5×10^{18} cm^{-3}. (a) Rapid thermal annnealing (RTA): (1050°C, 10s); (b) Furnace annealing (FA) and RTA: (850°C, 20 min and 1000°C, 10s) [After K. Rajendran et al., IEEE Trans Electron Dev., Vol. 48, 2001(2022–2031). © 2001 IEEE]

resulted in diffusion into the relaxed-Si layers on both sides of the diffused profiles as given in the above figures.

Figure 3.5 shows the B diffusivity as a function of strain, Ge fraction and temperature. B diffusivity decreases rapidly with the increase of strain as well as Ge fraction. By comparing with experimental values, the extracted (by using experiment and simulation) B diffusivity predicted a lower value (retardation). Figure 3.5(b) shows the extracted boron diffusivity as an function of depth. Here, the As-grown SIMS profile was used as an initial profile, before annealing at Ge fraction of 15 per cent. It appears that the two tails on both sides of each plot are the results of boron diffusivity in the Si region and the flat regions are in $Si_{1-x}Ge_x$ epitaxial layers. It is interesting to note that boron diffusivity has a comparatively lower value in the $Si_{1-x}Ge_x$ layer than in the adjacent Si regions. In addition, the graded Ge profile has a lower value of boron diffusivity compared to box-type profile.

Kuo *et al.* [16] showed that the diffusivity of B in $Si_{1-x}Ge_x$ was almost stress-independent. The B diffusivity in almost strain-free $Si_{0.80}Ge_{0.20}$ is approximately equal to B diffusivity in fully strained-$Si_{0.80}Ge_{0.20}$, which is about one order of magnitude lower than that in Si. The result indicates that boron diffusion in SiGe is predominantly a function of Ge content, rather than biaxial strain. The importance of Ge chemical content or microscopic strain has been suggested [16].

Following the work from References 17 and 20, a new model for the diffusion of boron in SiGe has been implemented in Silvaco SUPREM4 [22]. This new diffusion model treats SiGe as highly Ge-doped silicon and accounts for two effects: (1) boron diffusivity decreases exponentially as the Ge content increases and (2) the intrinsic carrier concentration increases linearly as the Ge content increases. The most

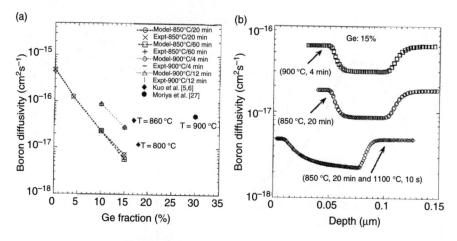

Figure 3.5   *(a) Extracted (simulation) and measured B diffusivity as a function of Ge fraction at different thermal budgets. (b) Extracted B diffusivity as a function of depth for box-type Ge profile for 900°C at 4 min and 850°C at 20 min and graded Ge profile with FA: 850°C at 20 min and 1100°C at 10 s [After K. Rajendran et al., IEEE Trans Electron Dev., Vol. 48, 2001(2022–2031). © 2001 IEEE]*

outstanding feature of the new model is the resulting boron pileup at the Si/SiGe inter-face. It was also shown that the diffused profiles in layers with uniform and graded Ge profiles differ significantly. Strain effects are considered negligible in this model. In this new diffusion model, SiGe layers are introduced by using the DEPOSIT statement to create a Ge-doped silicon layer.

3.3   Diffusion: P and As in strained-SiGe

While the diffusion of boron in SiGe is relatively well characterised [8, 9, 23], there is little information about n-type dopant diffusion in SiGe. The diffusion behaviour of ion-implanted arsenic and phosphorus in relaxed-$Si_{0.8}Ge_{0.2}$ has been investigated by Eguchi *et al.* [24]. Secondary ion mass spectrometry (SIMS) was used to obtain diffusion profiles in Si and SiGe samples. The 20 per cent enhancement for As and P effective diffusivities in SiGe were extracted by comparing measured SIMS data with simulated profiles from TSUPREM4 [25].

The effective diffusivity of As and P in SiGe under equilibrium conditions was found to be enhanced compared to that in Si. Both dopants exhibit enhanced diffusivities in SiGe compared to those in Si under equilibrium conditions. At high concentrations ($n = 5 \times 10^{19}$ cm^{-3}), the ratio of the effective diffusivity in SiGe relative to that in Si is found to be approximately 7 for arsenic, and roughly 2 for phosphorus in the temperature range from 900 to 1050 °C. However, under transient diffusion conditions, arsenic diffusion in SiGe is retarded while arsenic diffusion in Si is enhanced by the ion implant damage. The transient retarded diffusion of As in SiGe suggests that As diffuses primarily by a vacancy-mediated mechanism in SiGe. The absolute value of effective diffusivity of P in SiGe is roughly three times higher than that of As in SiGe. Therefore, to form shallow source/drains, As is a better choice for a n-type dopant in SiGe. The As diffusion model, implemented in TSUPREM4 was applied for modelling dopant diffusion in SiGe as in Si. As diffusion in Si is assumed to be based on neutral and single-negatively charged defects.

The effective diffusivity of As in Si under equilibrium conditions can be described as

$$D_{\text{eff}}^{\text{As}}(\text{Si}) = f_i \left[D_i^0 + D_i^- \left(\frac{n}{n_i(\text{Si})} \right) \right] + f_v \left[D_v^0 + D_v^- \left(\frac{n}{n_i(\text{Si})} \right) \right] \tag{3.2}$$

where n is the electron concentration and $n_i(\text{Si})$ is the intrinsic carrier concentration in Si. D_i^0 and D_i^- represent the contributions from neutral and single-negatively charged interstitial defects, respectively. D_v^0 and D_v^- represent the contributions from neutral and single-negatively charged vacancy defects. And the factors f_i and f_v represent the fractions of interstitial and vacancy-mediated diffusion, respectively. The P diffusion model in Si is assumed to be based on neutral, single- and double-negatively charged interstitial defects. The effective diffusivity of P under equilibrium conditions can be described as

$$D_{\text{eff}}^{\text{P}}(\text{Si}) = D_i^0 + D_i^- \left(\frac{n}{n_i(\text{Si})} \right) + D_i^- \left(\frac{n}{n_i(\text{Si})} \right)^2 \tag{3.3}$$

where D_i^- represents the contribution from double-negatively charged interstitial defects. These equations were modified to include appropriate intrinsic carrier concentration in SiGe, $n_i(\text{SiGe})$ and is given by

$$n_i(\text{SiGe}) = n_i(\text{Si}) \exp \left(\frac{\Delta E_g}{2kT} \right) \tag{3.4}$$

where ΔE_g between Si and relaxed-SiGe (Ge 20 per cent) is approximately 0.08 eV [26]. Hence, the diffusion models for As and P in SiGe under equilibrium conditions are described below:

$$D_{\text{eff}}^{\text{As}}(\text{SiGe}) = g_{\text{As}} \left[f_i \left\{ D_i^0 + D_i^- \left(\frac{n}{n_i(\text{SiGe})} \right) \right\} + f_v \left\{ D_v^0 + D_v^- \left(\frac{n}{n_i(\text{SiGe})} \right) \right\} \right] \tag{3.5}$$

$$D_{\text{eff}}^{\text{P}}(\text{SiGe}) = g_P \left[D_i^0 + D_i^- \left(\frac{n}{n_i(\text{SiGe})} \right) + D_i^- \left(\frac{n}{n_i(\text{SiGe})} \right)^2 \right] \tag{3.6}$$

In Equations (3.5) and (3.6), g_{As} and g_P are defined as enhancement factors for the diffusion coefficients. The values for each individual diffusivity (D_i^0, D_i^- etc.) used in (3.5) and (3.6) are the same as the Si values used in (3.2) and (3.3). g is varied in the simulations to fit the observed SIMS data in SiGe. Using the extracted enhancement factors g, and the intrinsic carrier concentration n_i, the effective diffusivities for As and P in SiGe under equilibrium conditions can be evaluated at any given value of n

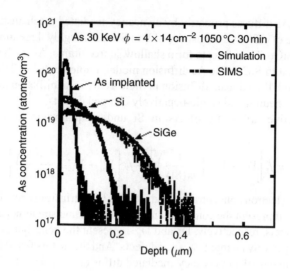

Figure 3.6   Arsenic SIMS profiles in Si and SiGe [After S. Eguchi et al., MIT Annual Report 2003]

using (3.5) and (3.6). Finally, r, ratio of the effective diffusivity of dopants in SiGe compared to that in Si is given by

$$r = D_{\mathrm{eff}}(\mathrm{SiGe})/D_{\mathrm{eff}}(\mathrm{Si}) \tag{3.7}$$

From the extracted diffusivities, r shows substantial magnification of D_{eff} in SiGe compared to that in Si. Figure 3.6 shows As profiles in Si and SiGe after 1050 °C, 30 min annealing. The As profiles in Si and SiGe are quite similar. However, after 1050 °C, 30 min annealing, the As diffusion in SiGe is enhanced compared to that in Si. In the temperature range from 900 to 1050 °C, the enhancement factor r is found to be roughly constant under equilibrium conditions and is about 7 for an average doping concentration $n = 5 \times 10^{19}\,\mathrm{cm}^{-3}$.

Figure 3.7 shows P profiles in Si and SiGe after 1050 °C, 30 min annealing. P diffusion in SiGe is observed to be enhanced compared to that in Si. For P, the extracted ratio r in the temperature ranges from 900 to 1050 °C is found to be roughly 2, (i.e., $r \sim 1.7$ at 1050 °C, $r \sim 2$ at 1000 °C and $r \sim 2.2$ at 900 °C). Figure 3.8 shows effective diffusivities of As and P in SiGe and Si under equilibrium, extrinsic conditions. In the plot, the diffusivities are evaluated at $n = 5 \times 10^{19}\,\mathrm{cm}^{-3}$. From the plot in Figure 3.8, in the temperature range from 900 to 1050 °C, under equilibrium, extrinsic conditions, the effective diffusivity of P in SiGe, $D_{\mathrm{eff}}^{\mathrm{P}}(\mathrm{SiGe})$ is approximately three times higher than the diffusivity of As in SiGe.

Figure 3.9 compares As profiles in Si and SiGe for an annealing temperature of 1000 °C and times of 1, 20, and 60 s. For annealing times of 20 and 60 s, As diffuses faster in SiGe compared to Si. However, the profile for As in SiGe after 1000 °C, 1 s is very similar to that in Si. These results demonstrate the possibility to form As ion implanted $\mathrm{n}^{+}/\mathrm{p}$ junctions in SiGe that are as shallow as those in Si. These

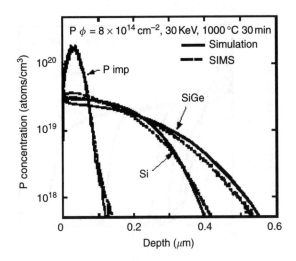

Figure 3.7   Phosphorus SIMS profiles in Si and SiGe [After S. Eguchi et al., MIT Annual Report 2003]

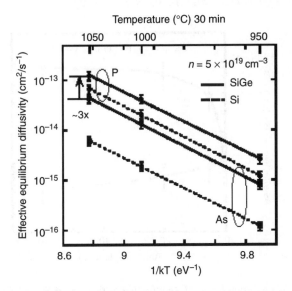

Figure 3.8   Comparison of effective diffusivity in Si and SiGe under equilibrium conditions [After S. Eguchi et al., MIT Annual Report 2003]

results also suggest that either As diffusion in Si is enhanced for short anneal times, or As diffusion in SiGe is dramatically retarded for short anneal times, compared to diffusion under equilibrium conditions.

Figure 3.10 shows the resulting time dependence of the extracted As effective diffusivities in Si and SiGe. It is observed that after relatively long time annealing,

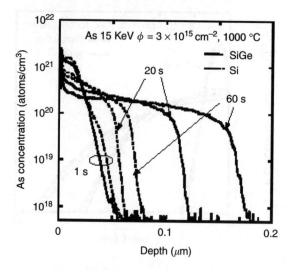

Figure 3.9   *Comparison of As profiles in Si and SiGe after 1000°C annealing [After S. Eguchi et al., MIT Annual Report 2003]*

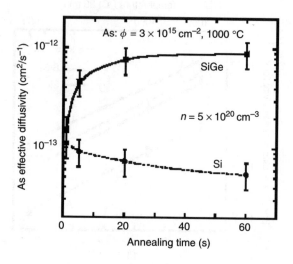

Figure 3.10   *Time dependence of As effective diffusivity in Si and SiGe for 1000°C annealing [After S. Eguchi et al., MIT Annual Report 2003]*

the As effective diffusivity in SiGe is larger than that in Si. For short anneal times, where the diffusion is dominated by the effects of ion implant damage, the As effective diffusivity in Si is observed to be enhanced. However, As effective diffusivities in SiGe are observed to be retarded for short anneal times, as shown in Figure 3.10. A possible explanation for the observed retarded transient diffusion of As in SiGe is that As diffuses primarily by vacancies in SiGe.

3.4 Dopant diffusion in SiGeC

Adding germanium to silicon allows for the adjustment of both bandgap and strain, making possible devices such as the heterojunction bipolar transistor and strained-Si MOSFETs. The out-diffusion of boron in $Si_{1-x}Ge_x$ has been discussed in Section 3.1. Low carbon concentrations can be used to suppress diffusion of dopants. This unique property has already found its first device application, the SiGeC heterojunction bipolar transistors [27, 28]. Introducing small amounts of substitutional carbon can adjust strain and bandgap, and has been shown to dramatically reduce the out-diffusion of boron and phosphorus atoms, allowing for increased control over dopant profiles. Thus, $Si_{1-x-y}Ge_xC_y$ alloys are of great interest for scaled Si-based devices.

Research on highly supersaturated, carbon containing alloys on silicon substrates started only a few years ago [29]. Meanwhile, knowledge has been accumulated on growth, strain manipulation, thermal stability, carbon effects on band structure and charge carrier transport [30–36]. Advanced SiGeC HBTs are fabricated in numerous different architectures [37–47], depending on whether they use selective or differential epitaxy, and self-alignment or quasi self-alignment. The selection of the architecture is required at the beginning of the development, while the assessment of whether the choice is optimum for the target applications can only be made after several technology optimisation cycles. Device performance figures of merit like f_T, f_{max} and NF_{min} are not always one-to-one correlated with the performance of basic circuit blocks. Especially when advanced materials or process steps are introduced, the lack of specific models or model coefficients might result in marginal impact of TCAD in the decision process and optimisation cycle.

Work on C-containing alloys has mainly been pursued at IHP in Germany since 1993. First grown with MBE, the process was later transferred to CVD. SiGeC HBTs from IHP now offer high-performance wireless solutions, with the potential to integrate digital and analog functions on a single chip [48–50]. They combine high performance with process simplicity, and hence low cost [51]. Both these features stem directly from the beneficial effects of adding small concentrations of C to SiGe.

Rucker and Heinemann [52] have shown that dopant diffusion coefficients in Si can be changed by more than one order of magnitude owing to alloying with Ge and C. The observed suppression of boron diffusion in C-rich layers was explained on the basis of coupled diffusion of Si point defects and carbon. A model for boron diffusion and segregation in SiGe heterostructures was also presented. The consequences of the unique diffusion properties of SiGeC alloys for heterojunction bipolar transistors were investigated by process and device simulation. It has been shown that by adding small amounts of carbon ($\sim$0.1–1.0 per cent) to either silicon or $Si_{1-x}Ge_x$, the diffusivity of boron and phosphorus atoms can be reduced by orders of magnitude below normal levels. This provides a mechanism by which diffusion can be controlled without altering processing conditions. By adding carbon to regions of a device where diffusion must be minimised, sharp profiles can be maintained even during high-temperature processing. This has been crucial in the continued scaling of high-performance HBTs; other potential applications include MOSFET channels and source/drains.

Several factors, however, limit the usefulness of carbon. Carbon has a very low solubility in silicon and $Si_{1-x}Ge_x$ and is therefore unstable. High levels of substitutional carbon, needed for the desired effect of reduced dopant diffusion, can be lost during thermal annealing to undesirable α-SiC precipitates. Carbon has also been shown to degrade the electrical properties of silicon and $Si_{1-x}Ge_x$ films in some cases. In addition, many effects of carbon are still not understood or have not been investigated. Possible atomic interactions between carbon and other dopants during thermal processing are largely unknown, and may have important technological implications, as it is essential that both the carbon and dopants remain on substitutional lattice sites and do not form defects. The exact mechanism by which carbon controls dopant diffusion, while a subject of many studies, is still not fully understood; in particular, whether carbon–dopant interactions are playing a role is not fully known. There has also been very little investigation of carbon in polycrystalline films.

Carbon controls boron spike diffusion in the base layer, which is critical to achieve high-speed HBTs [43, 53]. During the fabrication of transistors, attention should be given to low thermal budgets; however, the rapid thermal annealing (RTA) activation step used in typical BiCMOS technologies (typically 1000–1100 °C) modifies considerably the boron diffusion in SiGeC HBTs. Several workers [50, 54–59] have attempted to understand the diffusion mechanisms and modelling boron diffusion in SiGeC; however, major focus was on RTA or thermally annealed films in the temperature range of 700–900 °C.

Boron and carbon diffusion in SiGeC layers under different actual process conditions, typical for BiCMOS processes, has been studied in detail by Sibaja-Hernandez *et al.* [60]. Special attention was paid to the RTA temperatures in the range 1020–1100 °C. The impact of Si cap thickness, germanium profile, polySi emitter processes and atmosphere conditions in the RTA step on boron and carbon diffusion in SiGeC base layer was studied. The authors have derived model parameters for equilibrium and transient-enhanced diffusion conditions and incorporated in the process simulator TSUPREM4 [61] via User-Specified Equation Interface. The model and model parameters were validated for a 0.13 μm BiCMOS process. Following Reference 62, the boron diffusion model for SiGeC is implemented as [60]

$$\frac{\partial C_B}{\partial t} = -\nabla \cdot \left(-D_B \left[\nabla C_B + C_B \nabla \frac{N_T}{Q} + C_B \frac{2n}{Q} \nabla \ln(n_i) \right] \right) \tag{3.8}$$

where N_T and Q are the net doping concentration, N_A-N_D, and the total carrier concentration, $n + p$, respectively. The first term in (3.8) is the simple diffusion term (Fick's law), and the second and third terms are electric field contributions due to the dopant distribution and bandgap varying with Ge content, respectively [20]. The boron diffusivity is given by

$$D_B = \left[D_x + D_P \left(\frac{p}{n_i} \right) \right] \left[\exp \left(-\frac{\epsilon_{B}sx}{kT} \right) \right] \left[fI \left(\frac{I}{I^*} \right) + (1 - fI) \left(\frac{V}{V^*} \right) \right] \tag{3.9}$$

The first factor is the boron diffusivity in Si under extrinsic conditions. The second factor takes into consideration a linear dependence of the activation energy of the diffusion coefficient on the Ge content, x [10]. The characteristic strain in the plane of the SiGe layer (s) is modified with an energy per unit strain factor (ϵ_B). The third factor considers the dependency of boron diffusivity on vacancies (V), and silicon self-interstitials (I), where I^* and V^* denote the equilibrium interstitial and vacancy concentrations, respectively. The model also includes boron diffusion reduction due to trapping of boron atoms [20]. To simulate correctly the diffusion of impurities in carbon-doped devices, coupled diffusion of carbon and Si point defects has to be taken into account.

Some experimental Boron and carbon diffusion in SiGeC layers and simulation results under different process conditions, with RTA temperatures in the temperature range of 1000–1100 °C, are presented below. Good correlation between measured and simulated boron and carbon profiles (Figure 3.11) was obtained with the modified pre-factors and activation energies for carbon and Ge diffusivities, for different RTA temperatures in inert ambience. Figure 3.11 shows the impact of different RTA temperatures on B and C profiles for 150 nm Si cap. The model also agrees well with simulations under oxidising ambient conditions, where excess interstitials are injected, enhancing the boron diffusion. Default model coefficients [61, 63] and the

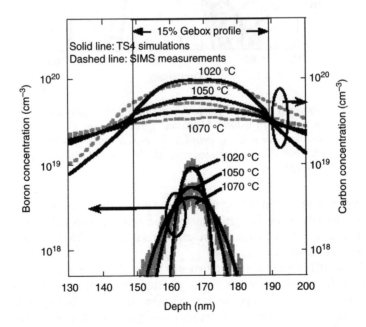

Figure 3.11   Experimental (dashed lines) and simulated (solid lines) boron and carbon profiles for different RTA temperatures. Samples with a 150 nm silicon cap were annealed for 10 s at temperatures of 1020, 1050 and 1070 °C [After A. Sibaja-Hernandez et al., Mat. Sci. Semicond. Proc., Vol. 8, 2005(115–120)]

Table 3.1   TSUPREM4 model coefficients

Parameter	TSUPREM4	Reference [60]
$D_{C_s}^{\text{eff}}$ (cm^2 s^{-1})	$1.90 \exp(-3.1/kT)$	$0.33 \exp(-3.1/kT)$
D_{C_i} (cm^2 s^{-1})	$0.44 \exp(-0.88/kT)$	$0.44 \exp(-0.88/kT)$
D_{Ge} (cm^2 s^{-1})	$0.037 \exp(-5.46/kT)$	$1.38 \times 10^5 \exp(-5.39/kT)$

Source: After A. Sibaja-Hernandez *et al.*, *Mat. Sci. Semicond. Proc.*, Vol. 8, 2005(115–120).

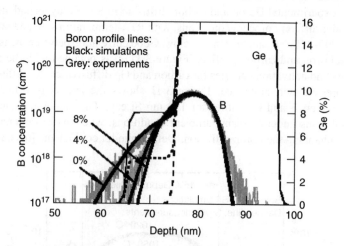

Figure 3.12   Impact of the Ge profile pedestal (8 per cent, 4 per cent and 0 per cent) on boron diffusion. Measured (grey – solid lines) and simulated (black – solid lines) profiles for boron under 1100°C spike anneal [After A. Sibaja-Hernandez et al., Mat. Sci. Semicond. Proc., Vol. 8, 2005(115–120)]

coefficients extracted from the experimental results are shown in Table 3.1. The germanium diffusivity is in good agreement with the measurements of Dorner *et al.* [64] and Hettich *et al.* [65], although the pre-exponential factor in TSUPREM4 is around 6–7 orders of magnitude lower.

In SiGeC HBTs generally using two-step Ge profiles, the current gain is tuned by varying the Ge concentration in the pedestal. The different germanium concentrations in the low-Ge layer impact the boron diffusion. Boron diffuses more for lower Ge concentrations resulting in a non-symmetric boron profile after the spike anneal. Figure 3.12 shows that variation of the Ge fraction in the pedestal is correctly simulated. The presence of a polyemitter modifies the boundary condition for the diffusion of interstitials and vacancies. Therefore, the model is found to be in good agreement for typical profiles of a 0.13-μm high-speed BiCMOS process (see Figure 3.13), using in situ As-doped polyemitter with 1100°C spike anneal.

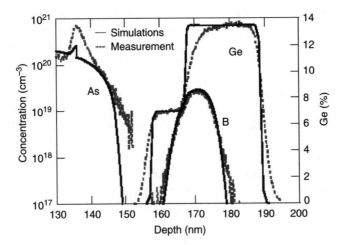

Figure 3.13   Measured and simulated profiles after 1100°C spike annealing for a typical 0.13 μm SiGeC BiCMOS process. Confirmation of process modelling accuracy is observed [After A. Sibaja-Hernandez et al., Mat. Sci. Semicond. Proc., Vol. 8, 2005(115–120)]

3.5   Oxidation: SiGe/SiGeC films

In order to use SiGe alloys in Si-based devices, it is of primary importance to study the oxidation of these alloys layers since it is a major step in Si technology. The most frequently used starting materials for the oxidation studies are strained-$Si_{1-x}Ge_x$ films with desired compositions. The films can be deposited with a variety of techniques such as ultra high vacuum chemical vapour deposition (UHVCVD) or molecular beam epitaxy (MBE). Moreover, since silicon oxidation has been studied for decades now, it is of interest to compare the basic physical phenomena involved in both silicon and SiGe oxidation.

Investigations performed so far using thermal oxidation have shown different behaviours depending on the oxidation conditions and alloy composition. In the case of wet oxidation, for Ge concentrations below 50 per cent, a pure SiO_2 layer is formed and a Ge pileup occurs at the SiO_2/SiGe interface [66–69]. Oxidation of polycrystalline $Si_{1-x}Ge_x$ films with different compositions, carried out in pyrogenic steam at 800 °C for various lengths of time, shows that the oxidation is enhanced by the presence of germanium, and the enhancement becomes more pronounced for the films with high germanium content. A mixed oxide in the form of either $(Si,Ge)O_2$ or SiO_2-GeO_2 is found at the sample surface if the initial $Si_{1-x}Ge_x$ contains more than 50 per cent of germanium. As high-pressure oxidation increases the oxidation rate at a relatively lower temperature, it has also been employed for the oxidation of SiGe [70].

In thermal oxidation (wet), an increase by a factor of about 2 of the oxidation rate in the linear regime of SiGe compared to Si is observed. As regards dry oxidation, the formation of a SiO_2 top layer with a Ge pileup at the interface has been reported [67, 70, 71]. However, no increase of the oxidation rate of SiGe alloys with respect to Si has been noted although a Ge-rich layer is present at the interface. A first explanation of increase in oxidation rate in wet oxidation has been given by Fathy *et al.* [68]. They assumed that during the linear regime of wet oxidation the oxidation rate is limited by the breaking of Si substrate bonds. Thus, the increase of the oxidation rate in the SiGe case can be explained by the fact that SiGe bonds are weaker than Si–Si bonds. LeGoues *et al.* [66] have explained the difference in oxidation rates in dry and wet conditions. They assumed that the Si interstitials formation is the limiting factor during the linear regime of wet oxidation but not during the linear regime of dry oxidation. Thus, the increase of oxidation rate, observed in the case of wet oxidation, can be explained by the fact that the Ge-rich interfacial layer would prevent interstitial formation by reducing the stress between the oxide and the substrate.

To avoid the problem of Ge segregation at the interface, which degrades gate oxide properties, attempts have been made to form a thin Si-cap layer on strained-SiGe films [72]. This thin Si-layer is supposed to be converted fully into SiO_2 during oxidation. However, a surface silicon cap layer of about 14 nm is found to have a significant impact on the oxidation of the $Si_{0.5}Ge_{0.5}$ films; it leads to the growth of about 115 nm thick SiO_2, which is about four times that of the SiO_2 resulting from the oxidation of the cap layer itself. On the $Si_{1-x}Ge_x$ films with only 30 per cent of germanium, the SiO_2 continues to grow after oxidation for 180 mins, resulting in 233-nm-thick SiO_2, which is about 2.4 times greater than the SiO_2 grown on (100) silicon substrates. Rejection of germanium results in piling up of germanium at the interface between the growing SiO_2 and the remaining $Si_{1-x}Ge_x$. Substantial inter-diffusion of silicon and germanium takes place in the remaining $Si_{1-x}Ge_x$.

Nevertheless, thermal oxidation processes requiring high temperatures (700–1000 °C) are not suitable when applied to strained-SiGe layers since they can undergo dislocations formation and degradation of electrical characteristics under such high-temperature treatments. Clearly, low thermal budgets are highly desirable for processing such strained layers. Low-temperature oxidation has been performed using electron cyclotron resonance plasma [73–75]. The results of such oxidation differ from that obtained with thermal oxidation and show that both Si and Ge are fully oxidised forming SiO_2 and GeO_2 without any Ge pileup at the oxide/substrate interface. Use of microwave plasma of oxygen [76] or N_2O [77], or ultraviolet radiation [78] can also enhance the oxidation rate. The $Si_{1-x}Ge_x$ alloys used for such oxidation studies are often thin films. They can be epitaxial (crystalline) [67, 70, 76, 77, 79–83], polycrystalline [80, 84] or amorphous [85].

3.5.1   Oxidation kinetics

The basic understanding of the kinetics of thermal oxidation in Si comes from the model proposed by Deal and Grove [86], which assumes that growth occurs by diffusion of oxidant through the oxide to the oxide/Si interface where it reacts with Si to form SiO_2. Although this model has been developed for thermal oxidation of

Si, most researchers use this model to interpret the oxidation kinetics of the oxidation of SiGe. A mathematical model of thermal oxidation of $Si_{1-x}Ge_x$ alloys has been reported by Hellberg *et al.* [87, 88]. The growth of SiO_2 is simulated in conjunction with the determination of silicon distribution in $Si_{1-x}Ge_x$ using numerical methods. The main feature of the model is the assumption of simultaneous oxidation of germanium and silicon when exposing the $Si_{1-x}Ge_x$ to an oxidising atmosphere. The enhanced oxidation of silicon in the presence of germanium is modelled as a result of the rapid oxidation of germanium followed by the quick reduction of GeO_2 by silicon. The growth of a mixed oxide in the form of either $(Si,Ge)O_2$ or SiO_2-GeO_2 only occurs when the supply of silicon to the $SiO_2/Si_{1-x}Ge_x$ interface is insufficient. A comparison is made between simulation and experiment for wet oxidation (in pyrogenic steam) of polycrystalline $Si_{1-x}Ge_x$ films. It is found that the model gives a good account of the oxidation process. Kinetic parameters, that is, interfacial reaction rate constant for oxidation of germanium and diffusion coefficient of silicon(germanium) in $Si_{1-x}Ge_x$, are extracted by fitting the simulation to the experiment. Thermal oxidation of $Si_{1-x}Ge_x$ proceeds as follows [88]:

1. When the $Si_{1-x}Ge_x$ surface is exposed to an oxidising atmosphere containing either O_2 or H_2O, the silicon atoms are preferentially oxidised forming SiO_2 while the germanium atoms pile up at the interface between the growing SiO_2 and the remaining $Si_{1-x}Ge_x$. This condition will prevail till the concentration x is above the critical value of x_e (typically 3.2×10^{-19} of the atomic concentration of silicon).

2. As the oxidation process proceeds, and when x is smaller than x_e, oxidation of germanium would take place resulting in simultaneous oxidation of silicon and germanium.

If one follows the formulation of the Deal–Grove model [86], the above model, however, does not work as the growth rate of SiO_2 should be directly proportional to the value of x. The value of x_e being on the order of 10^{-19} is a very demanding concentration criterion. The formation of mixed oxides for substrate concentrations with nearly 50 per cent silicon at 800–1000 °C [79, 81, 85, 87] shows quite clearly that this model does not apply. Germanium is known to be oxidised with an appreciable rate temperature below 450 °C, which is several hundred degrees lower than that for the oxidation temperature of silicon. The vast majority of oxidation studies on $Si_{1-x}Ge_x$, carried out in the temperature range of 500–1000 °C, show that a mixed oxide forms when the germanium concentration is above 50 per cent. There is not enough silicon at the interface to reduce all formed GeO_2 because of the limited flux of silicon.

Figure 3.14 shows schematically a one-dimensional model for the SiGe oxidation process. During the oxidation of $Si_{1-x}Ge_x$, one may distinguish three fluxes: J_{ox}, J'_{Si}, J'_{Ge}. J_{ox}, the flux of O_2 molecules through growing oxide, is given by

$$J_{ox} = D_{ox}\frac{c_0 - c_{iox}}{L_{ox}} \tag{3.10}$$

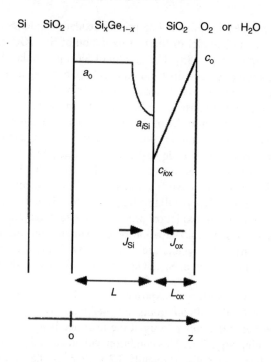

Figure 3.14   *A one-dimensional model schematically showing the various fluxes occurring during the oxidation of $Si_{1-x}Ge_x$ films on top of an SiO_2/silicon substrate. Note that the concentration gradient in the SiGe layer is steeper in the silicon-rich region than in the germanium-rich area near the interface. The constant slope of the oxygen concentration in the oxide results from the assumptions of (a) constant diffusion coefficient D_{ox}, and (b) the establishment of a steady state condition [After P.-E. Hellberg et al., J. Appl. Phys., Vol. 82, 1997(5779–5787)]*

where D_{ox} and c_0 are the diffusion coefficient and the solid solubility of O_2 in the oxide, respectively. c_{iox} is the concentration of O_2 at the $Si_{1-x}Ge_x$/oxide interface, and L_{ox} is the oxide thickness.

The second and third fluxes are those of silicon and germanium, respectively, at the $Si_{1-x}Ge_x$/oxide interface towards the oxide:

$$J'_{Si} = k_{Si}a_{iSi}c_{iox} \qquad (3.11)$$

$$J'_{Ge} = k_{Ge}(1 - a_{iSi})c_{iox} \qquad (3.12)$$

where k_{Si} and k_{Ge} are the rate constants for the reaction of silicon and germanium with O_2 to form SiO_2 and GeO_2, respectively. a_{iSi} is the silicon activity at the $Si_{1-x}Ge_x$/oxide interface. As $Si_{1-x}Ge_x$ is assumed to be an ideal solution, the silicon activity a is equal to the composition x. Under conditions of steady state the sum of the flux of silicon expressed by (3.11) and the flux of germanium expressed by (3.12) should be equal to the flux of O_2 expressed by (3.10):

$$J_{ox} = J'_{Si} + J'_{Ge} \tag{3.13}$$

By eliminating c_{iox}, the following is obtained from (3.10) to (3.13):

$$J_{ox} = \frac{[k_{Ge} + a_{iSi}(k_{Si} - k_{Ge})]c_0}{1 + (L_{ox}/D_{ox})[k_{Ge} + a_{iSi}(k_{Si} - k_{Ge})]} \tag{3.14}$$

In order to find the thickness of the oxide formed one should make use of an equation by relating the growth of the oxide to the flux of O_2 as

$$\frac{dL_{ox}}{dt} = \Omega J_{ox} \tag{3.15}$$

in which Ω is the molecular volume of either SiO_2 or GeO_2. As a first-order approximation, the molecular volume of GeO_2 is approximately equal to that of SiO_2. It should be noted that D_{ox}, c_0, and L_{ox} are quantities related to the mixed oxide containing both SiO_2 and GeO_2. During oxidation, the reduction reaction occurs, which has been studied by Prabkakaran *et al.* [89, 90]. They have shown that the cleavage of Ge-O bonds and the formation of Si-O bonds already take place around 200 °C. Thus, it can be anticipated that reaction proceeds rapidly at temperatures well above 200 °C. Therefore, as long as there is enough silicon to reduce GeO_2 at the interface, (3.11) and (3.12) together can account for the net flux of silicon towards the interface forming pure SiO_2:

$$J_{Si} = k_{eff}c_{iox} \tag{3.16}$$

with

$$k_{eff} = k_{Ge} + a_{iSi}(k_{Si} - k_{Ge}) \tag{3.17}$$

which can be regarded as the effective interface reaction rate constant. Thus, D_{ox}, c_0, and D_{ox} are now quantities only related to pure SiO_2. In order to find the oxide thickness as a function of time, the silicon activity at the $Si_{1-x}Ge_x$/oxide interface, a_{iSi}, has to be determined.

The diffusion process described below will allow the determination of the silicon activity, a, in the remaining $Si_{1-x}Ge_x$ film during oxidation. The silicon diffusion in $Si_{1-x}Ge_x$ obeys Fick's second law:

$$N\frac{\partial a}{\partial t} = \frac{\partial \phi_{Si}}{\partial z} = N\frac{\partial}{\partial z}D(x)\frac{\partial a}{\partial z} \tag{3.18}$$

in which N is the atomic concentration of silicon in $Si_{1-x}Ge_x$ (per cm^3), a is the silicon activity, t is the diffusion time, ϕ_{Si} is the flux of silicon and z is the coordinate (i.e., the depth in the $Si_{1-x}Ge_x$). The diffusion coefficient, $D(x)$ is a strong function of the silicon concentration in the $Si_{1-x}Ge_x$ film [91]. The simultaneous solution to (3.14), (3.15), and (3.18) is very complex and an analytic solution is difficult to find. Equation 3.18 is therefore solved with a finite difference method, with the first boundary condition at the $Si_{1-x}Ge_x$/SiO_2 interface being set by J_{Si} [Equation (3.16)].

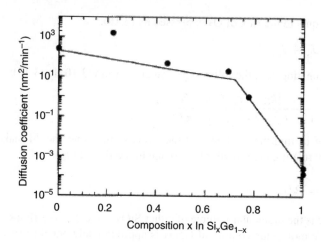

Figure 3.15   Diffusion coefficient D(x) of silicon and germanium in $Si_{1-x}Ge_x$ alloys. The solid points represent the experimental results in the literature and the solid lines represent the simple exponential functions that are found when modelling the oxidation of polycrystalline $Si_{1-x}Ge_x$ films [After P.-E. Hellberg et al., J. Appl. Phys., Vol. 82, 1997(5779–5787)]

It is assumed that no germanium diffuses into the formed SiO_2 under the oxidation conditions used.

McVay and DuCharme [91] have studied the diffusion of germanium in $Si_{1-x}Ge_x$ at temperatures typically around 1000 °C. Their results are extrapolated to 800 °C and are shown in Figure 3.15 together with other results of self-diffusion of germanium and diffusion of germanium in silicon [92]. Since large uncertainties have been found in the pre-exponential factors for the diffusion of germanium in $Si_{1-x}Ge_x$, large errors can be expected with the extrapolated results so they can only be used as guidelines for determination of the composition-dependent D(x). Figure 3.16 shows the growth of SiO_2 (pure) on various substrates at 800 °C. A comparison is made between the experimental [87] data (legends) and the simulation results (lines) generated by the mathematical model. It is apparent that the model gives a good account of the oxidation process.

3.5.2   Plasma oxidation of SiGe films

The oxide composition depends on the Ge concentration in the SiGe layers and varies with the oxidation time. In a 500 °C oxidation, it has been shown that the oxide first contains pure SiO_2 and the germanium piles up to form a pure Ge layer at the oxide/alloy interface [93]. Once the Ge interfacial layer reaches a critical thickness of about 12 Å, Ge starts to be embedded in the SiO_2 layer and forms microcrystals. For longer oxidation times, the Ge atoms start to be oxidised and the results are summarised in Figure 3.17. This transition between Ge pileup and oxidation has been observed for both wet thermal oxidation [94] and high-pressure dry oxidation [95].

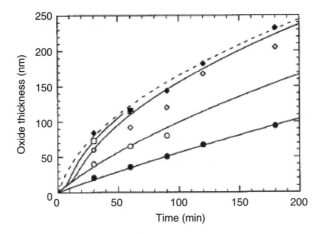

Figure 3.16   *Comparison between the experimental data (legends) and the simulation results (lines) of the growth of SiO₂ on various substrates: silicon
(100) (filled circles vs. continuous line), amorphous silicon (open circles vs. continuous line), Si₀.₇Ge₀.₃ with (open diamonds vs. continuous
line) and without (filled diamonds vs. broken line) the silicon cap layer
of 14 nm thickness, and Si₀.₅Ge₀.₅ without the cap layer (open squares
vs. continuous line). The simulation results were generated by using
the mathematical model (see text) [After P.-E. Hellberg et al., J. Appl.
Phys., Vol. 82, 1997(5779–5787)]*

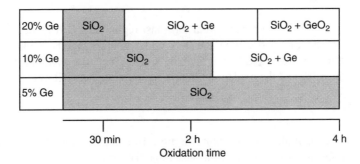

Figure 3.17   *Variation of the oxide composition with oxidation time after plasma-
assisted oxidation of SiGe layers at 500°C [After C. Tetelin et al.,
J. Appl. Phys., Vol. 83, 1998(2842–2846)]*

Tetelin *et al.* [96] have used the Deal–Grove (D-G) model to interpret the oxidation
kinetics of the plasma-assisted oxidation of SiGe. In this linear regime, the oxide
growth is limited by the interface kinetics. The interface reaction can be divided in
three steps: (1) breaking of the oxidant molecule, (2) breaking of a Si–Si or SiGe
substrate bond and (3) the relaxation of the interfacial stress between the oxide and
the substrate. For short oxidation times, the variation of the oxide thickness is linear

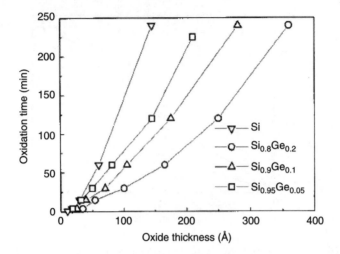

Figure 3.18   Oxidation kinetics of Si and SiGe layers after plasma-assisted oxi-
dation at 500°C [After C. Tetelin et al., J. Appl. Phys., Vol. 83,
1998(2842–2846)]

with the oxidation time. The relation between the oxidation time (t) and the oxide
thickness (d) can be written:

$$t = \frac{A}{B}d - \tau \tag{3.19}$$

where τ is a time constant for an initial ultra fast regime and B/A is the linear rate
constant. For longer oxidation time, the D-G model assumes that the oxide growth is
limited by the diffusion of oxidant through the oxide. Then, the relation between the
oxide thickness (d) and the oxidation time (t) can be written as follows:

$$t = \frac{d^2}{B} \tag{3.20}$$

where B is the parabolic rate constant. Figure 3.18 compares the rates of oxidation
of Si and SiGe for different Ge concentrations. These rates have about the same
magnitude of that observed during dry thermal oxidation at 800 °C. Nevertheless, the
results strongly differ from the case of dry oxidation because, for the temperature
studied here, the presence of Ge enhances the oxidation rate of Si. Figure 3.19 shows
that the oxidation kinetics of SiGe first follows a linear regime. In this regime, for
all Ge contents and all oxide thickness, the oxide formed is pure SiO_2. These results
show that during the linear regime the oxidation rate increases with the Ge content
of the SiGe layer. This can be explained either by interfacial stress relaxation or by
bond strength difference in Si and SiGe.

Figure 3.20 compares oxidation kinetics of Si (100) and Ge/Si (100) structures.
For the three types of Ge layer reported (4 and 9 Å amorphous Ge and 10 Å epitaxial
Ge), no difference in the oxidation rate has been observed. From these measurements,

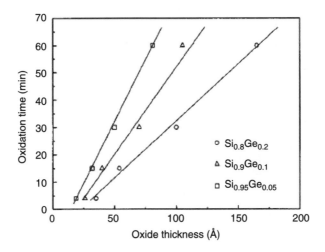

Figure 3.19   First linear regime of the oxidation kinetics of Si and SiGe layers at 500°C [After C. Tetelin et al., J. Appl. Phys., Vol. 83,
1998(2842–2846)]

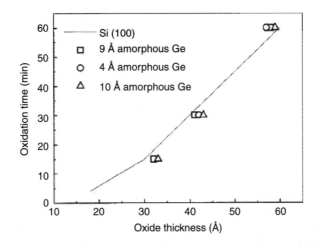

Figure 3.20   Comparison of oxidation rates of Si and Ge/Si (100) structures
at 500°C [After C. Tetelin et al., J. Appl. Phys., Vol. 83,
1998(2842–2846)]

it was concluded that unlike the case of wet thermal oxidation [66], the presence of
a pure Ge layer at the oxide/alloy interface has no effect on the oxidation kinetics.
Figure 3.21 shows the variation of the square of the oxide thickness with the oxidation
time. It is clear that for longer oxidation time, the plasma-assisted oxidation of SiGe
layers follows the parabolic regime described by Deal and Grove. The measured
oxidation rates for the different SiGe layers and the parabolic constant measured

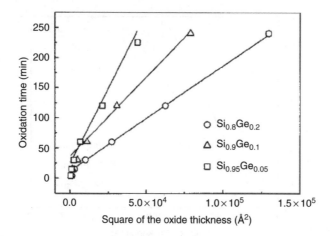

Figure 3.21   Parabolic regime of the oxidation kinetics of Si and SiGe layers at 500°C [After C. Tetelin et al., J. Appl. Phys., Vol. 83, 1998(2842-2846)]

Table 3.2   Linear rate constant and parabolic rate constant of the plasma-assisted oxidation of different SiGe layers [After C. Tetelin et al., J. Appl. Phys., Vol. 83, 1998(2842–2846)].

Sample	B/A (Å/min)	B(Å^2/min)
Si	1.27 ± 0.05	153.8 ± 4
Si$_{0.95}$Ge$_{0.05}$	1.35 ± 0.06	200 ± 8
Si$_{0.9}$Ge$_{0.1}$	1.95 ± 0.06	380 ± 32
Si$_{0.8}$Ge$_{0.2}$	2.70 ± 0.05	574 ± 23

from the oxidation kinetics are summarised in Table 3.2. These results show that the parabolic constant B increases with the Ge concentration in the SiGe layers. Other material properties of low-temperature plasma-grown oxides and qualitative model for the mechanism of oxide growth of SiGe that is consistent with published results for SiGe oxides grown with other oxidation systems have been reviewed by Riley and Hall [75].

3.6   Summary

In this chapter, a brief discussion on the nature of point defect-mediated diffusion, boron diffusion in silicon and SiGe, and reasons for strain relaxation in SiGe has been made. Because of the relationship between dopant diffusion and point defect diffusion, both the movement of point defects and dopants need to be modelled simultaneously. It has been shown that boron in strained, low Ge-composition SiGe layers diffuses primarily via an interstitial mediated mechanism. Mobile misfit dislocations can act as a strong interstitial sink but immobile dislocations appear to have very little effect on the point defect population. However, experiments should be performed to determine the segregation coefficient across the Si/SiGe interface as a function of germanium and dopant concentration. Further studies should also be taken up to find more closely the relationship between relaxation and interstitial absorption.

A moderate amount of boron pileup is observed at the Si/SiGe interface, and the diffusivity of boron is noticeably reduced within the SiGe boundary layers. The boron profile in the graded structure differs significantly. Very little pileup is seen at the SiGe interface owing to the low Ge content at that boundary. The boron diffusivity at the inner Si/SiGe interface is close to that in the silicon structure. Since most of the SiGe layers used in HBTs are metastable in nature, they will relax upon high temperature processing and produce dislocations at the Si/SiGe interface. It is important to isolate the effect of these parameters upon boron diffusivity.

TCAD models had been used to simulate the effects of boron diffusion in Si and $Si_{1-x}Ge_x$. From the models, it was found that boron diffusion is related more to Ge concentration rather than to strain. Model parameters for boron diffusion implemented both in the ATHENA and TSUPREM4 User-Specified Equation Interface have been extracted for SiGe/SiGeC layers for RTA temperatures in the 1020–1100 °C range. Good correlation between measured and simulated boron and carbon profiles under different process conditions were obtained for germanium and carbon diffusion coefficients.

A mathematical model of oxidation of $Si_{1-x}Ge_x$ alloys has been presented. A comparison between simulation and experiment was made, showing that the model gives a good account of the process of oxidation of $Si_{1-x}Ge_x$. While the diffusion of silicon and germanium is strongly dependent on the crystalline structure and the composition of the $Si_{1-x}Ge_x$ alloys, the interfacial reaction rate constants are mainly affected by the type of oxidant and the crystalline structure of the $Si_{1-x}Ge_x$.

The kinetics of plasma-assisted oxidation of SiGe layers at 500 °C has also been presented using the Deal and Grove model. In the linear regime, the oxidation rate increases with the Ge concentration in the layer. The comparison between the oxidation kinetics of different Ge/Si (100) structures and the oxidation kinetics of Si and SiGe allows one to conclude that unlike in the case of wet thermal oxidation, the limiting factor of the interface oxidation reaction is the breaking of substrate bonds in case of dry oxidation. There is still no explanation of the fact that the linear kinetics of the dry thermal oxidation of Si and SiGe are the same.

References

1 Z. Li, W. Zhang, D. Wang, Y. Chang, and Y. Sun, 'The base dopant out diffusion and the optimized setback layers in SiGe HBT', in *Proc. Solid-State and Integrated-Circuit Tech. Conf.*, pp. 596–599, 2001.

2 C. K. Maiti, and G. A. Armstrong, *Applications of Silicon-Germanium Heterostructure Devices*. Inst. of Physics Pub., UK, 2001.

3 E. J. Prinz, P. M. Garone, P. V. Schwartz, X. Xiao, and J. C. Sturm, 'The effect of base-emitter spacers and strain-dependent densities of states in $Si/Si_{1-x}Ge_x/Si$ heterojunction bipolar transistors', in *IEEE IEDM Tech. Dig.*, pp. 639–642, 1989.

4 E. J. Prinz, P. Garone, P. Schwartz, X. Xiao, and J. Sturm, 'The effect of base dopant outdiffusion and undoped $Si_{1-x}Ge_x$ junction space layers in $Si/Si_{1-x}Ge_x/Si$ heterojunction bipolar transistors', *IEEE Electron Device Lett.*, vol. EDL-12, pp. 42–44, 1991.

5 Z. A. Shafi, C. J. Gibbings, P. Ashburn, I. R. C. Post, C. G. Tuppen, and D. J. Godfrey, 'The importance of neutral base recombination in compromising the gain of Si/SiGe heterojunction bipolar transistors', *IEEE Trans. Electron Devices*, vol. 38, pp. 1973–1976, 1991.

6 Z. A. Shafi, P. Ashburn, I. R. C. Post, D. J. Robbins, W. Y. Leong, C. J. Gibbings, and S. Nigrin, 'Analysis and modeling of base currents of $Si/Si_{1-x}Ge_x$ heterojunction bipolar transistors fabricated in high and low oxygen content material', *J. Appl. Phys.*, vol. 78, pp. 2823–2829, 1995.

7 G. A. Armstrong, and T. C. Denton, 'HQUPETS – a two-dimensional simulator for heterojunction bipolar transistors', in *Proc. IMA Conf. on Semiconductor Modeling, Loughborough*, pp. 16–17, 1991.

8 P. Kuo, J. L. Hoyt, J. F. Gibbons, J. E. Turner, R. D. Jacowitz, and T. I. Kamins, 'Comparison of boron diffusion in Si and strained $Si_{1-x}Ge_x$ epitaixl layers', *Appl. Phys. Lett.*, vol. 62, pp. 612–614, 1993.

9 N. Moriya, L. C. Feldman, H. S. Luftman, C. A. King, J. Berk, and B. Freer, 'Boron diffusion in $Si_{1-x}Ge_x$ epitaxial layers', *Phys. Rev. Lett.*, vol. 71, pp. 883–886, 1993.

10 N. E. B. Cowern, P. C. Zalm, P. Van der Sluis, D. J. Gravesteijn, and W. B. de Boer, 'Diffusion in strained Si(Ge)', *Phys. Rev. Lett.*, vol. 72, pp. 2585–2588, 1994.

11 A. Antonelli and J. Bernholc, 'Pressure effects on self-diffusion in silicon', *Phys. Rev. B*, vol. 40, pp. 10643–10646, 1989.

12 K. Rajendran, and W. Schoenmaker, 'Simulation of boron diffusivity in strained $Si_{1-x}Ge_x$ epitaxial layers', *J. Appl. Phys.*, vol. 89, pp. 980–988, 2001.

13 D. X. Xu, C. J. Peters, J. P. Noed, S. J. Rolfe, and N. G. Turr, 'Control of an anomalous boron diffusion in the base of Si/SiGe/Si hetereojunction bipolar transistors using PtSi', *Appl. Phys. Lett.*, vol. 64, pp. 3270–3272, 1994.

14 P. Kuo, J. L. Hoyt, J. F. Gibbons, J. E. Turner and D. Lefforge, 'Effects of Si thermal oxidation on B diffusion in Si and strained $Si_{1-x}Ge_x$ layers', *Appl. Phys. Lett.*, vol. 67, pp. 706–708, 1995.

15 K. Rajendran, and W. Schoenmaker, 'Measurement and simulation of boron diffusion in strained $Si_{1-x}Ge_x$ epitaxial layers with a linearly graded germanium profile', *Solid State Electron.*, vol. 45, pp. 1879–1884, 2001.

16 P. Kuo, J. L. Hoyt, and J. F. Gibbons, 'Effects of strain on boron diffusion in Si and SiGe', *Appl. Phys. Lett.*, vol. 66, pp. 580–582, 1995.

17 K. Rajendran, W. Schoenmaker, S. Decoutere, and M. Caymax, 'Simulation of boron diffusion in strained $Si_{1-x}Ge_x$ epitaxial layers', in *Proc. SISPAD*, pp. 206–209, 2000.

18 K. Rajendran, W. Schoenmaker, S. Decoutere, R. Loo, M. Caymax, and W. Vandervorst, 'Measurement and simulation of boron diffusion in strained $Si_{1-x}Ge_x$ epitaxial layers', *IEEE Trans. Electron Devices*, vol. 48, pp. 2022–2031, 2001.

19 K. Rajendran, and W. Schoenmaker, 'Measurement and simulation of boron diffusivity in strained $Si_{1-x}Ge_x$ epitaxial layers,' in *Abs. Computational Electronics Symp.*, pp. 108–109, 2000.

20 R. F. Lever, J. M. Bonar, and A. F. W. Willoughby, 'Boron diffusion across silicon-silicon germanium boundaries', *J. Appl. Phys.*, vol. 83, pp. 1988–1994, 1998.

21 Avant! Corporation, Fremont, CA, *Taurus Process/Device, Ver. 98.4*, 1998.

22 Silvaco International, *Silvaco-ATHENA User's Manual*, 2005.

23 G. H. Loechelt, G. Tam, J. W. Steele, L. K. Knoch, K. M. Klein, J. K. Watanabe, and J. W. Christiansen, 'Measurement and simulation of boron diffusion in Si and strained $Si_{1-x}Ge_x$ epitaxial layers during rapid thermal annealing', *J. Appl. Phys.*, vol. 74, pp. 5220–5526, 1993.

24 S. Eguchi, C. W. Leitz, E. A. Fitzgerald, and J. L. Hoyt, 'Diffusion behavior of ion-implanted n-type dopants in silicon germanium', in *Proc. Mat. Res. Soc.*, vol. 686, pp. A1.7.1–A1.7.6, 2002.

25 SUPREM-3, *One-Dimensional Semiconductor Process Simulation, Technology Modeling Associates*, 1993.

26 R. Braunstein, A. R. Moore, and F. Herman, 'Intrinsic optical absorption in germanium-silicon alloys', *Phys. Rev.*, vol. 109, pp. 695–710, 1958.

27 D. Knoll, B. Heinemann, K.-E. Ehwald, B. Tillack, P. Schley, and H. J. Osten, 'Comparison of SiGe and SiGe:C heterojunction bipolar transistors', *Thin Solid Films*, vol. 369, pp. 342–346, 2000.

28 G. Lippert, H. J. Osten, K. Blum, R. Sorge, P. Schley, D. Kruger, and G. Fischer, 'Optimized processing for differentially moleculer beam epitaxy-grown SiGe(C) devices', *Thin solid film*, vol. 321, pp. 21–25, 1998.

29 H. J. Osten, *Carbon-Containing Layers on Silicon – Growth, Properties and Applications*. Trans-Tech Publications, Switzerland, 1999.

30 H. J. Osten, M. Kim, G. Lippert, and P. Zaumseil, 'Ternary SiGeC alloys: growth and properties of a new semiconducting material', *Thin Solid Films*, vol. 294, pp. 93–97, 1997.

31 H. J. Osten, B. Heinemann, D. Knoll, G. Lippert, and H. Rucker, 'Effects of carbon on boron diffusion in SiGe: Principles and impact on bipolar devices', *J. Vac. Sci. Technol. B*, vol. 16, pp. 1750–1753, 1998.

32 H. J. Osten, D. Knoll, and H. Rucker, 'Dopant diffusion control by adding carbon into Si and SiGe: Principles and device application', *Mat. Sci. and Eng. B*, vol. 87, pp. 262–270, 2001.

33 H. J. Osten, H. Rucker, J. P. Liu, and B. Heinemann, 'Wider latitude for sophisti-
    cated devices by incorporating carbon into crystalline Si or SiGe', *Microelectronic
    Eng.*, vol. 56, pp. 209–212, 2001.

34 H. J. Osten, 'MBE growth and properties of supersaturated, carbon-containing
    silicon/germanium alloys on Si(001)', *Thin Solid Films*, vol. 367, pp. 101–111,
    2000.

35 H. J. Osten, R. Barth, G. Fischer, B. Heinemann, D. Knoll, G. Lippert, *et al.*,
    'Carbon-containing group IV heterostructures on Si: Properties and device
    applications', *Thin Solid Films*, vol. 321, pp. 11–14, 1998.

36 B. Heinemann, D. Knoll, G. G. Fischer, P. Schley, and H. J. Osten, 'Comparative
    analysis of minority carrier transport in npn bipolar transistors with Si, $Si_{1-x}Ge_x$,
    and $Si_{1-y}C_y$ base layers', *Thin Solid Films*, vol. 369, pp. 347–351, 2000.

37 H. J. Osten, D. Knoll, B. Heinemann, and P. Schley, 'Increasing process margin in
    SiGe heteeterojunction bipolar technology by adding carbon', *IEEE Trans. Elec.
    Devices*, vol. 46, pp. 1910–1912, 1999.

38 D. Knoll, H. Rucker, B. Heinemann, R. Barth, J. Bauer, D. Bolze, *et al.*, 'HBT
    before CMOS, a new modular SiGe BiCMOS integration scheme', in *IEEE IEDM
    Tech. Dig.*, pp. 499–502, 2001.

39 B. Heinemann, D. Knoll, R. Barth, D. Bolze, K. Blum, J. Drews, *et al.*,
    'Cost-effective high-performance high-voltage SiGe:C HBTs with 100 GHz f_T
    and $BV_{CEO} \times f_T$ products exceeding 220 VGHz', in *IEEE IEDM Tech. Dig.*,
    pp. 348–351, 2001.

40 H. J. Osten, D. Knoll, B. Heinemann, H. Rucker, and K.-E. Ehwald, 'Carbon
    doped SiGe heterojunction bipolar transistor module suitable for integration in a
    deep submicron CMOS process', in *Asia-Pacific Microwave Conf*, pp. 757–762,
    2000.

41 I. M. Anteney, G. Lippert, P. Ashburn, H. J. Osten, B. Heinemann, G. J. Parker,
    et al., 'Characterization of the effectiveness of carbon incorporation in SiGe for
    the elimination of parasitic energy barriers in SiGe HBTs', *IEEE Electron Device
    Lett.*, vol. 20, pp. 116–118, 1998.

42 H. J. Osten, G. Lippert, D. Knoll, R. Barth, B. Heinemann, H. Rucker, and
    P. Schley, 'The effect of carbon incorporation on SiGe heterobipolar transistor
    performance and process margin', *IEEE IEDM Tech. Dig.*, pp. 803–806, 1997.

43 H. Rucker, B. Heinemann, R. Barth, D. Bolze, J. Drews, U. Haak, *et al.*,
    'SiGe:C BiCMOS technology with 3.6 ps gate delay', in *IEEE IEDM Tech. Dig.*,
    pp. 121–124, 2003.

44 B. Heinemann, R. Barth, D. Bolze, J. Drews, P. Formanek, O. Fursenko, *et al.*,
    'A complementary BiCMOS technology with high speed npn and pnp SiGe:C
    HBTs', in *IEEE IEDM Tech. Dig.*, pp. 117–120, 2003.

45 D. Knoll, B. Heinemann, K.-E. Ehwald, H. Rucker, B. Tillack, W. Winkler, and
    P. Schley, 'BiCMOS integration of SiGe:C heterojunction bipolar transistors', in
    IEEE BCTM Proc., pp. 162–166, 2002.

46 D. Knoll, K. E. Ehwald, B. Heinemann, A. Fox, K. Blum, H. Rucker, *et al.*,
    'A flexible, low-cost, high performance SiGe:C BiCMOS process with a one-
    mask HBT module', in *IEEE IEDM Tech. Dig.*, pp. 783–786, 2002.

47 B. Heinemann, H. Rucker, R. Barth, J. Bauer, D. Bolze, E. Bugiel, *et al.*, 'Novel collector design for high-speed SiGe:C HBTs', in *IEEE IEDM Tech. Dig.*, pp. 775–778, 2002.

48 H. J. Osten, D. Knoll, B. Heinemann, and B. Tillack, 'Carbon doping of SiGe heterobipolar transistors', in *Silicon Monolithic Integrated Circuits in RF Systems*, pp. 19–23, 1998.

49 H. J. Osten, D. Knoll, B. Heinemann, H. Rucker, and B. Tillack, 'Carbon doped SiGe heterojunction bipolar transistors for high frequency applications', in *IEEE BCTM Proc.*, pp. 109–116, 1999.

50 H. Rucker, B. Heinemann, D. Bolze, D. Knoll, D. Kruger, R. Kurps, *et al.*, 'Dopant diffusion in C-doped Si and SiGe: Physical model and experimental verification', in *IEEE IEDM Tech. Dig.*, pp. 345–348, 1999.

51 D. Knoll, B. Heinemann, H. J. Osten, K. E. Ehwald, B. Tillack, P. Schley, *et al.*, 'Si/SiGe:C heterojunction bipolar transistors in an epi-free well, single-polysilicon technology', in *IEEE IEDM Tech. Dig.*, pp. 703–706, 1998.

52 H. Rucker, and B. Heinemann, 'Tailoring dopant diffusion for advanced SiGeC heterojunction bipolar transistors', *Solid State Electron.*, vol. 44, pp. 783–789, 2000.

53 T. F. Meister, H. Schafer, K. Aufinger, R. Stengl, S. Boguth, R. Schreiter, *et al.*, 'SiGe bipolar technology with 3.9 ps gate delay', in *IEEE BCTM Proc.*, pp. 103–106, 2003.

54 R. F. Scholz, P. Werner, U. Gosele, and T. Y. Tan, 'The contribution of vacancies to carbon out-diffusion in silicon', *Appl. Phys. Lett.*, vol. 74, pp. 392–394, 1999.

55 J. L. Ngau, P. B. Griffin, and J. D. Plummer, 'Modeling the suppression of boron transient enhanced diffusion in silicon by substitutional carbon incorporation', *J. Appl. Phys.*, vol. 90, pp. 1768–1778, 2001.

56 R. Pinacho, P. Castrillo, M. Jaraiz, I. Martin-Bragado, J. Barbolla, H.-J. Gossmann, *et al.*, 'Carbon in silicon:modelling the diffusion and clustering mechanism', *J. Appl. Phys.*, vol. 92, pp. 1582–1587, 2002.

57 M. S. Carroll, and J. C. Sturm, 'Quantification of substitutional carbon loss from $Si_{0.998}C_{0.002}$ due to silicon self-interstitial injection during oxidation', *Appl. Phys. Lett.*, vol. 81, pp. 1225–1227, 2002.

58 M. S. Carroll, J. C. Sturm, E. Napolitani, D. De Salvador, M. Berti J. Stangl, *et al.*, 'Diffusion enhanced carbon loss from SiGeC layers due to oxidation', *Phys. Rev. B*, vol. 64, pp. 073308–073311, 2001.

59 N. E. B. Cowern, B. Colombeau, F. Roozeboom, M. Hopstaken H. Snijders, P. Meunier-Beillard, *et al.*, 'Diffusion suppression in silicon by substitutional C doping', in *Proc. ESSDERC*, pp. 203–206, 2002.

60 A. Sibaja-Hernandez, M. W. Xu, S. Decoutere, and H. Maes, 'TSUPREM-4 based modelling of boron and carbon diffusion in SiGeC base layers under rapid thermal annealing conditions', *Mat. Sci. Semicond. Proc.*, vol. 8, pp. 115–120, 2004.

61 Avant! Corporation, Fremont, CA, *Two-Dimensional Process Simulation Program, Tech. Rep. TSUPREM-4, Ver. 6.6*, 1998.

62 S. Decoutere and A. Sibaja-Hernandez, 'SiGeC HBTs : The TCAD challenge reduced to practice', *Mat. Sci. Semicond. Proc.*, vol. 8, pp. 283–288, 2005.

63 R. C. Newman, and J. Wakefield, 'Diffusion and precipitation of carbon in silicon', in *Metallurgy of semiconductor materials, vol. 15, (J. B. Schroeder, Ed.)*, pp. 201–208, Interscience, New York, 1962.

64 P. Dorner, W. Gust, B. Predel, and U. Roll, 'Investigations by SIMS of the bulk impurity diffusion of Ge in Si', *Philos. Mag. A*, vol. 49, pp. 557–571, 1984.

65 G. Hettich, H. Mehrer, and K. Maier, 'Tracer diffusion of ^{17}Ge and ^{31}Si in intrinsic and doped silicon', in *Defects and radiation effects in semiconductors, (J. H. Albany, Ed. Inst. Phys. Conf. Ser. 46)*, pp. 500–507, Institute of Physics, London, 1979.

66 F. K. LeGoues, R. Rosenberg, T. Nguyen, F. Himpsel, and B. S. Meyerson, 'Oxidation studies of SiGe', *J. Appl. Phys.*, vol. 65, pp. 1724–1728, 1989.

67 F. K. Legoues, R. Rosenberg, and B. S. Meyerson, 'Kinetics and mechanism of oxidation of SiGe: dry versus wet oxidation', *Appl. Phys. Lett.*, vol. 54, pp. 644–646, 1989.

68 D. Fathy, O. W. Holland, and C. W. White, 'Formation of epitaxial layers of Ge on Si substrates by Ge implantation and oxidation', *Appl. Phys. Lett.*, vol. 51, pp. 1337–1339, 1987.

69 O. W. Holland, C. W. White, and D. Fathy, 'Novel oxidation process in Ge^{+}-implanted Si and its effect on oxidation kinetics', *Appl. Phys. Lett.*, vol. 51, pp. 520–522, 1987.

70 D. C. Paine, C. Caragianis, and A. F. Schwartzman, 'Oxidation of Si$_{1-x}$Ge$_x$ alloys at atmospheric and elevated pressure', *J. Appl. Phys.*, vol. 70, pp. 5076–5084, 1991.

71 K. K. Liou, P. Mei, U. Gennser, and E. S. Yang, 'Effects of Ge concentration on SiGe oxidation behaviour', *Appl. Phys. Lett.*, vol. 59, pp. 1200–1202, 1991.

72 S. S. Iyer, P. M. Solomon, V. P. Kesan, A. A. Bright, J. L. Freeouf, T. N. Nguyen, and A. C. Warren, 'A Gate-quality dielectric system for SiGe Metal-Oxide-Semiconductor Devices', *IEEE Electron Device Lett.*, vol. EDL-12, pp. 246–248, 1991.

73 P. W. Li, and E. S. Yang, 'SiGe gate oxide prepared at low temperature in an electron cyclotron resonance plasma', *Appl. Phys. Lett.*, vol. 63, pp. 2938–2940, 1993.

74 M. Mukhopadhyay, S. K. Ray, C. K. Maiti, D. K. Nayak, and Y. Shiraki, 'Electrical properties of oxides grown on strained SiGe layer at low temperatures in a microwave oxygen plasma', *Appl. Phys. Lett.*, vol. 65, pp. 895–897 (see also Vol. 66(12), p.1566, March 20, 1995 issue), 1994.

75 L. S. Riley, and S. Hall, 'X-ray photoelectron spectra of low tempoerature plasma anodized Si$_{0.84}$Ge$_{0.16}$ alloy on Si(100)', *J. Appl. Phys.*, vol. 85, pp. 6828–6837, 1999.

76 M. Mukhopadhyay, S. K. Ray, C. K. Maiti, D. K. Nayak, and Y. Shiraki, 'Properties of SiGe oxides grown in a microwave oxygen plasma', *J. Appl. Phys.*, vol. 78, pp. 6135–6140, 1995.

77 M. Mukhopadhyay, S. K. Ray, D. K. Nayak, and C. K. Maiti, 'Ultrathin oxides using N_2O on strained $Si_{1-x}Ge_x$', *Appl. Phys. Lett.*, vol. 68, pp. 1262–1264, 1996.

78 V. Craciun, I. W. Boyd, A. H. Reader, W. J. Kersten, F. J. G. Hakkens, P. H. Oosting, *et al.*, 'Microstructure of oxidized layers formed by the low temperature UV assisted dry oxidation of strained $Si_{0.8}Ge_{0.2}$ on Si', *J. Appl. Phys.*, vol. 75, pp. 1972–1975, 1994.

79 J. Eugene, F. K. LeGoues, V. P. Kesan, S. S. Iyer, and F. M. d'Heurle, 'Diffusion versus oxidation rates in silicon-germanium alloys', *Appl. Phys. Lett.*, vol. 59, pp. 78–80, 1991.

80 S.-G. Park, W. S. Liu, and M.-A. Nicolet, 'Kinetics and mechanism of wet oxidation of GexSi1-x alloys', *J. Appl. Phys.*, vol. 75, pp. 1764–1770, 1994.

81 J. P. Zhang, P. L. F. Hemment, S. M. Newstead, A. R. Powell, T. E. Whall, and E. H. C. Parker, 'A Comparison of the behavior of $Si_{0.5}Ge_{0.5}$ alloys during dry and wet oxidation', *Thin Solid Films*, vol. 222, pp. 141–144, 1992.

82 J. Xiang, N. Herbots, H. Jacobsson, P. Ye, S. Hearne, and S. Whaley, 'Comparative study on dry oxidation of heteroepitaxial $Si_{1-x}Ge_x$ and $Si_{1-x-y}Ge_xC_y$ on Si(100)', *J. Appl. Phys.*, vol. 80, pp. 1857–1866, 1996.

83 D. K. Nayak, K. Kamjoo, J. S. Park, J. C. S. Woo, and K. L. Wang, 'Wet oxidation of GeSi strained layers by rapid thermal processing', *Appl. Phys. Lett.*, vol. 57, pp. 369–371, 1990.

84 H. Tsutsu, W. J. Edwards, D. G. Ast, and T. I. Kamins, 'Oxidation of polycrystalline-SiGe alloys', *Appl. Phys. Lett.*, vol. 64, pp. 297–299, 1994.

85 A. K. Rai, and S. M. Prokes, 'Wet oxidation of amorphous Si-Ge layer deposited on Si(001) at 800 and 900 °C', *J. Appl. Phys.*, vol. 72, pp. 4020–4025, 1992.

86 B. E. Deal, and A. S. Grove, 'General relationship for the thermal oxidation of silicon', *J. Appl. Phys.*, vol. 36, pp. 3770–3778, 1965.

87 P.-E. Hellberg, S.-L. Zhang, F. M. d'Heurle, and C. S. Petersson, 'Oxidation of silicon-germanium alloys. I. An experimental study', *J. Appl. Phys.*, vol. 82, pp. 5773–5778, 1997.

88 P.-E. Hellberg, S.-L. Zhang, F. M. d'Heurle, and C. S. Petersson, 'Oxidation of silicon-germanium alloys. II. A mathematical model', *J. Appl. Phys.*, vol. 82, pp. 5779–5787, 1997.

89 K. Prabhakaran, T. Ogino, T Scimeca, Y. Watanabe, and M. Oshima, 'Bonding partner change reaction in oxidation of Ge on Si(001): Observation of two step formation of SiO_2', *Appl. Phys. Lett.*, vol. 64, pp. 1839–1841, 1994.

90 K. Prabhakaran, T. Nishoka, K. Sumitomo, Y. Kobayashi, and T. Ogino, 'Oxidation of Ultrathin SiGe Layer on Si(001): Evidence for Inward Movement of Ge', *Jap. J. Appl. Phys.*, vol. 33, pp. 1837–1838, 1994.

91 G. L. McVay, and A. R. DuCharme, 'Diffusion of Ge in SiGe Alloys', *Phys. Rev. B*, vol. 9, p. 627, 1974.

92 W. Frank, U. Gosele, H. Mehrer, and A. Seeger, *Diffusion in Crystalline Solids*. Academic, New York, 1984.

93 C. Tetelin, X. Wallart, L. Vescan, and J. P. Nys, 'Plasma assisted oxidation of SiGe layers at 500 °C: interface characterization', *Appl. Surf. Sci.*, vol. 104/105, pp. 385–391, 1996.

94 W. S. Liu, E. W. Lee, M-A. Nicolet, V. Arbet-Engels, K. L. Wang, N. M. Abuhadba, *et al.*, 'Wet oxidation of GeSi at 700 °C', *J. Appl. Phys.*, vol. 71, pp. 4015–4018, 1992.

95 E. C. Frey, N. Yu, B. Patnaik, N. R. Parikh, M. L. Swanson, and W. K. Chu, 'Transition between Ge segregation and trapping during high-pressure oxidation of Ge_xSi_{1-x}/Si', *J. Appl. Phys.*, vol. 74, pp. 4750–4755, 1993.

96 C. Tetelin, X. Wallart, J. P. Nys, L. Vescan, and D. J. Gravesteijn, 'Kinetics and mechanism of low temperature atomic oxygen-assisted oxidation of SiGe layers', *J. Appl. Phys.*, vol. 83, pp. 2842–2846, 1998.

Chapter 4
Strain-engineered MOSFETs

In the field of microelectronics, the planar Si metal-oxide-semiconductor field-effect transistor (MOSFET) is perhaps the most important invention. It started in 1928 when J.E. Lilienfeld proposed the concept of field-effect conductivity modulation and the MOSFET [1]. With the discovery of silicon dioxide (SiO_2) passivation for the Si semiconductor system by Atalla in 1958, the modern Si MOSFET era started [2]. Since then MOSFET performance has improved at a dramatic rate owing to gate length scaling and has become the dominant technology for integrated circuits.

During the early years of transistor scaling, Gordon Moore predicted that the number of transistors in a chip will increase exponentially [3, 4], which is now known as Moore's law. However, according to Moore himself, 'no exponential is forever' [5]. Many of the exponential trends are approaching limits that require new means to circumvent to continue the historic rate of progress. Accordingly, new and more fundamental barriers must be confronted in the coming decades – the fact that the technology is approaching atomic dimensions. For example, Figure 4.1 shows the increase in computer performance benefiting from both faster transistors and an exponentially increasing transistor budget. Figure 4.2 shows the increasing power dissipation of microprocessor chips, despite all attempts to decrease operating voltage (see Figure 4.3) and minimise leakage currents.

Key to continuing the historic trends, the transistor itself must evolve from the planar structure, generally used today. The key challenge for continuing transistor scaling below a gate length of 30 nm is the growth in off-state leakage current such as has been observed in an experimental prototype 10 nm planar device [6].

However, future progress can continue with new materials and device structures. Several ideas have been put forward to decrease leakage and continue speed improvements, including fully depleted silicon-on-insulator [7], as well as double-gate [8] and triple-gate [9] structures. Figure 4.4 shows directions in which the standard planar transistor is evolving. Even the crystal structure of the silicon has been modified by introducing strain into the lattice to increase carrier mobility and, hence, transistor performance [10].

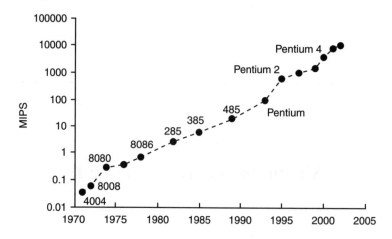

Figure 4.1   Processor performance (MIPS) [After G. E. Moore, Proc. ISSCC, 2003(20–23). © 2003 IEEE]

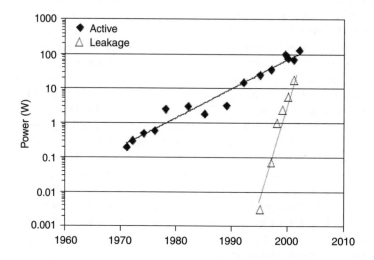

Figure 4.2   Processor power (W) – active and leakage [After G. E. Moore, Proc. ISSCC, 2003(20–23). © 2003 IEEE]

In this chapter, the key scaling limits are identified for MOS transistors (see Table 4.1), and methods for improving device performance are discussed. For improved short channel effects, creation of shallow source/drain extension (SDE) profiles, the use of retrograde and halo well profiles to improve leakage characteristics and the effect of scaling the gate oxide thickness are discussed in detail. Experimental data and simulations are used to show that although conventional scaling of junction depths is still possible, increased resistance for junction depths below 30 nm results in

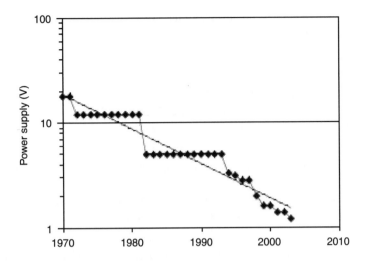

Figure 4.3 *Processor supply voltage [After G. E. Moore, Proc. ISSCC, 2003(20–23). © 2003 IEEE]*

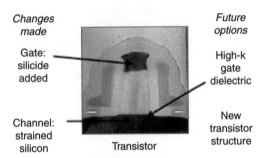

Figure 4.4 *New materials and device structures extending transistor scaling [After G. E. Moore, Proc. ISSCC, 2003(20–23). © 2003 IEEE]*

Table 4.1 *The key scaling limits for conventional MOS transistors [After S. E. Thompson et al., Lecture notes.]*

Feature	Limit	Reason
Gate length	$0.10\ \mu m$	leakage (I_{off})
Channel length	$0.06\ \mu m$	leakage (I_{off})
Oxide thickness	2.3 nm	leakage (I_{gate})
Junction depth	30 nm	resistance (R_{sde})
Channel doping	$V_t = 0.25$ V	leakage (I_{off})
SDE under diffusion	15 nm	resistance (R_{inv})

performance degradation. Fundamental trade-offs and scaling trends in engineering these effects are analysed.

4.1   Scaling issues

Shrinking the conventional MOSFET beyond the 90-nm technology node requires innovations to circumvent barriers due to the fundamental physics that constrain the conventional MOSFET. The limits most often cited are

- quantum-mechanical tunnelling of carriers through the thin gate oxide;
- tunnelling of carriers from source to drain, and from drain to the body of the MOSFET;
- control of the density and location of dopant atoms in the MOSFET channel and source/drain region to provide a high on-off current ratio;
- the increase in subthreshold slope.

These fundamental limits have led to pessimistic predictions of the imminent end of technological progress for the Si semiconductor industry [11–16]. The key scaling limits for conventional Si MOS transistors are shown in Table 4.1. Because of these limits, new solutions need to be developed for continued transistor scaling. Possible device performance improvement opportunities are shown in Table 4.2.

Small geometry effects become more severe than the fundamental limits of switching owing to quantum fluctuations, energy equipartition or thermal fluctuations. Short

Table 4.2   Device performance improvement opportunities [After H.-S. P. Wong, Lecture notes.]

Source of improvement	Parameters affected	Method
Charge density	S (inverse subthreshold slope) Q_{inv} at a fixed off-current	double-gate FET low operating temperature
Carrier transport	mobility (μ_{eff}) carrier velocity ballistic transport	strained-Si high-mobility substrates reduced mobility degradation factors shorter channel length
Ensuring device scalability to a shorter channel length	generalised scale length channel length	via electrostatic control sharp doping profiles halo/pocket implants high gate capacitance
Parasitic resistance	R_{ext}	extended/raised source/drain low-barrier Schottky contact
Parasitic capacitance	C_{jn} C_{gd}, C_{gs}, C_{gb}	SOI double-gate FET

channel effect due to the charge sharing along the transistor channel is more severe, as it is strongly dependent on the gate/channel coupling capacitance, the junction depth, the channel length and the doping concentration. This effect is the main limitation to minimal design rule and it can be of the order of threshold voltage itself, if low threshold voltage values are reached. Another CMOS challenge is to achieve symmetry of threshold voltage for n- and p-channel devices to yield ideal transfer CMOS inverter characteristics, thereby providing an acceptable trade-off between performance and standby leakage.

The use of n^+-poly gate for n-MOSFETs and p^+-poly gate for p-MOSFETs theoretically allows threshold voltage adjustment by enhancement of the well dopant concentration. However, this solution suffers essentially from B penetration into SiO_2 coming from the p^+-doped gate. Nitrided SiO_2 is a way to limit, without avoiding, this effect. However, the creation of trapping centres in the oxide or at the SiO_2/Si interface will decrease carrier mobility. The use of midgap gate material offers an attractive possibility of a single material for both n- and p-MOSFETs. However, the flat-band voltage is shifted by approximately 0.55 eV. It could lead to too high an absolute value of V_t, if one does not reduce or compensate the surface concentration. The metal gate integration of metal gate technology with conventional processing is not straightforward. The popularity of silicon gate technology comes from the self-alignment of source and drain on the gate. The process sequence has to be modified in order to allow a metal gate integration. Alternative approaches such as the damascene gate [17] or the replacement gate have been proposed to avoid the source and drain integration problem. For sub-0.10 μm devices, the following are the main challenging issues that cannot be avoided:

- Direct tunnelling through SiO_2 when the sub-2.5 nm thickness is reached. This increases the contribution of the leakage component to power consumption. Nevertheless, SiO_2 operating in the direct tunnelling current regime has been demonstrated to be usable down to a thickness of 1.4 nm without affecting device reliability [18–20].
- High doping levels in the channel reaching more than 5×10^{18} cm^{-3} will enhance reverse tunnelling current in source and drains up to values of 1 A/cm^2 under 1 V [21].
- Reverse short-channel effect due to lateral non-uniform doping resulting in V_{th} variation when channel length varies.
- Drain-induced threshold voltage shift due to change in drain-induced barrier lowering (DIBL) for long channel devices when the halo-implant influence on the channel diminishes.
- Poly-depletion is getting significant for ultrathin gate oxide, which accounts for about an 8-nm increase in equivalent oxide thickness (EOT) for most devices.
- High field in the drain to gate causes band-to-band tunnelling due to high junction doping and abrupt junctions.
- Proximity effect of shallow trench isolation (STI) stress reduces electron mobility but increases hole mobility, thus affecting I_{dsat}.
- Statistical dopant fluctuations.

The dopant fluctuations become very severe for geometries lower than 50 nm. Moreover, the discrete nature of dopant distribution can also give rise to devices characteristics asymmetry [22]. Special attention needs to be paid to these issues because the number of dopants in the channel of a MOSFET tends to decrease with scaling. The random dopant placement in the volume of the MOSFETs channel by the ion-implantation technique must be considered.

In the following, possible solutions to overcome the physical limitations that one could encounter in a classical scaling scenario through the different aspects of MOS devices optimisation are discussed:

1. gate, channel and substrate engineering;
2. source and drain engineering;
3. gate dielectric engineering.

Recently, a special issue has been brought out by the editors of the *IEEE Transactions on Electron Devices* to highlight some of the new innovations in strain-engineered MOSFETs and to survey their current progress and potential [23]. Interested readers may refer to this issue, which focuses on front-end materials such as strained-Si, SiGe and processes, as well as advanced transistor structures.

4.2   Mobility-enhanced substrate engineering

An engineered substrate is a material that can be manufactured and introduced into the conventional silicon processing, resulting in products that are unique and could not have been created on Si substrates. High carrier mobility MOSFETs and the integration of photonic materials like GaAs are some examples in which new ICs can be created using new materials platforms or substrates [24, 25]. Uniaxial process-induced strain is being adopted in all 90-, 65- and 45-nm high-performance logic technologies. Uniaxial strain offers large performance improvement at low cost and minimally increased manufacturing complexity and is scalable to future technology nodes.

In nanoscale MOSFETs, biaxial strain has been the conventional method to strain the MOSFET channel. Recent studies in the field of uniaxial process-induced strain have revealed significant advantages over its biaxial counterpart [26, 27]. Starting at the 90-nm technology generation, process-induced strained-Si is one such material change that has been widely adopted [10, 28–30]. At present, strained-Si offers performance gains much larger than any other new material options. Three techniques are being used to introduce strain:

1. tensile and/or compressive capping layer post salicide;
2. local epitaxial films grown in the source and drain regions;
3. capping layers on top of the polySi gate before source/drain anneal for stress memorisation of the polySi gate, and the combination of these various techniques is additive.

Figure 4.5 shows the feature size progress during the last 30 years and highlights the new materials needed to continue the scaling trends. A possible method

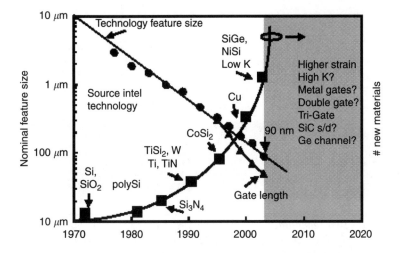

Figure 4.5   Technology and gate feature size and new materials needed vs. time [After S. E. Thomson et al., IEEE Trans. Semicond. Manufacturing, Vol. 18, 2005(26–36)]

for overcoming the limitations of conventional Si CMOS is the enhancement of the carrier transport properties in the transistors channel region through the use of new materials and/or the modification of the channel material properties. As VLSI technology uses various materials, it is important to understand how the different materials used and their patterning affects the stress present in nanoscale transistors. A complete understanding of these stress effects and their interaction with the intentionally or otherwise strained-Si layer is needed to ensure an optimum and stable process performance. Besides the intentionally added strain (substrate strain), the device and process-induced strain also have to be carefully controlled to obtain maximum device performance.

Two basic approaches exist for implementing strain in MOSFETs: a global approach, where biaxial stress is introduced across the entire substrate (substrate strain), and a local approach, where uniaxial stress is engineered into the device by means of epitaxial layers and/or high-stress nitride capping layers (process-induced stress). Two techniques to introduce stress or strain are shown in Figure 4.6. The difference between the uniaxial process stress and biaxial substrate strain is that process strain is present in both the n^+-polySi gate and Si channel while biaxial stress strains only the Si channel. For biaxial stress, a undesirably large n-channel threshold voltage shift has been reported [31–33] which is due to strain-induced bandgap narrowing [34], and is larger for biaxial than uniaxial tensile stress.

Currently, the most important area of engineered substrates is strained- or relaxed-SiGe/Si. The lattice mismatch between Si and Ge is about 4.18 per cent and offers a host of materials and devices that can be fabricated in the Si CMOS or bipolar processing, such as low-power or high-frequency CMOS and also the integration of III-V photonics with Si. Most of the initial work on strained-Si has focused on biaxial

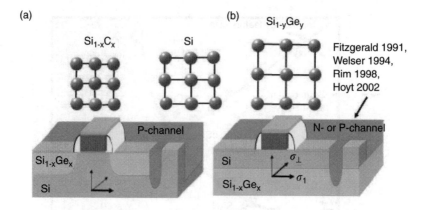

Figure 4.6   Two techniques to introduce stress or strain. (a) uniaxial stress and (b) biaxial strain [After S. E. Thompson, IEEE IDEAS Symp. Proc., 2005(14–16). © 2005 IEEE]

global strain using a wafer-based approach and using a thin strained-Si layer on a thick relaxed-SiGe virtual substrate. Uniaxial process-induced stress (as opposed to biaxial) is being pursued now because larger hole mobility enhancement can be achieved at low strain and because it results in significantly smaller stress-induced n-channel MOSFET threshold voltage shift [35]. For first- and second-generation strained-Si MOSFETs the industry has adopted process-induced uniaxial stress [28, 29, 36].

Various methods have been proposed to induce the desired stress in the channel, such as the use of a SiGe virtual substrate [37], tensile films [38] or mechanical force [39]. The strain in the channel region can be obtained by the optimisation of the stress introduced by the individual process steps (process-induced stress, PSS) and can as well be implemented by using strained-Si substrates (substrate-induced, SS). It is important to make sure that the combined effect of the three stress components (along x, y and z axes) is still positive and different contributions do not cancel each other out. Accurate process/stress models need to be developed to incorporate both the intentional and unintentional stress sources in order to maximise device performance. Major progress has been made in new engineered substrate materials and their potential needs to be examined for microelectronic applications with a view to (1) new materials needing to be incorporated to continue downscaling and (2) performance advantage. In the following, a brief review on the latest developments in mobility-enhanced substrate engineering driven by IC industry requirements is given and examples of silicon heterostructure substrates of interest for applications in heterostructure CMOS (HCMOS) are shown.

4.2.1   Orientation-dependent mobility engineering

It is well known that hole mobility is more than doubled on (110) silicon substrates with current flow direction along ⟨110⟩ [40, 41] compared with conventional

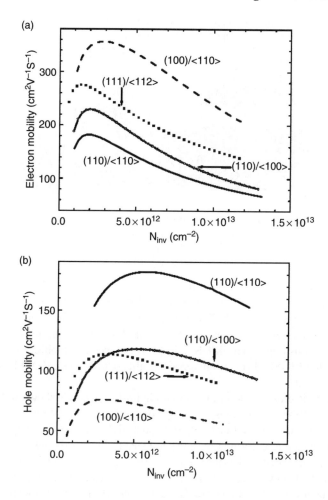

Figure 4.7 Carrier mobility in the inversion layer with various substrate orientations. Electron mobility (a) is highest on (100) substrate while hole
mobility (b) is highest on (110) substrate with current flow along ⟨110⟩
direction. The gate oxides (about 2 nm thick) were grown simultaneously
in N_2O ambient at 800°C in convention furnace [After M. Yang et al.,
IEEE IEDM Tech. Dig., 2003(453–456). © 2003 IEEE]

(100) substrates. The carrier mobility dependence on surface orientation is shown
in Figure 4.7. The electron mobility is found to be the highest on (100) substrates.
Uniaxial process-induced strain is generally applied either parallel (longitudinal) or
perpendicular (transverse) to the direction of MOSFET current flow.

   Although the hole mobility on (110) surface is about two times as high as that
on (100) surface [40, 42–44], the electron mobility of (110) n-MOSFETs is much
lower and about one-half of that of (100) counterpart [40, 42, 45]. This is a serious

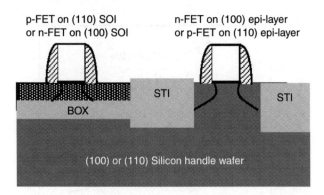

Figure 4.8   Schematic cross-section of CMOS on a hybrid-substrate with p-FET on (110) surface orientation and n-FET on (100) surface orientation [After M. Yang et al., IEEE IEDM Tech. Dig., 2003(453–456). © 2003 IEEE]

problem for the enhancement of CMOS current drive. As a result, it is important to develop a new device technology with enhanced electron mobility in the (110)-surface CMOS. Strained-SOI n-MOSFETs, fabricated on (110)-surface, show an electron mobility enhancement of 23 per cent along the (001) direction, against that of the (110)-surface unstrained-bulk MOSFETs [45]. An electron mobility ratio of (110) strained-SOI n-MOSFETs to the universal mobility of (100) bulk MOSFETs increases up to 81 per cent.

A schematic cross-section of CMOS on hybrid substrate is shown in Figure 4.8, including two types of structures: (1) p-FET on (110) SOI and n-FET on (100) silicon epitaxial layer, and (2) n-FET on (100) SOI and p-FET on (110) silicon epitaxial layer. The hybrid substrate is formed by layer transfer technique through wafer bonding (Figure 4.9). Interested readers may refer to an excellent review on hybrid-orientation technology (HOT) by Yang *et al.* [46]. The authors describe a novel planar silicon CMOS structure where n-FETs are fabricated on (100) surface orientation Si and p-FETs on (110) surface orientation on the same die, through wafer bonding and selective silicon epitaxy.

4.2.2   Mobility enhancement by process-induced stress

The influence of externally applied stress on the electrical performance of integrated circuits has been an active area of research for almost 30 years. STI can create large stresses due to thermal mismatch, oxide growth and trench fill, and the sharp corners at the top and bottom of the trenches are major contributors to the stress behaviour. A major concern when scaling local oxide isolation (LOCOS) and SIT structures is the build-up of localised stress near the isolation edge, often leading to dislocation formation and degrading device characteristics.

In the front-end process, STI is a major source of stress in the MOSFET channel. The proximity and amount of the stress in the silicon substrate limits the density of

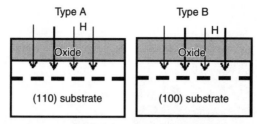

Step 1: Buried oxide formation and H I/I.

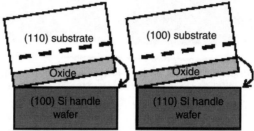

Step 2: Flip-bond implanted wafer to handle wafer
with different orientation.

Step 3: Wafer split and SOI thinning

Figure 4.9 *Process flow to form the hybrid-substrates (type A and type B) [After M. Yang et al., IEEE IEDM Tech. Dig., 2003(453–456). © 2003 IEEE]*

ICs, and when too much stress is exerted in the silicon, it will yield by releasing dislocations that lead to higher leakage currents and degraded device performance. Thin films such as silicon nitride, silicon dioxide and polySi are frequently encountered in device fabrication; these films contain intrinsic stresses as a result of the deposition process.

The advantage of uniaxial stress in Si process flow is that compressive strain is introduced into the p-type and tensile strain in the n-MOSFETs to improve both the electron and hole mobilities on the same wafer. Three state-of-the-art techniques for introducing uniaxial stress in the Si channel are described below. Transmission emission micrograph (TEM) micrographs of 45 nm p- and n-MOSFETs are shown in Figure 4.10. In the first approach [29], reported by Intel, Texas Instruments, Applied Materials and recently by IBM [47], TSMC, and Freescale [48], a local epitaxial SiGe film is grown in the source and drain regions and introduces uniaxial stress into the Si channel. The process flow consists of the following steps: the Si source and drain

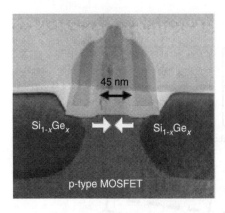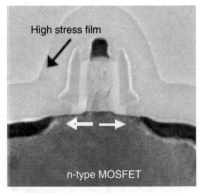

Figure 4.10   Transistor cross-section of 45-nm gate length strained-Si transistors used at 90-nm technology node [After S. E. Thompson, IEEE IDEAS Symp. Proc., 2005(14–16). © 2005 IEEE]

are etched, creating an Si recess, and then SiGe (for p-channel) or SiC (for n-channel) is epitaxially grown in the source and drain. This creates primarily a uniaxial compressive or tensile stress, respectively, in the channel of the MOSFET. For 17 per cent Ge, 500–900 MPa of channel stress is created, depending on the proximity of the SiGe to the channel. Impressive 60–90 per cent drive current enhancements on short devices (~35 nm) have been demonstrated, which offer greater device performance than alternative Si device enhancement concepts, such as multigates or high-k dielectrics.

To fabricate the strained-Si, in post-spacer etch, a Si recess etch is inserted followed by selective epitaxial $Si_{1-x}Ge_x$ deposition [see Figure 4.11(a)]. The silicon etch is blocked from n-channel devices and polySi gates. The Si recess etch removes 100 nm vertically and 70 nm laterally from the p-channel source and drain. Next, epitaxial $Si_{1-x}Ge_x$ is grown in the source and drain [see Figure 4.11(b)].

The $Si_{1-x}Ge_x$ growth is targeted to reduce the external resistance. The mismatch in the $Si_{1-x}Ge_x$ to Si lattice causes the smaller lattice constant Si channel to be under compressive strain. The next process flow is conventional except for the salicide [see Figure 4.11(c)]. $Si_{1-x}Ge_x$ in the source and drain requires extensive changes to the salicide since Ge inhibits the $CoSi_2$ transition to the low-resistivity disilicide phase [33]. To solve this problem, NiSi instead of $CoSi_2$ is used. Nickel silicide requires extensive changes since all formation and post-formation process steps need to be less than 500 °C.

After salicide formation, longitudinal uniaxial tensile strain is introduced into the n-MOSFET by engineering the tensile stress and thickness of the Si nitride-capping layer [see Figure 4.11(d)]. The thickness of the capping layer is approximately 80 nm. The capping layer introduces longitudinal tensile and compressive out-of-plane (z) stress [38, 49]. Florida object oriented three-dimensional finite element process simulator (FLOOPS) [50, 51], is a very useful technology computer aided design

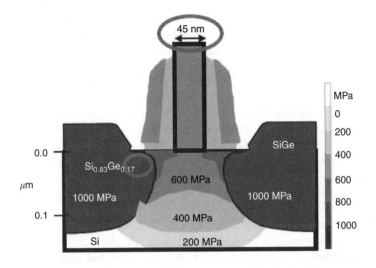

Figure 4.11   Stress contours for 45 nm gate length transistor with $Si_{0.83}Ge_{0.17}$ in the source and drain [After S. E. Thompson, IEEE IDEAS Symp. Proc., 2005(14–16). © 2005 IEEE]

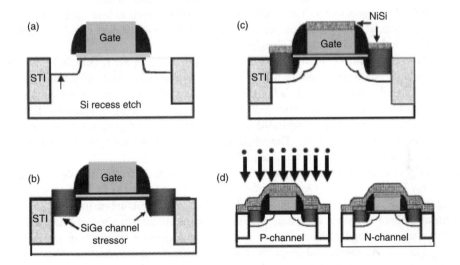

Figure 4.12   (a–d) Process-induced strained-Si formation process flow [After S. E. Thompson, IEEE IDEAS Symp. Proc., 2005(14–16). © 2005 IEEE]

(TCAD) aid to calculate the channel stress for both n- and p-channel MOSFETs. For a nominal 45-nm gate length transistor used at the 90-nm technology node, ~500 MPa of uniaxial longitudinal compressive stress is introduced into the p-channel MOSFET in the inversion layer, as demonstrated by stress contours in Figure 4.12.

The use of $Si_{1-x}Ge_x$ in the p-channel source and drain and a tensile capping film, and capping film strain relaxation off the p-MOSFET, allows independent targeting of the n- and p-type MOSFET Si channel strain. The channel stresses are predominantly uniaxial, which is desirable since uniaxial stress offers many advantages in electrical performance over biaxial stress, as will be discussed later. Stress-induced tensile capping films are adopted at the 90-nm technology generation and can improve n-channel device saturated drive current 10–15 per cent [28,30]. However, the tensile stress from the capping layer needs to be relaxed from the p-channel device since it causes significant degradation [28,30,38,49]. There are several techniques to almost completely neutralise the capping layer strain; one is the use of a Ge implant and masking layer.

A second, less complex technique for introducing strain in the MOSFET is the use of a tensile and/or compressive capping layer [52,53]. The capping films are introduced either as a permanent layer post salicide, as discussed here, or as a sacrificial layer before source and drain anneal, as will be discussed next. A dual capping layer approach enabled by the creation of very high-compressive and high tensile stress SiN has been reported [54]. SiN layers with more than 2 GPa of tensile stress and more than 2.5 GPa of compressive stress have been developed by Applied Materials. IBM, AMD and Fujitsu [52,53] have reported a CMOS architecture in which high-tensile and high-compressive Si nitride layers are selectively deposited on n-MOSFET and p-MOSFET, respectively. This dual-stress liner (DSL) architecture creates longitudinal uniaxial tensile and compressive stress in the Si channel to simultaneously improve both n- and p-channel transistors.

The process flow consists of a uniform deposition of a highly tensile Si_3N_4 liner post silicidation over the entire wafer, followed by patterning and etching the film off p-channel transistors. Next, a highly compressive SiN layer is deposited, and this film is patterned and etched from n-channel regions. High-stress compressive films can induce channel stress comparable in magnitude to the first-generation embedded SiGe in the source and drain. The advantages of DSL flow over epitaxial SiGe are reduced process complexity and integration issues.

Higher channel strain is possible by increased strain in the (1) nitride-capping layer, (2) epitaxial $Si_{1-x}Ge_x$ by higher Ge concentration or (3) fabrication of the epitaxial $Si_{1-x}Ge_x$ closer to the Si channel. More complex structures to integrate strain are also possible, such as epitaxial SiC in the source and drain of n-channel MOSFETs, that may offer higher channel strain than the capping-layer approach or strained-$Si_{1-x}Ge_x$ in the source and drain for uniaxial tensile strain introduction. There are also numerous other techniques and process steps to introduce strain such as high-strain capping layers introduced before polySi gate crystallisation [55], high-stress shallow-trench isolation and silicide [56].

In the near term for future technology generations, combinations of all these techniques will be used. This will place additional restrictions and requirements on new structures and materials. For example, for alternate device structures and materials such as FinFET, tri-gate and high-k gate dielectrics to be competitive with strained planar CMOS, strain or some other mobility-enhancing technique is needed in these structures as well [57].

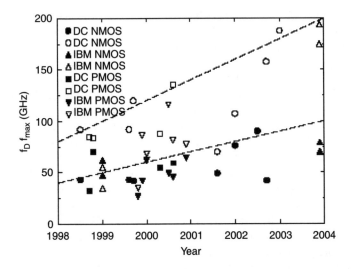

Figure 4.13 Cut-off frequencies f_T (filled symbols) and f_{max} (open symbols) of strained-Si FETs [After V. Palankovski et al., Asia Pacific Microwave Conference Abstr. and Proc., 2004]

4.2.3 Mobility enhancement by substrate-induced strain

Through the conventional scaling of MOSFET devices via gate oxide thickness, SDE, junction depths and gate lengths have enabled MOSFET gate lengths to be reduced from 10 μm in the 1970s to 0.1 μm or below now. The push to scale the conventional MOSFET continues to show remarkable progress [58]. Figure 4.13 summarises reported values for the cut-off frequencies f_T and f_{max}. In order to maintain this rapid rate of improvement, aggressive engineering of the source/drain and channel regions is required. It is also important to research in the area of materials compatible with Si technology and device structures for improving the speed of VLSI circuits. SiGe has emerged as a promising material because of its electrical and material properties.

Currently, SiGe has the major impact on silicon device technology primarily because of its electrical properties and also because of its other material properties. The areas of major impact include bipolar/BiCMOS technology and CMOS technology. There has been tremendous progress to grow device quality SiGe films. The initial growth technique used was Si MBE but this technique was replaced by high-quality chemical vapour deposition (CVD) techniques such as UHVCVD. The ability to in situ dope SiGe films with B and P and incorporate C in the film has greatly extended the use of the film and the performance levels achieved with it. The most mature and well-known example application of SiGe is SiGe BiCMOS technology, where it is employed in manufacturing processes [59, 60]. Because of the application potential of SiGe BiCMOS in wireless, wired, storage, test and other areas SiGe revenues are expected to grow at a rate of over 30 per cent per year.

SiGe applications in hetero-FET technologies has also been demonstrated. SiGe has been used to advantage for extrinsic as well as intrinsic applications. However, there are no qualified production technologies using SiGe in FET structures. For the CMOS technology, although the SiGe channel has been used to enhance the performance of p-MOS, a desired improvement of the complimentary n-MOS transistors is not achievable with SiGe. Oxides grown on SiGe have poor characteristics so there has been some concern about the quality of its use as a gate material where it is deposited on the gate oxide. Some initial reliability studies have revealed an enhanced degradation in polySiGe gate devices compared to polySi gate devices when the polySiGe was deposited on the gate oxide [61].

SiGe has been used to enhance the intrinsic conduction in FET structures using bandgap engineering. Several suggested structures for a CMOS type of process may be grouped into three categories: (1) pseudomorphic compressively strained-SiGe channels [62], (2) tensile strained surface channel [37, 63, 64] and (3) modulation doped buried channel MODFETs [65]. For these three classes of devices the primary aspects of each are summarised in Table 4.3.

There are two major extrinsic application areas of SiGe in hetero-FET structures: that of replacing the gate polySi with gate polySiGe and using it for raised source/drains. There are three advantages for using it as a gate material: (1) lower thermal budget than silicon [66], (2) superior boron dopant activation [67] and (3) a variable workfunction p-type gate [68]. The second extrinsic application raised source/drain, which is currently being used in CMOS technology. The fundamental problem is realising a shallow junction contact to the FET channel with a low resistance contact typically done by forming a silicide. Advanced scaled devices

Table 4.3   Comparison of various CMOS SiGe FET structures [After D. L. Harame et al., Appl. Surf. Sci., Vol. 224, 2004(9–17).]

Parameter	Pseudomorphic compressive strained-SiGe	Tensile strained-Si surface channel	Modulation-doped strained-SiGe channel
Substrate	Silicon	SiGe on relaxed buffer	SiGe on relaxed buffer
RMS cross-hatch (nm)	none	2–10	2–10
Threading dislocations (cm^{-2})	negligible	10^5–10^7	10^5–10^7
Stable at 1000 °C	Yes	Yes	No
Layer structure	simple	moderate	complex
p/n-channel	buried/surface	surface/surface	buried/buried
p/n μ enhancement	(1.5–2.0)/1.0	(1.4–1.8)/2.0	~5/3
p/n g_m enhancement	(1.2–1.6)/1.0	(1.4–1.7)/1.6	~2.2–1.3
Projected stage delay improvement (%)	10–20	30–40	100

require shallow junctions that are problematic in achieving low resistance. A raised S/D structure prior to S/D implant enables a shallower junction contact to the device. Using SiGe in the raised source/drain can reduce the thermal budget of the selective epitaxial growth [69] and reduce the contact resistance of the silicide/semiconductor interface [70]. Though Co has been shown to be problematic for use with SiGe, it has been demonstrated that Ni and Pt form good low-resistance silicides on SiGe layers [71].

It is well known that due to lattice mismatch, an epitaxial ultrathin Si layer grown pseudomorphically on relaxed-SiGe layer on Si experiences a biaxial compressive (tensile) strain, provided that the layer thickness is below the critical thickness [59]. This strain leads to a modification of both the conduction and valence bands. It lifts the degeneracy of the light and heavy hole bands and lowers the spin-orbit band resulting in reduction of inter-band scattering and improvement of hole mobility. In order to investigate and design strained-Si device structures, it is necessary to model the carrier mobilities in these devices. In the following, the theoretical and experimental mobility reported to describe the doping and material composition dependence for the strained-Si material are discussed. Reported enhancements with local and global strain engineering techniques are shown in Table 4.4.

The presence of the Ge atoms is responsible in modifying the band structure of strained-$Si_{1-x}Ge_x$ layer. In SiGe, Ge composition may change from 0.0 to 1.0, and this change is non-linear in nature, giving rise to Si like band structure when $x < 0.85$ and Ge like band structure when $x > 0.85$. As is well known, the strain in the silicon layer causes the six-fold degenerate valleys of the silicon conduction band minimum to split into two groups: two lowered valleys with the longitudinal effective mass axis perpendicular to the interface, and four raised valleys with the longitudinal mass

Table 4.4   Reported enhancements with local and global strain engineering techniques

Manufacturer	Strain type	L_g (nm)	NMOS I_{dsat} enhancement	PMOS I_{dsat} enhancement	References
Intel	local	45	10%	25%	[28]
AMD	global	25	25%	n/a	[32]
IBM/Sony/Toshiba/ AMD/Chartered	local on SOI	45	11%	20%	[53]
AMD	local on SOI	40	13%	20%	[72]
Freescale/TSMC/ SOITEC	global on SOI	150	28%	12%	[73]
IBM	global	67	35%	n/a	[74]
Intel	global	140	24%	n/a	[75]
Toshiba	global	40	n/a	19%	[76]
TSMC	global	60	15%	n/a	[77]

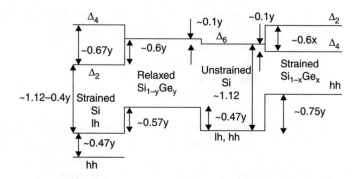

Figure 4.14   Bandgap alignment (in eV) of strained-Si relative to relaxed-Si$_{1-y}$Ge$_y$, strained-Si$_{1-x}$Ge$_x$ and Si [After V. Palankovski et al., Asia Pacific Microwave Conference Abstr. and Proc., 2004]

axis parallel to the interface. The combination of a lighter effective mass and reduced intervalley scattering gives rise to higher electron mobility and originates spectacular electron velocity overshoot.

For Ge < 0.85, compressively straining SiGe leads to splitting of the six-fold degenerate Δ_6-valleys in Si into two-fold degenerate Δ_2 valleys higher in energy and four-fold degenerate Δ_4 valleys lower in energy. Figure 4.14 shows the band alignment of strained-Si on SiGe relative to relaxed-SiGe and unstrained-Si. The figure shows the strain-induced splitting of the conduction and valence bands, together with the band edge discontinuities, as a function of the germanium content y in the SiGe layer. The higher in-plane effective electron mass of Δ_4 valleys leads to a reduction of the electron mobility for strained-SiGe. In the case of tensile strained-Si, the direction of motion of the splitting is reversed with the Δ_4 valley moving lower in energy and Δ_4 higher. The lower in-plane effective mass of electrons in the Δ_2 valleys and the reduction of inter-valley phonon scattering lead to an enhanced electron mobility. In the calculations a linear dependence of the discontinuities on y has been assumed, which gives a good agreement with reported data [60].

In Si-heterostructures, the mobility enhancement factor is defined as the ratio between the extracted mobility from strained-Si MOSFETs and the extracted mobility in conventional Si MOSFETs processed in same run. Figure 4.15(a) shows the mobility enhancement ratio for electrons as a function of the Ge content y in the SiGe buffer layer. The figure compares experimental data reported by various research groups. As can be seen in the figure, the enhancement of the electron mobility increases gradually with the Ge content *y* for *y* < 0.2 and tends to saturate for higher values. It is noted that the electron mobility enhancement of more than 50 per cent is observed in a wide range of effective fields (up to 2 MV/cm) and doping concentrations (up to 6×10^{18} cm^{-3}). Similarly, Figure 4.15(b) shows the mobility enhancement ratio for holes as a function of the Ge content of the Si$_{1-y}$Ge$_y$ buffer

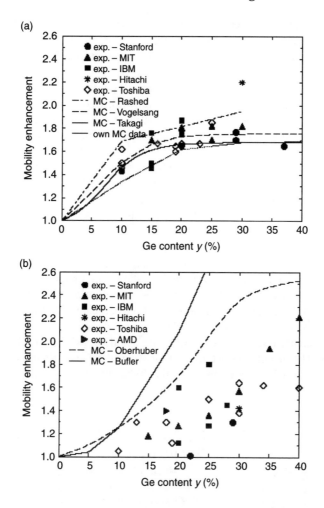

Figure 4.15 *Mobility enhancement ratio for (a) electrons and (b) holes as a function of the Ge content y in the Si$_{1-y}$Ge$_y$ buffer layer [After V. Palankovski et al., Asia Pacific Microwave Conference Abstr. and Proc., 2004]*

layer. The figure compares experimental data from various research groups. It can be seen that the enhancement of the hole mobility increases gradually with the Ge content y for $y \leq 0.4$.

Transport properties of strained-Si and SiGe layers have been theoretically inves-tigated using Monte Carlo calculations [78–80] or near equilibrium solutions to the Boltzmann equation [81]. To enable predictive simulations using TCAD tools a reliable set of models for the Si/SiGe material system is required. A com-prehensive set of strain-dependent models for parameters such as the low-field, high-field and the surface mobility, energy relaxation time and carrier life times for TCAD purposes is necessary. For this, appropriate models for the band structure parameters and deformation potentials must be used. Pseudopotential calculations for the transport properties of strained-Si or SiGe have been reported in References 81 and 82.

Monte Carlo simulation, which accounts for alloy scattering and the splitting of the anisotropic conduction band valleys due to strain [83], in combination with an accurate ionised impurity scattering model [84], allows one to obtain results for strained-Si for the complete range of donor and acceptor concentrations and Ge contents in the $Si_{1-y}Ge_y$ buffer layer. Strain effects on the device characteristics have been studied by Monte Carlo device simulation [85]. Use of the same functional form for Si to fit the doping dependence of the in-plane and the perpendicular mobility component for $y = 0$ and $y = 1$ (Si and strained-Si on Ge) has been reported [86]. Other possible approaches are to use analytical models [87] or tabulated Monte Carlo data in a device simulator [88].

Figure 4.16 shows the in-plane (parallel) and the perpendicular minority electron mobility in strained-Si as a function of y in the $Si_{1-y}Ge_y$ buffer at 300 K for different acceptor doping concentrations. Figure 4.17 shows the minority electron mobility in strained-$Si_{1-x}Ge_x$ as a function of acceptor concentration N_A in comparison with Monte Carlo simulation data both for in-plane and perpendicular directions. It is seen that the electron mobility in perpendicular (vertical) direction is enhanced for typical base doping concentrations (above 10^{18} cm^{-3}) and typical Ge content, $x \leq 0.2$ in SiGe HBTs. The model parameters used for strained-SiGe on Si and for strained-Si on relaxed-SiGe at 300 K are summarised in Table 4.5.

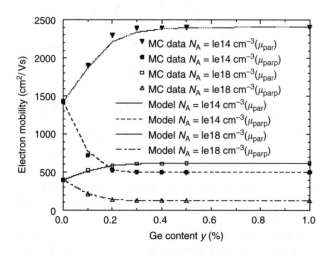

Figure 4.16   Minority electron mobility in strained-Si as a function of the Ge content y in the $Si_{1-y}Ge_y$ buffer layer for different acceptor doping concentrations [After V. Palankovski et al., Asia Pacific Microwave Conference Abstr. and Proc., 2004]

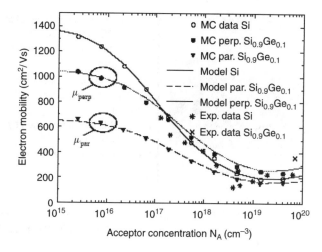

Figure 4.17 *Minority electron mobility in $Si_{1-x}Ge_x$ as a function of N_A and x: comparison with measurements and Monte Carlo simulation data for in-plane and perpendicular directions [After V. Palankovski et al., Asia Pacific Microwave Conference Abstr. and Proc., 2004]*

Table 4.5 *Parameter values for the majority/minority electron mobility in strained-SiGe and Si at 300 K [After V. Palankovski et al., Asia Pacific Microwave Conference Abstr. and Proc., 2004.]*

Parameter	Si	Ge (on Si)	SSi($\mu_\parallel$)	Strained-Si($\mu_\perp$)
μ_n^L(cm^2/Vs)	1430	560	2420	502
μ_{maj}^{mid} (cm^2/Vs)	44	80	95	20
μ_{maj}^{hi} (cm^2/Vs)	57	59	123	25
μ_{min}^{mid} (cm^2/Vs)	141	124	232	49
μ_{min}^{hi} (cm^2/Vs)	218	158	315	62
α	0.65	0.65	0.65	0.65
β	2.0	2.0	2.0	2.0

4.3 Channel engineering

The simplest SiGe CMOS structure is the pseudomorphic compressively strained-SiGe channel. Since the band offsets are almost entirely in the valence band, the holes are confined away from the interface and the mobility is significantly improved. Threshold voltage implants in the channel may degrade the mobility. At high over-drive, a parallel surface channel conduction path exists that further degrades the

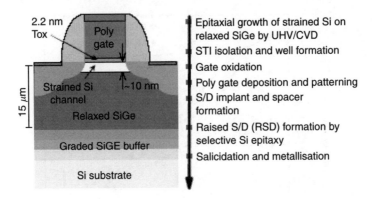

Figure 4.18   *Strained-Si CMOS device cross-section and process flow in a 0.13-μm process [After K. Rim et al., Proc. VSLI Tech. Symp., 2002(98–99). © 2002 IEEE]*

overall mobility. With the rapid improvement in CMOS performance and the lack of improvement in the n-FET, these devices have not found widespread use.

The replacement of the channel material by strained-Si, which utilizes an underlying relaxed-SiGe layer for its functioning, renders a solution to the problem since it leads to enhancement of both the electron and hole mobilities. Strained-Si/SiGe FETs exhibit superior performance for radio frequency (RF) applications. The strained-Si CMOS and the MODFET CMOS structures require a relaxed-SiGe buffer layer. The structure and process of a 0.13 μm strained-Si CMOS fabricated on tensile strained-Si epitaxially grown on a relaxed-SiGe/SiGe buffer layer [89] is shown in Figure 4.18. The three substrate regions included in the strained-Si hetero-FET devices are (1) tensile strained-Si device, (2) relaxed-SiGe and (3) SiGe buffer layers.

The SiGe buffer layer gradually increases the Ge content (15–30 per cent Ge) until the layer relaxes by means of a high density of misfit dislocations. The high-defect density in the SiGe buffer layers is shown in Figure 4.19. The relaxed-SiGe is a high-quality layer with low (10^5–10^7 cm^{-2}) defect density on which a thin tensile strained-Si layer is grown. The asymmetric strain in the silicon changes the band structure and splits both the conduction and valence bands, increasing the mobility of both electrons and holes [37, 90]. The band alignment is also changed such that electrons are confined to the strained-Si layer but holes may also be confined to a parasitic buried channel at the relaxed-SiGe layer interface as shown in Figure 4.20. Strained-Si CMOS is the most promising SiGe hetero-FET device candidate because of its similarity with conventional CMOS processing, since it is a material innovation.

The improvement in mobility is a function of the Ge content achieved in the relaxed layer with electron mobility plateauing after 20 per cent and hole mobility continuing to improve up to 40 per cent Ge. The electron mobility of strained-Si FETs is improved by 70 per cent on relaxed $Si_{0.8}Ge_{0.2}$ buffer layers and hole mobility is improved by 70 per cent on relaxed-$Si_{0.6}Ge_{0.4}$ buffer layers [33]. Significantly, these mobility enhancements of 50 per cent have been shown for E_{eff} up to 2 MV/cm^{-1} and

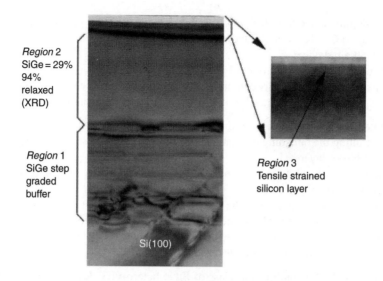

Figure 4.19   *TEM micrographs of tensile strained-Si quantum well. Note the three regions; (1) SiGe step graded buffer, (2) relaxed-SiGe and (3) device layers [After K. Rim, Application of Silicon-Based Heterostructures to Enhamced Mobility Metal-Oxide-Semiconductor Field-Effect Transistors, PhD Thesis, Stanford University, 1999]*

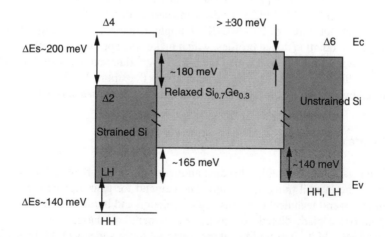

Figure 4.20   *Band alignment in strained-Si CMOS [After K. Rim, Application of Silicon-Based Heterostructures to Enhamced Mobility Metal-Oxide-Semiconductor Field-Effect Transistors, PhD Thesis, Stanford University, 1999]*

channel doping above 6×10^{18} cm^{-3} [64], which easily covers the range of channel doping (up to 6×10^{18} cm^{-3}), and electric fields (up to 1 MV/cm) found in CMOS down to 50 nm devices.

At moderate Ge content of $\sim$15 per cent enhancements in I_{ON} of 15–20 per cent have also been experimentally demonstrated. There are several challenges in realising these benefits in conventional CMOS processes [91], which include:

1. material quality of the epitaxial layers;
2. controlling the threading dislocations from the high Ge content (20–30 per cent) SiGe buffer layers;
3. thermal conductivity $\sim$15$\times$ lower in 20 per cent SiGe, which reduces the performance enhancements;
4. process issues arising from the presence of the SiGe layer (e.g., dopant diffusion (B reduction, as enhancement), isolation, oxidation and silicidation of high Ge content SiGe layers). In spite of these difficulties, strained-Si CMOS looks promising, particularly for nodes beyond 90 nm.

The MODFET is the third in class of SiGe hetero-FET devices that have been proposed for very low-noise high f_T devices for RF/analog applications. Very high mobilities can be achieved with modulation doping and buried undoped channels. For example, mobilities as high as 2800 cm^2/Vs can been achieved in n-type doped Si/SiGe MODFETs [92], an increase of 3–5 times, compared to unstrained Si n-MOSFETs. The hole mobility of compressively strained layers on SiGe shows even greater enhancement compared to unstrained Si. Si$_{0.2}$Ge$_{0.8}$-channel quantum wells grown on relaxed-Si$_{0.7}$Ge$_{0.3}$ have produced mobilities over 1000 cm^2/Vs, while Ge quantum wells grown on Si$_{0.4}$Ge$_{0.6}$ have yielded room temperature hole mobilities around 1800 cm^2/Vs [93]. The later value is about ten times higher than the hole mobility in Si p-MOSFETs for the same carrier density. Process integration is more difficult for a true CMOS MODFET implementation, which requires both p- and n-type modulation doped profiles, which makes the approach less attractive. An implementation for CMOS in which the n-FET is modulation doped and the p-FET has a very high Ge content buried channel above the strained-Si channel has been suggested by Armstrong [94]. Owing to complexity, this approach is not optimal.

4.4   Gate engineering

MOSFET gate engineering has become an increasingly important technology component in the overall transistor design. The demand for gate engineering has been driven by several technical concerns. Gate depletion and boron penetration through thin gate oxide place directly opposing requirements on the gate engineering for advanced MOSFETs. The traditional gate material for n- and p-MOSFETs is heavily doped n-type (n$^+$-polySi). The n$^+$-polySi gate enables the n-MOSFET to operate in a surface-channel mode while the p-MOSFET operates in a buried-channel mode. To have more balanced devices and improved short-channel effects, the p-MOSFET should also operate in the surface-channel mode and therefore should have a p$^+$-polySi gate. However, for scaling in the deep sub-micron and beyond 100 nm, the effects of gate depletion and boron penetration become the major concern in devices incorporating p$^+$-polySi gates. The problem of B penetration in p$^+$-polySi-gated devices would obviously be aggravated in advanced CMOS technologies with ultrathin SiO$_2$ (<3 nm).

In general, the gate electrode should be doped very heavily in order to decrease the gate depletion effects. For nanoscale devices, if an insufficient doping level in the gate is obtained owing to the use of a low-thermal budget and/or low implant dose to avoid B penetration, the polySi gate depletion effect (PDE) will become unacceptable. Figure 4.21(a) shows the poly gate depletion due to a relatively low-doped gate. However, the high amount of dopant density in the poly gate causes a high-density charge layer at the polysilicon gate–oxide interface when gate depletion occurs [see Figure 4.21(b)]. These parasitic charges behave as charge centres for scattering to the carriers in the channel region. This degrades the drain current.

Process simulation results of p- and n-MOSFETs with various gate doping profiles using TSUPREM4 are shown in Figure 4.22. The goal of the process simulations is to simulate devices for a fixed threshold voltage (V_t). The simulated p-MOS devices have a gate doping of 1×10^{20} cm^{-3} (p$^+$ gate), down to 5×10^{13} cm^{-3} (p-gate and n-gate). However, such low doping concentrations are not useful, because of very

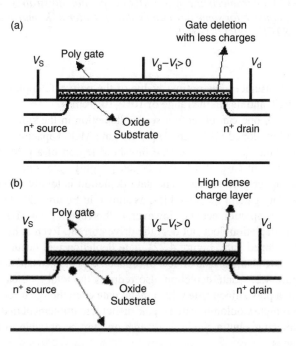

Figure 4.21   Schematic diagrams of (a) polySi gate-depletion due to a relatively low-doped gate and (b) high-density charge layer at polySi-gate oxide interface [After R. Komaragiri et al., Proc. SAFE, 2004(704–709)]

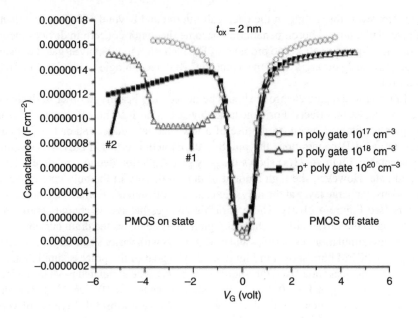

Figure 4.22   C-V characteristics of a n-MOS capacitor illustrating gate depletion (adjusted for same flat-band voltage) [After R. Komaragiri et al., Proc. SAFE, 2004(704–709)]

high poly gate resistance. The optimal value of the poly gate doping depends upon the channel length and lies in the range of 10^{17}–10^{19} cm^{-3}.

In Figure 4.22, one can observe a severe depletion in the simulated C-V characteristics of a low-doped (10^{17} cm^{-3}) p-poly gate MOS capacitor. The gate starts depleting at around $V_g = -1$ volt (the threshold region of a p-MOS) in the low doped p-type case. In case of a highly doped p^{+}-poly gate, one can see the gate depletion at a higher gate voltage. The gate depletion is totally absent in the case of an n-doped poly gate in negative bias, as shown in Figure 4.22. Only at positive gate bias, the gate depletion occurs. However, in this region the p-MOSFET is turned off, so that gate depletion does not degrade drive current. Even at high doping concentrations (p^{+} doped gate, Figure 4.22), gate depletion will occur by introducing more parasitic gate charges. The effects of reduced capacitance and drain current are shown in Figure 4.22. Gate depletion also reduces the total gate capacitance. The replacement of a poly silicon gate with a metal gate even though it removes the poly depletion and remote Coulomb scattering, is difficult to implement in dual workfunction gate architectures but a single workfunction may be possible in fully-depleted SOI technologies.

The previously described limitations have given rise to interest in alternative materials for both the gate electrode and gate dielectric. The threshold voltage instability is due to shifts in the flat-band voltage through both the gate workfunction, which would be lower than that of a degenerate gate, and the charges induced and/or introduced in the gate oxide and at the interfaces. This problem would be exacerbated in nanoscale devices where channel dopant fluctuation further worsens the situation. However, these problems may be alleviated by introducing nitrogen in the oxide and/or the gate, as will be shown later. It is worth to note that the tunnelling current increases exponentially with decreasing oxide thickness, and for 2 nm oxide this current can be as high as $1–10$ A/cm^2 [95].

Recently the introduction of a metal gate with a mid-gap workfunction alongside high-k dielectric for the gate insulator has been proposed. In relation to Si/SiGe/Si MOSFETs, such high-k dielectrics may be favourable in terms of increasing the transconductance while still maintaining lower power-delay product compared to conventional Si devices. Examples of alternative metal gate electrodes are damascene metals W/TiN or Al/TiN [96], copper [97] and tantalum [98]. Such metal gates can indeed be used to eliminate gate depletion and avoid the problems generated by B penetration. This, however, yields a buried-channel device, which suffers from more short-channel effects than the surface channel device, but with better channel mobility. However, the thermal budget needed may be incompatible with standard CMOS processing for some metals.

As discussed above, the B penetration effects in polySi-gated devices that shift the threshold voltage of the MOSFET is one of the consequences. Poly-Si$_{1-x}$Ge$_x$ gate material is compatible with standard Si processing techniques. It is suitable in dual p$^+$/n$^+$ gate technology and has the potential of not only suppressing the effects of B penetration and gate depletion but more importantly the ability to adjust the threshold voltage [99,100]. The diffusivity of B in polySi$_{1-x}$Ge$_x$ is smaller than in polySi [101] and as such helps in suppressing B penetration through the oxide because B can be activated at lower thermal budgets with even better activation levels. Moreover, the polySi$_{1-x}$Ge$_x$ material has also been shown to benefit from an improved resistivity, compared to polySi, owing to the presence of Ge [100]. The tunable workfunction of the polySi$_{1-x}$Ge$_x$ gate can be used to adjust the threshold voltage, while engineering the channel doping profile can be used to control short channel effects [102].

Depending on the Ge content in the gate, the valence band edge can be raised up because Ge has a smaller energy gap than Si and therefore the polySi$_{1-x}$Ge$_x$ would have an energy gap in between that of Si and Ge. For a degenerate p$^+$-polySiGe gate material, the Fermi level would approximately be at the valence band edge. That is how Ge can be used to modify the workfunction of the gate that is, adjustment of the threshold voltage of a MOS device. Gate depletion effect, which is due to insufficient implant dose and/or insufficient activation, would also be suppressed in devices incorporating polySi$_{1-x}$Ge$_x$ gates. As a result, one can obtain a much lower threshold voltage instability due to effect of B penetration and gate depletion. Therefore, by employing a p$^+$-polySi$_{1-x}$Ge$_x$ gate, not only the threshold voltage is adjusted by changing the Ge content in the gate to match that of a n-MOSFET, but also the threshold shifts that are caused by B penetration and gate depletion effects are suppressed.

The advantages of gate engineering may be achieved together with the channel engineering, as both physical characteristics will affect the nominal threshold voltage value given by the well-known expression for n-MOSFETs

$$V_t = V_{FB} + 2\Phi_F + \frac{Q_B}{C_{ox}} \qquad (4.1)$$

where

$$V_{FB} = \Phi_{ms} - \frac{Q_{ox}}{C_{ox}} \qquad (4.2)$$

and V_t is the threshold voltage, V_{FB} the flat-band voltage, Φ_F the distance from Fermi level to the intrinsic Fermi level, Q_B the depletion charge, Φ_{ms} is the metal-semiconductor work function, Q_{ox} is the oxide charge density, and C_{ox} is the unit area capacitance of the gate insulator. From the above expressions, V_t depends on the gate material as well as the doping concentration in the channel.

Low threshold voltage values may be achieved by (1) adjusting gate insulator thickness and (2) tuning surface doping concentration as low as possible. High-quality gate insulator and subthreshold characteristics optimisation require modifications of the device vertical structure and low thermal budget. An alternative way to give low threshold voltage is to use SOI, where a fully depleted architecture can yield subthreshold slope approaching an ideal 60 mV/decade by using ultra thin films [103, 104]. Practically, ultrathin SOI films are difficult to control and use of partially depleted SOI has been more prevalent [105]. Lim *et al.* [106] have reported on the scaling of complementary threshold voltages for fully depleted SOI n- and p-channel transistors using polySiGe as gate material. The authors have shown that a carefully optimised polySiGe gate, in a FDSOI transistor allows better control of threshold voltage than a conventional design, which requires a very highly doped layer. Via simulation, it has been shown that a p^+-polySi$_{0.4}$Ge$_{0.6}$ gate can yield a similar workfunction to a mid-gap metal, but at the same time offer much greater compatibility with conventional silicon processing.

As Si$_{1-x}$Ge$_x$ is one of the few materials that offers the possibility of a mid-gap gate material that is fairly compatible to the standard Si CMOS processing, it is now being considered seriously. A comparison of the energy levels of bulk-Si, Si$_{1-x}$Ge$_x$ and Ge [107] are shown in Figure 4.23. The electron affinities of Si and Ge are

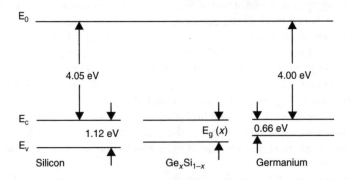

Figure 4.23   Energy levels of Si and Ge. For Si$_{1-x}$Ge$_x$ the valence band level will be in between the values for Si and Ge

similar, but because of the difference in bandgap the valence band edge of Ge lies more than 0.5 eV closer to the free electron energy than is the case for Si. The bandgap of $Si_{1-x}Ge_x$ has a value in between those of Ge and Si, but it is not a linear interpolation and the exact bandgap changes when the layers are strained [108]. For heavily doped material, as is the case for gate material, the Fermi level lies close to the conduction band edge for n-type doping and close to the valence band for p-type doped materials. For heavily p-type doped SiGe the Fermi level will lie in between that of Si and Ge at an energy that would lie in the forbidden gap for polySi gates – hence the expression mid-gap gate material. Replacing p^+-polySi by p^+-polySiGe gates will in this way give the desired workfunction shift, leading to more negative threshold voltages. When in future generations single p^+ doped gates will be used for CMOS, the workfunction difference between n^+-polySi and p^+-polySiGe will cause V_t to shift towards more positive values.

Workfunction engineering by exploiting the difference in valence band edge can in principle be applied for all Ge contents. However, a Fermi energy level close to midgap can only be achieved with very high Ge percentages, where problems are encountered with process compatibility. For example, the melting point of SiGe is a more or less linear interpolation between that of pure Si and pure Ge. High-temperature steps, like rapid thermal annealing (RTA) would not be possible if the Ge percentage becomes too high. Also alloys with a high Ge content will behave more Ge-like, which could pose a problem with certain chemical cleaning or etching steps. Some reports state that the process compatibility will extend to up to 60 per cent Ge, but for every single step that would have to be carefully investigated.

Figure 4.24 shows how the gate workfunction decreases with the Ge content [109]. Since the gate workfunction can be reduced by approximately 0.45 eV for a Ge content

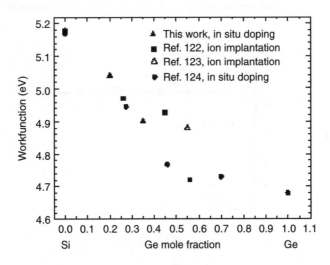

Figure 4.24   The workfunction of B-doped p^+-polySi$_{1-x}$Ge$_x$ gate material as a function of Ge content [After M. Y. A. Yousif et al., Solid-State Electron., Vol. 45, 2001(1931–1937)]

of about 50 per cent, this offers the possibility to engineer the threshold voltage from the gate material, rather than from the substrate material. As evident from this figure, the values obtained for the workfunction of in situ B-doped p^+-polySi$_{1-x}$Ge$_x$ films with $x = 0$, 0.2 and 0.35 are in excellent agreement with reported results. For the case of B/BF^{+2}-implanted samples, one may notice some discrepancies in the workfunction around 35–60 per cent Ge content. A reduction of more than 0.40 eV in the workfunction can be seen in p^+-polySiGe films with 56 per cent Ge and strain is confirmed to take place in such films [108]. At a Ge content of 45 per cent also a difference of about 150 meV between the workfunction of in situ doped films [110] and ion-implanted films can be observed. Also note that the effect of 35 per cent Ge is more than that of 45 per cent Ge in the case of ion-implanted gates.

Gate oxide thickness scaling has been instrumental in controlling short channel effects as MOS gate dimensions have been reduced. In order to ensure that short channel effects are not aggravated to an extent where the gate electrode loses control over the channel, thereby decreasing the threshold voltage, it is absolutely imperative that the gate oxide be shrunk to 2 nm or below, for a 100-nm MOSFET. This in turn leads to unacceptable levels of gate leakage due to direct quantum mechanical tunnelling. Gate oxide thickness is approximately linearly scaled with channel length to maintain the same amount of gate control over the channel to ensure good short channel behaviour. Figure 4.25 shows the electrical channel length divided by gate oxide thickness for Intel's process technologies over the past 25 years. Each data point represents a process technology, developed approximately every three years, which was used to fabricate Intel's leading-edge microprocessors.

With gate oxide thickness of the order of 2 nm or less for a 90-nm logic technology in order to achieve better short-channel margin, direct tunnelling of carriers in between the channel region and the gate electrode becomes unacceptably high.

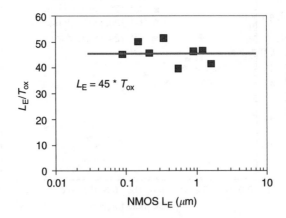

Figure 4.25   Channel length divided by gate oxide thickness vs. channel length [After S. E. Thompson et al., Lecture notes]

A comprehensive account of the different tunnelling current modelling techniques from a historical perspective may be found in Reference 111.

4.5   Strained-engineered CMOS technology

Process-induced strain engineering, along with aggressively scaled devices has been the key feature for enhancing the leading-edge device performance for high-speed 90-nm logic technology [29, 112]. Optimised strain-engineered 45-nm gate length CMOS technology, offering high-performance n-MOSFETs with no impact on p-MOSFETs devices and with minimum manufacturing complexity, has been announced [30]. Through aggressive scaling of the gate length and gate oxide thickness (1.2 nm EOT), high-performance 45-nm logic devices have seamlessly been integrated into an existing 90-nm logic technology [112].

Other front-end process technology features include shallow abrupt SDE, halo implants, deep source/drain implants and spike activation anneal. The silicon channel has been strained from both sides by STI and contact etch stop layer. Device performance has been improved by optimising the stress effects. Spacer and the subsequent thermal cycle were optimised to reduce dopant transient enhanced diffusion (TED) and improve the junction abruptness. Optimised strain engineering obtained via both the trench isolation and contact etch stop nitride film has enabled the high-performance devices, which are amongst the best reported so far.

It has been shown that the silicon channel can be strained from both STI and contact etch stop layer. Device performance can be improved by optimising the stress effects. Conventional contact etch stop nitride films have tensile stress in the range of 0 to 7 Gdyne/cm^2. When the tensile stress is increased to 14 Gdyne/cm^2, the resulting film in which a more gentle subthreshold slope results in a higher mobility and a lower intercept indicating lower external parasitic resistance, gives better n-MOSFET drive current. Optimisation of the nitridation process with ultrathin gate dielectric has led to achieve high drive current without increasing gate leakage, degrading channel mobility or causing large V_t shift.

The mobility enhancement in p-MOSFETs with SiGe source/drain has been studied [113, 114]. The shape of the SiGe source/drain is engineered to vary the compressive stress in the channel from 200 MPa to 1.5 GPa. The depth of the SiGe recess varies from 20 to 120 nm. The source/drain elevation above the channel level varies from 20 to 80 nm. The source/drain recess is obtained by either anisotropic or isotropic etching. Ten-micrometre wide transistors are fabricated with standard orientation and polySi gate lengths down to 35 nm. Two diffusion distances from the polySi gate to STI are used: 0.5 and 10 μm. Experimental p-MOSFETs with $Si_{0.8}Ge_{0.2}$ source/drain have been reported [113]. A TEM of a typical transistor is shown in Figure 4.26.

The channel stress is calculated using 2D process simulation, where all intentional and unintentional stress sources, as well as the stress evolution during the entire process flow, are taken into account [115]. The distribution of the longitudinal stress component σ_{xx} in a typical case is shown in Figure 4.27. The stress distribution

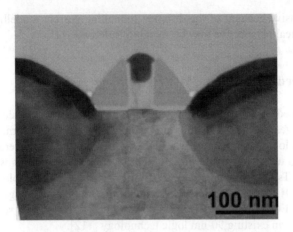

Figure 4.26   TEM image of a 35-nm p-MOSFET [After L. Smith et al., IEEE Electron Dev. Lett., Vol. 26, 2005(652–654). © 2005 IEEE]

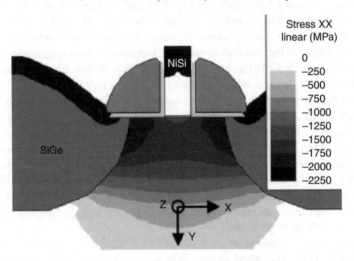

Figure 4.27   Simulated stress distribution for the device structure shown in Figure 4.26 [After L. Smith et al., IEEE Electron Dev. Lett., Vol. 26, 2005(652–654). © 2005 IEEE]

was uniform, varying by less than 4 per cent throughout the channel. The following analysis is based on the simulated longitudinal σ_{xx} and transverse σ_{zz} stresses that are averaged along the channel at a depth of 2 nm from the channel surface.

The transverse stress component is always compressive and consistently about 18.5 per cent of the longitudinal stress component. Such relatively high compressive transverse stress is typical for transistors with wide channels. As the channel gets narrower, the transverse stress reduces and can even become tensile [116]. Since the effects of the longitudinal and the transverse stress components on the hole mobility are comparable in magnitude [35], it is important to consider both of these in-plane components when estimating the stress-enhanced mobility.

The mobility enhancements observed in purely uniaxial stress via wafer bending experiments and more complicated stress patterns have been compared by Smith *et al.* [117]. The authors have proposed an effective channel stress as $\sigma_{eff} = 0.8\sigma_{zz} - \sigma_{xx}$. This effective stress takes into account the mobility degradation due to compressive transverse stress. The σ_{zz} weight of 0.8 is extracted from wafer bending data as a ratio of the transverse to longitudinal piezoresistance coefficients. For the wafer bending experiments, the effective channel stress is identical to σ_{xx} because of zero transverse stress. Figure 4.28 shows the low-field mobility enhancement extracted along with the enhancement derived from longitudinal wafer bending.

Figures 4.29 and 4.30 show subthreshold characteristics and I_{ds}-V_{gs} for the high speed n- and p-MOSFET devices with gate length of 45-nm. n- and p-MOSFET drive current at $|V_{ds}| = |V_{gs}| = 1.0$ V are 1010 and 400 $\mu A/\mu m$ at 100 nA/μm off-state leakage and 45-nm nominal channel length. Good halo and extension implant designs resulted in subthreshold slope of 85 mV/dec and low drain-induced barrier lowering of 100 mV/V.

Figures 4.31 and 4.32 show the effect of STI proximity to the gate edge in the device performance. The stress in p-MOSFETs increases current when the gate to

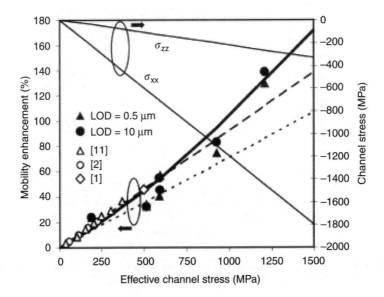

Figure 4.28   *Hole mobility enhancement and simulated stress components as a function of the effective channel stress. For the mobility enhancement, measured data are shown as symbols, the dotted line shows bulk piezoresistance, the dashed line shows best-fit piezoresistance and the solid line shows the mobility model (see text) [After L. Smith et al., IEEE Electron Dev. Lett., Vol. 26, 2005(652–654). © 2005 IEEE]*

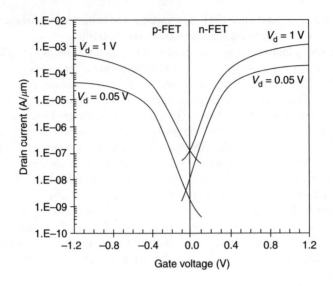

Figure 4.29   Subthreshold behaviour for 45-nm devices [After V. Chan et al., IEEE IEDM Tech. Dig., 2003(77–80). © 2003 IEEE]

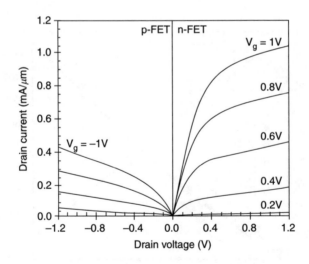

Figure 4.30   Output characteristics for 45-nm devices [After V. Chan et al., IEEE IEDM Tech. Dig., 2003(77–80). © 2003 IEEE]

STI edge distance becomes shorter, but n-MOSFET current is decreased. When the channel width is reduced, n-MOSFETs are less sensitive to the stress effect; however, on the contrary, p-MOSFETs become more sensitive to it. Several devices, namely high-speed (HS), high-V_t (HVT) and super-high-V_t (SHVT), covering a wide off-state leakage current range of 1–100 nA/μm at V_{dd} of 1 V have been realised in the 45-nm

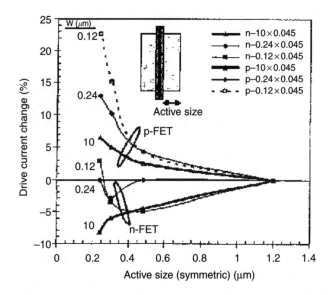

Figure 4.31   Shallow trench isolation (STI) mechanical stress effect (different active width) on the n- and p-MOSFET drive current change with different device width [After V. Chan et al., IEEE IEDM Tech. Dig., 2003(77–80). © 2003 IEEE]

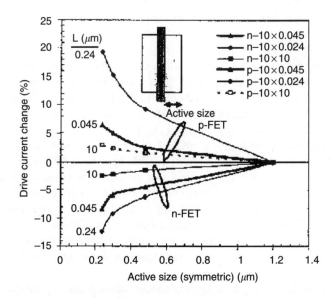

Figure 4.32   STI mechanical stress effect on different device length [After V. Chan et al., IEEE IEDM Tech. Dig., 2003(77–80). © 2003 IEEE]

technology. The nominal unloaded ring oscillator speed of the three devices are 7.0, 9.0 and 10.8 ps/stage at 1 V V_{dd} and 45 nm L_{nom}, respectively.

Several interesting observations are as follows:

1. The stress in different channel length devices will have different effects. The drive current enhancement in long channel devices with high tensile stress is negligible.
2. As the gate length is scaled down, the electron mobility enhancement becomes significant but reduces for shorter gate length, although the current improvement in linear region remains high.
3. High halo dose and high parasitic resistance can offset the current enhancement by the local strain benefits.

4.6   Layout engineering

It has been shown that the induced channel stress and the consequent drive current increase are dependent on various processing factors, such as recess depth, SiGe overgrowth and the channel direction [113]. The impact of process-induced strain on electrical behaviour necessitates increasing complexity in TCAD modelling. STI is a dominant source of mechanical stress variations in MOSFET channel, and is dependent on MOSFET geometry variations, such as active area size and shape. The stress generated by shallow-trench isolation can alter the drive current of n- and p-MOSFETs by up to 20 per cent depending on the length of diffusion [118]. p-MOS transistors with epitaxial SiGe in the S/D regions are widely regarded as a viable and manufacturable way to enhance the performance up to 35 per cent [28].

Also, the drive current is not only related to gate length and width but also to the exact layout of the individual transistor. As MOSFET dimensions shrink, the stress becomes dependent also on the width of the STI surrounding the transistor. The question remains how this technology is applicable for next-generation nodes with the active area dimensions scaled down to the decananometre range. A similar problem arises with the intentional application of stress by depositing SiGe layers beneath the source and drain regions [29, 113, 119].

Layout dependence of n- and p-MOSFETs have been studied by several workers [120]. Stress simulations and mobility models have been calibrated and verified for test structures with SiGe source/drain as a stressor. The numerical results show that variations of 15 per cent in drive currents and of 44 per cent in hole mobility due to layout-induced stress variations can occur. For 45 nm gate length p-MOS transistors with large active areas, measurements and simulations show a 65 per cent increase in drive current, but this improvement may be seriously degraded when transistor dimensions such as the source-drain length and the device width are further scaled [121].

In the following, the impact of layout on SiGe stress-enhanced p-MOSFET performance, where numerical simulation is compared with measurements on test structures comprising isolated and nested transistors, are discussed. The isolated transistor is surrounded by STI, whereas the nested transistor is bounded by multiple polySi gates

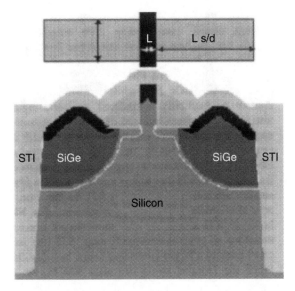

Figure 4.33   *Simulated 2D cross-section of an isolated 35-nm p-MOS. Layout is shown on the top [After V. Moroz et al., Proc. SISPAD, 2005(143–146). © 2005 IEEE]*

and doped SiGe source and drains. The characteristic dimensions are the channel length L, width W, source and drain length $L_{s/d}$ and the STI width.

Simulation results performed using Taurus-Process and Taurus-Device (2D/3D) TCAD tools that contain an extensive set of stress-related features have been reported [115]. The authors used a model for the stress-induced enhancement of the low-field hole mobility that was extracted for compressive uniaxial stress up to a level of 1.5 GPa [117]. Figure 4.33 shows the simulated 2D cross-section of a typical isolated 35 nm p-MOS transistor, and Figure 4.34 shows a typical nested 35 nm p-MOS. The corresponding layouts are shown in the figures as the top inserts. The white lines shown in the silicon regions in both figures represent the simulated P/N junctions.

The simulated distribution of the stress component along the channel direction for the isolated transistors with 0.4 μm $L_{s/d}$ is shown in Figure 4.35. Figure 4.36 shows measured and simulated stress-induced mobility enhancements as a function of $L_{s/d}$. The mobility enhancement increases with $L_{s/d}$ owing to the increasing size of SiGe stressor. At small $L_{s/d}$, transistors with nested gates exhibit higher mobility enhancement than the isolated transistors owing to the absence of STI relaxer. For large $L_{s/d}$ cases, $(L_{s/d} > 1~\mu$m), the mobility enhancements saturate at the same level for both isolated and nested transistors. When the $L_{s/d}$ is scaled down to 200 nm, the minimum size for the 90-nm technology node, the mobility enhancement is 30$\times$ less than the saturation value. There is a small but noticeable effect of the width of the transistor. As the width increases the stress increases and reaches the saturation level as shown in Figure 4.37.

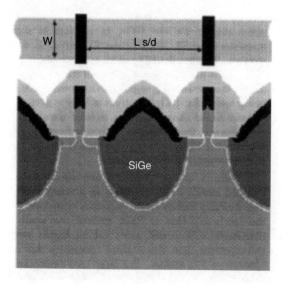

Figure 4.34   Simulated 2D cross-section of a nested 35-nm p-MOS. Layout is shown on the top [After V. Moroz et al., Proc. SISPAD, 2005(143–146). © 2005 IEEE]

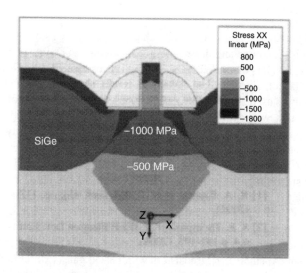

Figure 4.35   Simulated stress distribution of the normal stress component along the channel direction for an isolated p-MOS transistor with 0.4 μm $L_{s/d}$ [After V. Moroz et al., Proc. SISPAD, 2005(143–146). © 2005 IEEE]

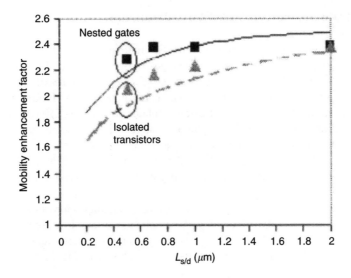

Figure 4.36 *Mobility enhancement of isolated and transistors with nested gates as a function of $L_{s/d}$. Measured data is shown as symbols, and simulation results are shown as lines [After V. Moroz et al., Proc. SISPAD, 2005(143–146). © 2005 IEEE]*

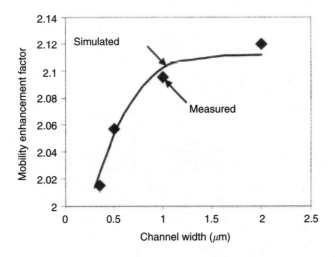

Figure 4.37 *Mobility enhancement of isolated transistors as a function of channel width. Measured data is shown as symbols, and simulation results are shown as lines [After V. Moroz et al., Proc. SISPAD, 2005(143–146). © 2005 IEEE]*

4.7   Strain-engineered MOSFETs: simulation

Although the transition from conventional Si-processing to strained-Si/SiGe MOS-
FETs may appear to be only a substrate change, but due to doping differences between
Si and SiGe and other material parameters, channel re-engineering is necessary
to achieve successful sub-100 nm devices using heterolayers. Figure 4.38 shows
the typical strained-Si and strained-Si-on-insulator (strained-SOI) MOSFETs, using
strained-Si/SiGe heterostructures. The advantage of these device structures using
Si/SiGe heterointerfaces against the devices using process-induced strain like capping layers, STI, silicides and so on is the capability to introduce much larger and
more uniform strain in the channels, expected for tight distributions in the strain
variation.

In the following, an example of integrated modelling approach, from dopant
implantation and thermal processing to device simulation, is presented following
Thean *et al.* [122], which allows one to optimise short-channel effects and parasitic
S/D resistance for scaled strained-Si devices as shown in Figure 4.39. The SiGe

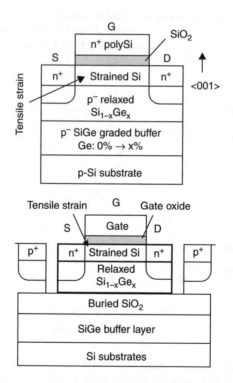

*Figure 4.38   Schematic diagrams of typical device structure of strained-Si MOS-
FETs fabricated on bulk-Si substrates and strained-Si-on-insulator
(Strained-SOI) MOSFETs [After T. Mizuno et al., IEEE Trans. Electron
Dev., Vol. 48, 2001(1612–1618). © 2001 IEEE]*

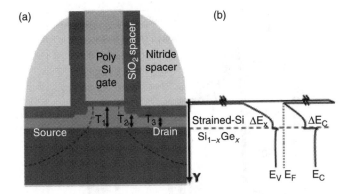

Figure 4.39 (a) *A schematic of gate and channel structures with exaggerated recessed features that illustrate typical differences in strained-Si thicknesses (T1, T2 and T3) under the gate and spacer regions. (b) The energy band diagram showing the type-II heterostructure in the vertical direction under the gate [After A. V.-Y. Thean et al., Proc. SISPAD, 2003(195–198). © 2003 IEEE]*

substrate is modelled by doping Si to alloying concentrations corresponding to the Ge mole fraction, x, where the mole fraction-dependent strained-Si and SiGe bandgaps are according to Rieger *et al.* [82]. A drift-diffusion approach used in DESSIS [123] has been applied to model the carrier transport, where the drive current enhancement is calibrated to experiments through detailed matching of mobility and transconductance behaviour. Coupled dopant diffusion processes during thermal steps as described in DIOS [124], implementation of the pair-diffusion model where dopant-defect pairs and unpaired point defects are the mobile species, were used in simulation.

The device simulation included a model of type-II heterostructure at the strained-Si/SiGe interface (see Figure 4.39). The strained-Si/SiGe conduction and valence band offsets, ΔE_c and ΔE_v, respectively, were taken from Reference 82. Simulations based on hydrodynamic models were also performed as comparisons. Band-to-band tunnelling according to Schenk [125] is included to account for junction leakage, in addition to diode current increase arising from reduced bandgaps of strained-Si and SiGe. Dopant implantations were simulated by two-dimensional Monte Carlo method and analytic implant tables with nuclear and electronic-stopping of SiGe calibrated to secondary ion mass spectrometry (SIMS) data for the device (see Figure 4.39). The transient enhanced diffusion was simulated by the addition of point defects from implant damage. After process simulation, the final structure was checked against transmission electron microscopy images of the devices [see Figure 4.39(a)].

Figures 4.40 and 4.41 show the SIMS and simulated As profiles after 1025 °C and 900 °C anneals, respectively, for SiGe with 25 per cent Ge. It is seen that high-temperature anneal enhances the diffusion of As in SiGe. The diffusion can be sensitive to the presence of the heterostructure and Ge distribution, leading to very different doping profiles when the diffusion occurs in the absence of the strained-Si

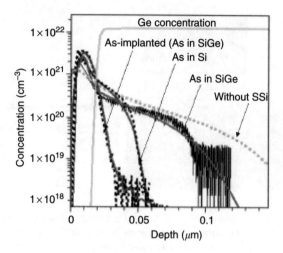

Figure 4.40   Diffusion of As in Si and strained-Si/SiGe heterostructure after a 10 s anneal at 1025°C (solid: simulated, dashed: SIMS) [After A. V.-Y. Thean et al., Proc. SISPAD, 2003(195–198). © 2003 IEEE]

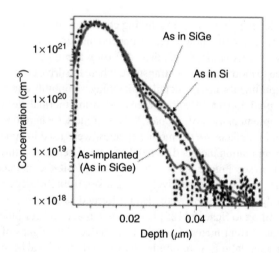

Figure 4.41   Diffusion of As in Si and strained-Si/SiGe heterostructure after a 900°C 10 s thermal anneal (solid: simulated, dashed: SIMS) [After A. V.-Y. Thean et al., Proc. SISPAD, 2003(195–198). © 2003 IEEE]

layer (see Figure 4.40). When the anneal temperature is lowered to 900 °C, dopant diffusion in SiGe becomes comparable to that of Si (see Figure 4.41). Figure 4.42 shows the B doping profiles after the same annealing condition. In this case, boron diffusion in the SiGe substrate is retarded when the anneal temperature is high and becomes comparable to the Si case when temperature is lowered. Figure 4.42 also

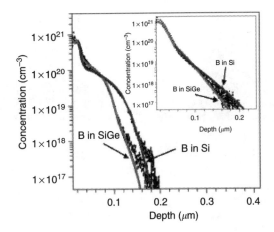

Figure 4.42   *Diffusion of B in Si and strained-Si/SiGe heterostructure after a 10 s high temperature anneal (solid: simulated, dashed: SIMS). Inset: Dopant diffusion when anneal temperature is lowered to 900 °C [After A. V.-Y. Thean et al., Proc. SISPAD, 2003(195–198). © 2003 IEEE]*

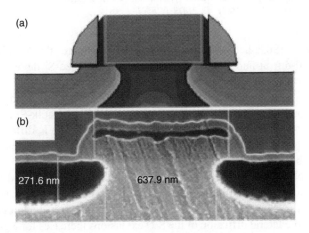

Figure 4.43   *(a) Simulated junction profile of a long strained-Si/SiGe device before S/D optimisation. (b) Scanning electron microscopy of a dopant-selective etched device showing the junction profile due to the strained-Si/SiGe heterostructure [After A. V.-Y. Thean et al., Proc. SISPAD, 2003(195–198). © 2003 IEEE]*

shows the junction profile of a strained-Si device that received S/D implants meant for optimised Si devices.

Figure 4.43(a) shows the simulated junction profile of a long strained-Si/SiGe device before S/D optimisation and Figure 4.43(b) shows the scanning electron microscopy of a dopant-selective etched device showing the junction profile due

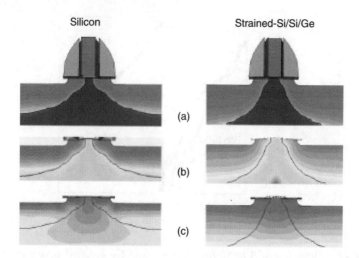

Figure 4.44   *(a) Net doping concentration delineating the junction profiles of small NMOS devices, (b) contour plots of As doping and (c) Contour plots showing the B halos [After A. V.-Y. Thean et al., Proc. SISPAD, 2003(195–198). © 2003 IEEE]*

to the strained-Si/SiGe heterostructure. Figure 4.44 compares the As S/D and B halo profiles for 50-nm Si and strained-Si devices. The strained-Si devices show more abrupt halo profiles due to retarded B diffusion in SiGe and deeper junctions due to enhanced As diffusion. The differences in density of states, bandgap and band offsets contribute to a V_t difference between strained-Si and bulk-Si devices. From capacitance-voltage measurement of long-channel devices, the V_t difference is determined to be approximately 150 mV. The long-channel V_t of the strained-Si devices was found to be very sensitive to the ΔE_c and ΔE_v values.

From simulation, it has been shown that the thickness of the strained-Si layer is an important factor in determining the S/D profiles and consequently the short-channel behaviour of the device. It is seen that when the strained-Si thickness is reduced below 100 Å, the lateral diffusion of the S/D extensions increases for a given implant and thermal anneal conditions. Figure 4.45 shows the contour plot of the junction and halo profiles of two strained-Si devices with different strained-Si thickness that underwent the same implant and anneal processes. The physical channel defined by the S/D junctions severely limits the gate length scaling by promoting the early onset of punchthrough and short channel effects.

The enhanced lateral diffusion leads to a reduction in physical channel length, defined by the separation between the S/D extensions [see Figure 4.45(a)]. As the strained-Si thickness is decreased, the junctions come closer together. This is due largely to the fact that a larger portion of the S/D As dopants is located in the SiGe when the strained-Si is thin and more dopants undergo enhanced diffusion in the presence of Ge. In addition, the B halo distributions in the two devices are also markedly different [see Figures 4.45(b) and (d)] owing to different amount of retardation experienced by

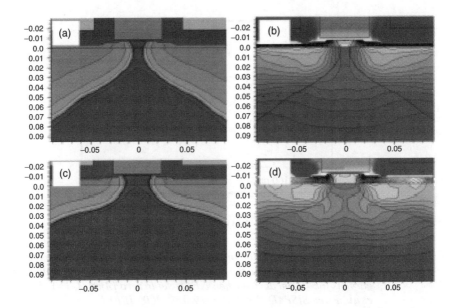

Figure 4.45   *(a) Net doping profile of strained-Si device with strained-Si thickness of 70 Å, (b) the boron halo profile of the strained-Si device with 70 Å thick strained-Si, (c) net doping profile of strained-Si device with strained-Si thickness of 100 Å and (d) the boron halo profile of the strained-Si device with 100 Å thick strained-Si [After A. V.-Y. Thean et al., Proc. SISPAD, 2003(195–198). © 2003 IEEE]*

the dopants when the strained-Si thickness changes. Thus, careful implant condition to avoid excessive lateral diffusion is necessary so that the short channel effect can be improved in sub-100 nm strained-Si devices, even with high-temperature anneals.

Figure 4.46 shows a plot of the drain-induced barrier lowering as a function of strained-Si thickness. As the strained-Si thickness is thinned from 100 Å to 50 Å, DIBL increases rapidly for the 70-nm device, owing to enhanced lateral diffusion of the S/D extensions. However, the changes in DIBL is negligible for the slightly longer device (L_g = 100 nm) for the same reduction in strained-Si thickness. The 70- and 100-nm devices demonstrate the same DIBL when the strained-Si layer thickness is greater than 100 Å, indicating that enhanced lateral diffusion is minimised for strained-Si thicker than 100 Å. It has been observed that short channel behaviour of the devices is not only influenced by the location of the strained-Si/SiGe interface but also the band offset at the heterojunction. When the conduction band offset ΔE_c is reduced to zero, DIBL increases slightly, especially when the strained-Si is thinned, indicating that the band structure of the heterointerface can help to reduce short channel effect. The role of strained-Si layer thickness with respect to dopant and device behaviour has been investigated for 45–50 nm devices. Figure 4.47 compares the simulated

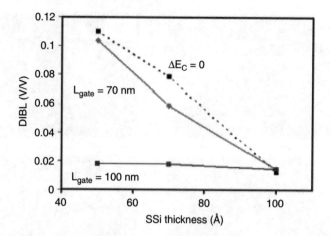

Figure 4.46   DIBL metric as a function of strained-Si thickness. Black curve is for a $L_g = 100$ nm device and the grey curve is for a $L_g = 70$ nm; the dashed curve is for 70-nm device when ΔE_c is set to zero [After A. V.-Y. Thean et al., Proc. SISPAD, 2003(195–198). © 2003 IEEE]

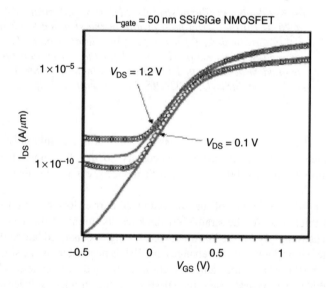

Figure 4.47   Drain current vs. gate voltage characteristic of a 50-nm strained-Si device. (symbols: measurement and solid: simulation) [After A. V.-Y. Thean et al., Proc. SISPAD, 2003(195–198). © 2003 IEEE]

and measured characteristics of a $L_g = 50$ nm strained-Si device with good short-channel control, exhibiting a DIBL of only 52 mV/V. Well-behaved short channel characteristics are observed from simulation for strained-Si thickness of 60 Å under the gate.

4.8 Summary

In this chapter, some of the engineering solutions to boost the performance of the conventional Si MOSFET by exploiting the favourable electrical properties of the strained-Si and $Si_{1-x}Ge_x$ materials and their compatibility with standard CMOS technology are addressed. Both biaxial and uniaxial stress provide significant enhanced mobility to improve MOSFET performance. As an in-depth understanding of how stress affects the semiconductor fabrication processes, stress simulations for strained-Si have been discussed to understand the effects of strain on carrier mobility, drive current and threshold voltage shift. The importance of an integrated TCAD approach to design sub-100-nm strained-Si/SiGe devices has been discussed. The impact of layout characteristic lengths on stress-enhanced transistor performance is discussed. It may be noted that $Si_{1-x}Ge_x$ may be used in three different regions of the MOSFET device:

1. as a quantum well to engineer the channel of the p-MOSFETs;
2. to engineer the gate work function of the transistor;
3. as a virtual substrate for tensile strained-Si MOSFETs.

References

1 J. E. Lilienfeld, 'Method and apparatus for controlling electric currents', U. S. Patent 1,745,175, 1925.
2 M. M. Atalla, M. Tannenbaum, and E. J. Scheibner, 'Stabilization of silicon surface by thermally grown oxides', *Bell Syst. Tech. J.*, vol. 38, pp. 749–783, 1959.
3 G. E. Moore, 'Cramming more components onto integrated circuits', *Electronics*, vol. 38, pp. 114–117, 1965.
4 G. E. Moore, 'Progress in Digital Integrated Electronics', in *IEEE IEDM Tech. Dig.*, pp. 11–13, 1975.
5 G. E. Moore, 'No exponential is forever: but "Forever" can be delayed!', in *Proc. ISSCC*, pp. 20–23, 2003.
6 B. Doyle, R. Arghavani, D. Barlage, S. Datta, S. Doczy, J. Kavalieros, *et al.*, 'Transistor elements for 30 nm physical gate lengths and beyond', *Intel Technol. J.*, vol. 6, pp. 42–54, 2002.
7 R. Chau, J. Kavalieros, B. Doyle, A. Murthy, N. Paulsen, D. Lionberger, *et al.*, 'A 50 nm depleted-substrate CMOS transistor (DST)', in *IEEE IEDM Tech. Dig.*, pp. 621–624, 2001.
8 X. Huang, W. Lee, C. Kuo, D. Hisamoto, L. Chang, J. Kedzierski, *et al.*, 'Sub 50-nm FinFET: PMOS', in *IEEE IEDM Tech. Dig.*, pp. 67–70, 1999.
9 J.-T. Park and J. P. Colinge, 'Multiple-gate SOI MOSFETs: device design guidelines', *IEEE Trans. Electron Devices*, vol. 49, pp. 2222–2229, 2002.
10 S. Thompson, N. Anand, M. Armstrong, C. Auth, B. Arcot, M. Alavi, *et al.*, 'A 90 nm logic technology featuring 50 nm strained silicon channel transistors,

7 layers of Cu interconnects, low k ILD and 1 μm^2 SRAM cell', in *IEEE IEDM Tech. Dig.*, pp. 61–64, 2002.

11 J. Meindl, 'Low power microelectronics – retrospect and prospect', *Proc. IEEE*, vol. 83, pp. 619–635, 1995.

12 Y. Taur, D. A. Buchanan, W. Chen, D. J. Frank, K. E. Ismail, S. H. Lo, *et al.*, 'CMOS scaling into the nanometer regime', *IEEE Proc.*, vol. 85, pp. 486–504, 1997.

13 H.-S. P. Wong, D. J. Frank, P. M. Solomon, C. H. J. Wann, and J. J. Wesler, 'Nanoscale CMOS', *Proc. IEEE*, vol. 87, pp. 537–570, 1999.

14 R. W. Keyes, 'Fundamental limits of silicon technology', *Proc. IEEE*, vol. 89, pp. 227–239, 2001.

15 J. D. Plummer and P. B. Griffin, 'Material and process limits in silicon VLSI technology', *Proc. IEEE*, vol. 89, pp. 240–258, 2001.

16 P. A. Packan, 'Pushing the limits', *Science*, vol. 285, pp. 2079–2080, 1999.

17 A. Yagishita, T. Saito, K. Nakajima, S. Inumiya, Y. Akasaka, Y. Ozawa, *et al.*, 'High performance metal gate MOSFETs fabricated by CMP for 0.1 μm regime', in *IEEE IEDM Tech. Dig.*, pp. 785–788, 1998.

18 H. Iwai, 'Future semiconductor manufacturing-challenges and opportunities', in *IEEE IEDM Tech. Dig.*, pp. 11–16, 2004.

19 H. Sasaki, M. Ono, T. Yoshitomi, T. Ohguro, S. Nakamura, M. Saito, *et al.*, '1.5 nm direct-tunneling gate oxide Si MOSFET's', *IEEE Trans. Electron Devices*, vol. 43, pp. 1233–1242, 1996.

20 C. Caillat, S. Deleonibus, G. Guegan, S. Tedesco, B. Dal'zotto, M. Heitzmann, *et al.*, '65 nm physical gate length NMOSFETs with heavy ion implanted pockets and highly reliable 2 nm thick gate oxide for 1.5 V operation', in *IEEE VLSI Tech. Symp. Dig.*, pp. 89–90, 1999.

21 Y. Taur, Y. J. Mii, D. J. Frank, H. S. Wong, D. A. Buchanan, S. J. Wind, *et al.*, 'CMOS scaling into the 21st century: 0.1 μm and beyond', *IBM J. Res. and Devices*, vol. 39, pp. 245–260, 1995.

22 H.-S. Wong, and Y. Taur, 'Three-dimensional "atomistic" simulation of discrete random dopant distribution effects in sub-0.1 μm MOSFET's', in *IEEE IEDM Tech. Dig.*, pp. 705–708, 1993.

23 IEEE, 'Special issue on nonclassical Si CMOS devices and technologies: Extending the roadmap', *IEEE Trans. Electron Devices*, vol. 53, pp. 941–1095, 2006.

24 P. R. Chidambaram, C. Bowen, S. Chakravarthi, C. Machala, and R. Wise, 'Fundamentals of silicon material properties for successful exploitation of strain engineering in modern CMOS manufacturing', *IEEE Trans. Electron Devices*, vol. 53, pp. 944–964, 2006.

25 E. A. Fitzgerald, 'Engineered substrates and their future role in microelectronics', *Mat. Sci. and Eng. B*, vol. 124–125, pp. 8–15, 2005.

26 S. E. Thompson, G. Sun, Y. S. Choi, and T. Nishida, 'Uniaxial-process-induced strained-Si: Extending the CMOS roadmap', *IEEE Trans. Electron Devices*, vol. 53, pp. 1010–1020, 2006.

27 S. E. Thompson, 'Strained Si and the future direction of CMOS', in *IEEE IDEAS Symp. Proc.*, pp. 14–16, 2005.

28 T. Ghani, M. Armstrong, C. Auth, M. Bost, P. Charvat, G. Glass, *et al.*, 'A 90nm high volume manufacturing logic technology featuring novel 45nm gate length strained silicon CMOS', in *IEEE IEDM Tech. Dig.*, pp. 978–980, 2003.

29 S. E. Thompson, M. Armstrong, C. Auth, S. Cea, R. Chau, G. Glass, *et al.*, 'A logic nanotechnology featuring strained-silicon', *IEEE Electron Device Lett.*, vol. 25, pp. 191–193, 2004.

30 V. Chan, R. Rengarajan, N. Rovedo, W. Jin, T. Hook, P. Nguyen, *et al.*, 'High speed 45 nm gate length CMOSFETs integrated into a 90 nm bulk technology incorporating strain engineering', in *IEEE IEDM Tech. Dig.*, pp. 77–80, 2003.

31 J.-S. Goo, Q. Xiang, Y. Takamura, F. Arasnia, E. N. Paton, P. Besser, *et al.*, 'Band offset induced threshold variation in strained-Si nMOSFETs', *IEEE Electron Device Lett.*, vol. 24, pp. 568–570, 2003.

32 J. Goo, Q. Xiang, Y. Takamura, H. Wang, J. Pan, F. Arasnia, *et al.*, 'Scalability of strained-Si nMOSFETs down to 25 nm gate length', *IEEE Electron Device Lett.*, vol. 24, pp. 351–353, 2003.

33 K. Rim, J. Chu, H. Chen, K. A. Jenkins, T. Kanarsky, K. Lee, *et al.*, 'Characteristics and device design of sub-100 nm strained si n- and p-MOSFETS', in *IEEE VLSI Tech. Symp. Dig.*, pp. 98–99, 2002.

34 J. G. Fossum, and W. Zhang, 'Performance projections of scaled CMOS devices and circuits with strained Si-on-SiGe channels', *IEEE Trans. Electron Devices*, vol. 50, pp. 1042–1049, 2003.

35 S. E. Thompson, G. Sun, K. Wu, J. Kim, and T. Nishida, 'Key differences for process-induced uniaxial vs. substrate-induced biaxial stressed Si and Ge channel MOSFETs', in *IEEE IEDM Tech. Dig.*, pp. 221–224, 2004.

36 M. D. Giles, M. Armstrong, C. Auth, S. M. Cea, T. Ghani, T. Hoffmann, *et al.*, 'Understanding stress enhanced performance in Intel 90nm CMOS technology', in *IEEE VLSI Tech. Symp. Dig.*, pp. 118–119, 2004.

37 C. K. Maiti, L. K. Bera, S. S. Dey, D. K. Nayak, and N. B. Chakrabarti, 'Hole mobility enhancement in strained-Si p-MOSFETs under high vertical fields', *Solid State Electron.*, vol. 41, pp. 1863–1869, 1997.

38 A. Shimizu, K. Hachimine, N. Ohki, H. Ohta, M. Koguchi, Y. Nonaka, *et al.*, 'Local mechanical-stress control (LMC): A new technique for CMOS-performance enhancement', in *IEEE IEDM Tech. Dig.*, pp. 433–436, 2001.

39 A. Lochtefeld, and D. A. Antoniadis, 'Investigating the relationship between electron mobility and velocity in deeply scaled NMOS via mechanical stress', *IEEE Electron Device Lett.*, vol. 22, pp. 591–593, 2001.

40 H. S. Momose, T. Ohguro, K. Kojima, S. Nakamura, and Y. Toyoshima, '110 GHz cut-off frequency of ultra-thin gate oxide p-MOSFETs on (110) surface-oriented Si substrate', in *IEEE VLSI Tech. Symp. Dig.*, pp. 156–157, 2002.

41 M. Yang, E. P. Gusev, M. Ieong, O. Gluschenkov, D. C. Boyd, K. K. Chan, *et al.*, 'Performance dependence of CMOS on silicon substrate orientation for ultrathin oxynitride and HfO$_2$ gate dielectrics', *IEEE Electron Device Lett.*, vol. 24, pp. 339–341, 2003.

42 T. Sato, Y. Takeishi, and H. Hara, 'Mobility anisotropy of electrons in inversion layers on oxidized silicon surface', *Phys. Rev. B*, vol. 4, pp. 1951–1960, 1971.

43 M. Kinugawa, M. Kakumu, and J. Matsunaga, 'Submicron 3D surface-orientation-optimized CMOS technology', in *IEEE VLSI Tech. Symp. Dig.*, pp. 17–18, 1986.

44 S. Sugawa, I. Ohshima, H. Ishino, Y. Saito, M. Hirayama, and T. Ohmi, 'Advantage of silicon nitride gate insulator transistor by using microwave-excited high-density plasma for applying 100 nm technology node', in *IEEE IEDM Tech. Dig.*, pp. 817–820, 2001.

45 S. Takagi, A. Toriumi, M. Iwase, and H. Tango, 'On the universality of the inversion layer mobility in Si MOSFETs: Part I: Effects of surface orientation', *IEEE Trans. Electron Devices*, vol. 41, pp. 2363–2368, 1994.

46 M. Yang, V. W. C. Chan, K. K. Chan, L. Shi, D. M. Fried, J. H. Stathis, *et al.*, 'Hybrid-orientation technology (HOT): Opportunities and challenges', *IEEE Trans. Electron Devices*, vol. 53, pp. 965–978, 2006.

47 Q. Ouyang, M. Yang, J. Holt, S. Panda, H. Chen, H. Utomo, *et al.*, 'Investigation of CMOS devices with embedded sige source/drain on hybrid orientation substrates', in *IEEE VLSI Tech. Symp. Dig.*, pp. 28–29, 2005.

48 D. Zhang, B. Y. Nguyen, T. White, B. Goolsby, T. Nguyen, V. Dhandapani, *et al.*, 'Embedded SiGe S/D PMOS on thin body SOI substrate with drive current enhancement', in *IEEE VLSI Tech. Symp. Dig.*, pp. 26–27, 2005.

49 S. Ito, H. Namba, K. Yamaguchi, T. Hirata, K. Ando, S. Koyama, *et al.*, 'Mechanical stress effect of etch-stop nitride and its impact on deep submicron transistor design', in *IEEE IEDM Tech. Dig.*, pp. 247–250, 2000.

50 M. E. Law, and S. M. Cea, 'Continuum based modeling of silicon integrated circuit processing: An object oriented approach', *Comp. Mat. Sci.*, vol. 12, pp. 289–308, 1998.

51 S. Cea, and M. Law, 'Multidimensional nonlinear viscoelastic oxidation modeling', *Simulat. Semicond. Device Process.*, vol. 6, pp. 139–142, 1995.

52 S. Pidin, T. Mori, K. Inoue, S. Fukuta, N. Itoh, E. Mutoh, *et al.*, 'A novel strain enhanced CMOS architecture using selectively deposited high tensile and high compressive silicon nitride films', in *IEEE IEDM Tech. Dig.*, pp. 213–216, 2004.

53 H. S. Yang, R. Malik, S. Narasimha, Y. Li, R. Divakaruni, P. Agnello, *et al.*, 'Dual stress liner for high performance sub-45 nm gate length SOI CMOS manufacturing', in *IEEE IEDM Tech. Dig.*, pp. 1075–1077, 2004.

54 A. A. Bayati, L. Washington, L.-Q. Xia, M. Balseanu, Z. Yuan, M. Kawaguchi, *et al.*, 'Production processes for inducing strain in CMOS channels', *Semiconductor Fabtech.*, vol. 26, pp. 84–88, 2005.

55 C.-H. Chen, T. L. Lee, T. H. Hou, C. L. Chen, C. C. Chen, J. W. Hsu, *et al.*, 'Stress memorization Technique (SMT) by selectively strained-nitride capping

for sub-65nm high-performance strained-Si device application', in *IEEE VLSI Tech. Symp. Dig.*, pp. 56–57, 2004.

56 A. Steegen, M. Stucchi, A. Lauwers, and K. Maex, 'Silicide induced pattern density and orientation dependent transconductance in MOS transistor', in *IEEE IEDM Tech. Dig.*, pp. 497–500, 1999.

57 M. Yang, M. Ieong, L. Shi, K. Chan, V. Chan, A. Chou, *et al.*, 'High performance CMOS fabricated on hybrid substrate with different crystal orientations', in *IEEE IEDM Tech. Dig.*, pp. 453–456, 2003.

58 R. Chau, J. Kavalieros, B. Roberds, R. Schenker, D. Lionberger, D. Barlage, *et al.*, '30 nm physical gate length CMOS transistors with 1.0 ps n-MOS and 1.7 ps p-MOS gate delays', *IEEE IEDM Tech. Dig.*, pp. 45–48, 2000.

59 C. K. Maiti, and G. A. Armstrong, *Applications of Silicon-Germanium Heterostructure Devices*. Inst. of Physics Pub., UK, 2001.

60 C. K. Maiti, N. B. Chakrabarti, and S. K. Ray, *Strained Silicon Heterostructures: Materials and Devices*. The IEE, UK, 2001.

61 K. Uejima, T. Yamamoto, and T. Mogami, 'Highly reliable poly-SiGe/ amorphous-Si gate CMOS', in *IEEE IEDM Tech. Dig.*, pp. 445–448, 2000.

62 S. Subbanna, V. P. Kesan, M. J. Tejwani, P. J. Restle, D. J. Mis, and S. S. Iyer, 'Si/SiGe p-channel MOSFETs', in *Proc. VLSI Technol. Symp.*, pp. 103–104, 1991.

63 D. K. Nayak, K. Goto, A. Yutani, J. Murota, and Y. Shiraki, 'High-mobility strained-Si PMOSFETs', *IEEE Trans. Electron Devices*, vol. 43, pp. 1709–1715, 1996.

64 J. L. Hoyt, H. M. Nayfeh, S. Eguchi, I. Aberg, G. Xia, T. Drake, *et al.*, 'Strained silicon MOSFET technology', in *IEEE IEDM Tech. Dig.*, pp. 23–26, 2002.

65 S. Verdonckt-Vandebroek, E. F. Crabbe, B. S. Meyerson, D. L. Harame, P. J. Restle, J. M. C. Stork, *et al.*, 'High-mobility modulation-doped grades SiGe-Channel p-MOSFET's', *IEEE Electron Device Lett.*, vol. EDL-12, pp. 447–449, 1991.

66 T.-J. King, and K. C. Saraswat, 'Low-temperature ($\leq$550 °C) fabrication of poly-Si thin-film transistors', *IEEE Electron Device Lett.*, vol. 13, pp. 309–311, 1992.

67 T.-J. King, and K. C. Saraswat, 'A low-temperature ($\leq$ 550°C) silicon-germanium MOS thin-film transistor technology for large-area electronics', in *IEEE IEDM Tech. Dig.*, pp. 567–570, 1991.

68 T.-J. King, J. R. Pfiester, and K. C. Saraswat, 'A variable-work-function polycrystalline-$Si_{1-x}Ge_x$ gate material for submicrometer CMOS technologies', *IEEE Electron Device Lett.*, vol. 13, pp. 533–535, 1991.

69 T. Uchino, T. Shiba, K. Ohnishi, A. Miyauchi, M. Nakata, Y. Inoue, *et al.*, 'A raised source/drain technology using in situ P-doped SiGe and B-doped Si for 0.1-μm CMOS ULSIs', in *IEEE IEDM Tech. Dig.*, pp. 479–482, 1997.

70 H. Huang, K. Chen, C. Chang, L. Chen, G. Huang, and T. Huang, 'Reduction of source/drain series resistance and its impact on device performance for PMOS transistors with raised $Si_{1-x}Ge_x$ source/drain', *IEEE Electron Device Lett.*, vol. 21, pp. 448–450, 2000.

71 M. C. Ozturk, J. Liu, H. Mo, and N. Pesovic, 'Advanced $Si_{1-x}Ge_x$ source/drain and contact technologies for sub-70 nm CMOS', in *IEEE IEDM Tech. Dig.*, pp. 375–378, 2002.

72 M. Horstmanna, D. Greenlawa, Th. Feudela, A. Weia, K. Frohberga, G. Burbacha, *et al.*, 'Sub-50 nm gate length SOI transistor development for high performance microprocessors', *Mat. Sci. and Eng. B*, vol. 114–115, pp. 3–8, 2004.

73 A. V. Y. Thean, T. White, M. Sadaka, L. McCormick, M. Ramon, R. Mora, *et al.*, 'Performance of super-critical strained-Si directly on insulator (SC-SSOI) CMOS based on high-performance PD-SOI technology', in *IEEE VLSI Tech. Symp. Dig.*, pp. 134–135, 2005.

74 K. Rim, S. Koester, M. Hargrove, J. Chu, P. M. Mooney, J. Ott, *et al.*, 'Strained Si NMOSFETs for high performance CMOS technology', in *IEEE VLSI Tech. Symp. Dig.*, pp. 59–60, 2001.

75 S. Datta, G. Dewey, M. Doczy, B. S. Doyle, B. Jin, J. Kavalieros, *et al.*, 'High mobility Si/SiGe strained channel MOS transistors with HfO_2/TiN gate stack', in *IEEE IEDM Tech. Dig.*, pp. 653–656, 2003.

76 T. Sanuki, A. Oishi, Y. Morimasa, S. Aota, T. Kinoshita, R. Hasumi, *et al.*, 'Scalability of strained silicon CMOSFET and high drive current enhancement in the 40nm gate length technology', in *IEEE IEDM Tech. Dig.*, pp. 65–68, 2003.

77 H. C.-H. Wang, Y.-P. Wang, S.-J. Chen, C.-H. Ge, S. M. Ting, J.-Y. Kung, *et al.*, 'Substrate-strained silicon technology: Process integration', in *IEEE IEDM Tech. Dig.*, pp. 61–64, 2003.

78 Th. Vogelsang, and K. R. Hofmann, 'Electron transport in strained Si layers on $Si_{1-x}Ge_x$ substrates', *Appl. Phys. Lett.*, vol. 63, pp. 186–188, 1993.

79 L. E. Kay, and T. W. Tang, 'Monte Carlo calculation of strained and unstrained electron mobilities in $Si_{1-x}Ge_x$ using an improved an ionized-impurity model', *J. Appl. Phys.*, vol. 70, pp. 1483–1488, 1991.

80 F. M. Bufler, P. Graf, S. Keith, and B. Meinerzhagen, 'Full band Monte Carlo investigation of electron transport in strained Si grown on $Si_{1-x}Ge_x$ substrate', *J. Appl. Phys.*, vol. 70, pp. 2144–2146, 1997.

81 M. V. Fischetti, and S. E. Laux, 'Band structure, deformation potentials, and carrier mobility in strained Si, Ge, and SiGe alloys', *J. Appl. Phys.*, vol. 88, pp. 2234–2252, 1996.

82 M. M. Rieger, and P. Vogl, 'Electronic-band parameters in strained $Si_{1-x}Ge_x$ alloys on $Si_{1-y}Ge_y$ substrates', *Phys. Rev. B*, vol. 48, pp. 14276–14287, 1993.

83 S. Smirnov, H. Kosina, and S. Selberherr, 'Investigation of the electron mobility in strained $Si_{1-x}Ge_x$ at high Ge composition', in *Proc. SISPAD*, pp. 29–32, 2002.

84 H. Kosina, and G. Kaiblinger-Grujin, 'Ionized-impurity scattering of majority electrons in silicon', *Solid State Electron.*, vol. 42, pp. 331–338, 1998.

85 F. M. Bufler, and W. Fichtner, 'Scaling and strain dependence of nanoscale strained-Si p-MOSFET performance', *IEEE Trans. Electron Devices*, vol. 50, pp. 2461–2466, 2003.

86 V. Palankovski, S. Dhar, H. Kosina, and S. Selberherr, 'Improved carrier transport in strained Si/SiGe devices', in *Asia Pacific Microwave Conference, Abstr. and Proc.*, 2004.

87 J. B. Roldan, F. Gamiz, P. Cartujo-Cassinello, P. Cartujo, J. E. Carceller, and A. Roldan, 'Strained-Si on $Si_{1-x}Ge_x$ MOSFET mobility model', *IEEE Trans. Electron Devices*, vol. 50, pp. 1408–1411, 2003.

88 F. Gamiz, J. B. Roldan, H. Kosina, and T. Grasser, 'Improving strained-Si on $Si_{1-x}Ge_x$ deep submicron MOSFETs performance by means of a stepped doping profile', *IEEE Trans. Electron Devices*, vol. 48, pp. 1878–1884, 2001.

89 K. Rim, J. L. Hoyt, and J. F. Gibbons, 'Transconductance enhancement in deep submicron strained-Si n-MOSFETs', *IEEE IEDM Tech. Dig.*, pp. 707–710, 1998.

90 K. Rim, J. L. Hoyt, and J. F. Gibbons, 'Fabrication and analysis of deep submicron strained-Si N-MOSFET's', *IEEE Trans. Electron Devices*, vol. 47, pp. 1406–1415, 2000.

91 K. Rim, 'Strained Si surface channel MOSFETs for high-performance CMOS technology', in *IEEE ISSCC Tech. Dig.*, pp. 116–117, 2001.

92 K. Ismail, 'Si/SiGe high-speed field-effect transistors', in *IEEE IEDM Tech. Dig.*, pp. 509–512, 1995.

93 S. J. Koester, R. Hammond, and J. O. Chu, 'Extremely high transconductance $Ge/Si_{0.4}Ge_{0.6}$ p-MODFET's grown by UHV-CVD', *IEEE Electron Device Lett.*, vol. 21, pp. 110–112, 2000.

94 M. A. Armstrong, D. A. Antoniadis, A. Sadek, K. Ismail, and F. Stern, 'Design of Si/SiGe heterojunction complementary metal-oxide semiconductor transistors', in *IEEE IEDM Tech. Dig.*, pp. 761–764, 1995.

95 D. A. Buchanan, 'Scaling the gate dielectric: materials, integration, and reliability', *IBM J. Res. Develop.*, vol. 43, pp. 245–264, 1999.

96 A. Yagishita, T. Saito, K. Nakajima, S. Inumiya, Y. Akasaka, Y. Ozawa, *et al.*, 'High performance damascene metal gate MOSFETs for 0.1 μm regime', *IEEE Trans. Electron Devices*, vol. 47, pp. 1028–1034, 2000.

97 Y. Ma, D. R. Evans, T. Nguyen, Y. Ono, and S. T. Hsu, 'Fabrication and characterization of sub-quarter-micron MOSFET's with a copper gate electrode', *IEEE Electron Device Lett.*, vol. 20, pp. 254–255, 1999.

98 T. Ushiki, M. Yu, Y. Hirano, H. Shimada, M. Morita, and T. Ohmi, 'Reliable tantalum-gate fully-depleted-SOI MOSFET technology featuring low-temperature processing', *IEEE Trans. Electron Devices*, vol. 44, pp. 1467–1472, 1997.

99 M. Y. A. Yousif, M. Friesel, M. Willander, P. Lundgren, and M. Caymax, 'On the performance of in situ B-doped P^+-poly-$Si_{1-x}Ge_x$ gate material for nanometer scale MOS technology', *Solid State Electron.*, vol. 44, pp. 1425–1429, 2000.

100 T. J. King, J. R. Pfriester, J. D. Scott, J. P. McVittie, and K. C. Saraswat, 'A polycrystalline SiGe gate CMOS technology', in *IEEE IEDM Tech. Dig.*, pp. 253–256, 1990.

101 P.-E. Hellberg, A. Gagnor, S.-L. Zhang, and C. S. Petersson, 'Boron doped polycrystalline Si_xGe_{1-x} films: Dopant activation and solid solubility', *J. Electrochem. Soc.*, vol. 144, p. 3968, 1997.

102 Y. V. Ponomarev, P. A. Stolk, C. Salm, J. Schmitz, and P. H. Woerlee, 'High-performance deep submicron CMOS technologies with polycrystalline-SiGe gates', *IEEE Trans. Electron Devices*, vol. 47, pp. 848–855, 2000.

103 S. Cristoloveanu, 'Silicon on insulator technologies and devices: from present to future', *Solid State Electron.*, vol. 45, pp. 1403–1411, 2001.

104 K. D. Hobart, F. J. Kub, M. E. Twigg, G. G. Jernigan, and P. E. Thompson, 'Ultra-cut: a simple technique for the fabrication of SOI substrates with ultra-thin (<5 nm) silicon films', in *IEEE SOI Conf. Proc.*, pp. 145–146, 1998.

105 E. Leobandung, E. Barth, M. Sherony, S. Lo, R. Schulz, W. Chu, *et al.*, 'High performance 0.18 μm SOI CMOS technology', in *IEEE IEDM Tech. Dig.*, pp. 679–682, 1999.

106 T. C. Lim, N. Jankovic, and G. A. Armstrong, 'Caling of fully depleted SOI MOSFETs with p^+-poly SiGe gates', in *Proc. ISIC*, pp. 231–234, 2004.

107 S. M. Sze, *Physics of Semiconductor Devices*. John Wiley & Sons, New York, 2nd ed., 1981.

108 T. J. King, J. P. McVittie, K. C. Saraswat, and J. R. Pfiester, 'Electrical properties of heavily doped polycrystalline silicon-germanium films', *IEEE Trans. Electron Devices*, vol. 41, pp. 228–232, 1994.

109 M. Y. A. Yousif, O. Nur, and M. Willander, 'Recent critical issues in $Si/Si_{1-x}Ge_x/Si$ heterostructure FET devices', *Solid State Electron.*, vol. 45, pp. 1931–1937, 2001.

110 P. Hellberg, S. Zhang, and C. S. Petersson, 'Work function of boron-doped polycrystalline Si_xGe_{1-x} films,' *IEEE Electron Device Lett.*, vol. 18, pp. 456–458, 1997.

111 A. Chatterjee, R. A. Chapman, K. Joyner, M. Otobe, S. Hattangady, M. Bevan, *et al.*, 'CMOS metal replacement gate transistors using tantalum Pentoxide gate insulator', *IEEE IEDM Tech. Dig.*, pp. 777–780, 1998.

112 S.-F. Huang, C. Lin, Y. Huang, T. Schafbauer, M. Eller, Y. Cheng, *et al.*, 'High performance 50 nm CMOS devices for microprocessor and embedded processor core applications', in *IEEE IEDM Tech. Dig.*, pp. 237–240, 2001.

113 F. Nouri, P. Verheyen, L. Washington, V. Moroz, I. De Wolf, M. Kawaguchi, *et al.*, 'A systematic study of trade-offs in engineering a locally strained pMOSFET', in *IEEE IEDM Tech. Dig.*, pp. 1055–1058, 2004.

114 L. Smith, V. Moroz, G. Eneman, P. Verheyen, F. Nouri, L. Washington, *et al.*, 'Exploring the limits of stress-enhanced hole mobility', *IEEE Electron Device Lett.*, vol. 26, pp. 652–654, 2005.

115 V. Moroz, N. Strecker, X. Xu, L. Smith, and I. Bork, 'Modeling the impact of stress on silicon processes and devices', *Mat. Sci. Semicond. Processing.*, vol. 6, pp. 27–36, 2003.

116 V. Moroz, X. Xu, D. Pramanik, F. Nouri, and Z. Krivokapic, 'Analyzing strained-silicon options for stress-engineering transistors', *Solid State Technol.*, vol. July, pp. 49–52, 2004.

117 L. Smith, V. Moroz, G. Eneman, P. Verheyen, F. Nouri, L. Washington, *et al.*, 'Exploring the limits of stress-enhanced hole mobility', *IEEE Electron Device Lett.*, vol. 26, pp. 652–654, 2005.

118 R. A. Bianchi, G. Bouche, and O. Roux-dit-Buisson, 'Accurate modeling of trench isolation induced mechanical stress effects on MOSFET electrical performance', in *IEEE IEDM Tech. Dig.*, pp. 117–120, 2002.

119 P. R. Chidambaram, B. A. Smith, L. H. Hall, H. Bu, S. Chakravarthi, Y. Kim, *et al.*, '35% drive current improvement from recessed-SiGe drain extensions on 37 nm gate length PMOS', in *IEEE VLSI Tech. Symp. Dig.*, pp. 48–49, 2004.

120 V. Moroz, G. Eneman, P. Verheyen, F. Nouri, L. Washington, L. Smith, *et al.*, 'The impact of layout on stress-enhanced transistor performance', in *Proc. SISPAD*, pp. 143–146, 2005.

121 G. Eneman, P. Verheyen, R. Rooyackers, F. Nouri, L. Washington, R. Degraeve, *et al.*, 'Layout impact on the performance of a locally strained PMOSFET', in *IEEE VLSI Tech. Symp. Dig.*, pp. 22–23, 2005.

122 A. V.-Y. Thean, A. L. Barr, T. R. White, Z.-H. Shi, B.-Y. Nguyen, C.-L. Liu, *et al.*, 'Computer aided design of sub-100nm strained-Si/Si$_{1-x}$Ge$_x$ NMOSFET through integrated process and device simulations', in *Proc. SISPAD*, pp. 195–198, 2003.

123 DESSIS-ISE, Release 8.5, 'Integrated Systems Engineering AG, Zurich, Switzerland', 2003.

124 ISE Integrated Systems Engineering AG., Zurich, Switzerland, *Tech. Rep. DIOS-ISE, Release 6.0*, 1999.

125 A. Schenk, 'Rigorous theory and simplified model of the band-to-band tunneling in silicon', *Solid State Electron.*, vol. 36, pp. 19–34, 1993.

Chapter 5

SOI MOSFETs

Silicon-on-insulator (SOI) refers to the first engineered semiconductor substrate where a buried layer of silicon dioxide is created beneath a thin layer of silicon. After a long development history for around 30 years, SOI technology has now joined the microelectronics roadmap and has started to be used as the mainstream technology in CMOS LSIs. The advantages of SOI devices over conventional silicon devices are manifold and have been summarised in many references [1,2]. Circuits based on SOI technology are extremely attractive for two basic reasons: (1) enhanced performance (higher speed, lower power dissipation) and (2) extended scalability.

SOI devices intrinsically have very low junction capacitance, where the buried oxide (BOX) can yield a three-fold reduction. The phenomenon of CMOS latchup is eliminated in SOI, as there exists no path for current flow to the substrate. SOI devices eliminate the body-effect since their body potential is not tied to ground. SOI devices have a body factor close to unity and hence this reduces their subthreshold slope, an attractive feature for switching applications. SOI technology is compatible with conventional silicon technology. In principle, by suitable process modifications, a circuit can be realised using fewer process steps. Scaling in SOI is easy, as the BOX thickness need not be scaled, so only the dimensions of the device need to be scaled. However, SOI devices suffer also from some drawbacks:

1. Floating body effect in partially depleted SOI MOSFETs are prone to 'kink' effect in d.c. circuits and drain-current overshoot in switching circuits.
2. There is higher off-current due to a forward biased body-source junction when the drain voltage is high.
3. Fully depleted devices require ultra-thin films that are difficult to realise.
4. High quality of buried oxide-silicon interface is vital.

However, the drawbacks of SOI devices are largely overshadowed by the significant advantages in performance the SOI technology has to offer, particularly for a low power.

Initial success has been achieved in using SOI substrate for mainstream CMOS technology. IBM has built and tested SOI-based chips that have produced 20–25 per cent cycle time and 25–35 per cent improvement over equivalent bulk CMOS technology. The ability of SOI as a low power source originates from the fact that SOI circuits can operate at a lower voltage than the equivalent bulk technology. At present SOI applications are mostly for microprocessors [3], but other possibilities are expanding from high speed/low-power logic LSIs [4, 5] to radio frequency (RF)/analog LSIs [6, 7], DRAMs [8–10], very low-voltage logic LSIs [11, 12] and system-on-chip (SoC) [13]. Excellent reviews of SOI materials and technology, evolution of devices and future trends can be found in References 14–16.

As far as the reliability is concerned, SOI MOSFETs are extremely robust to transient radiation effects. SOI devices offer a significant improvement in the soft-error rate, since the presence of the BOX significantly reduces the volume susceptible to ionising radiation. For example, in a circuit exposed to radiations, most of the electron-hole pairs are generated in the thick silicon substrate and do not affect the performance of SOI devices. This is why SOI technologies have initially been dedicated to the niche of radiation-hard circuits.

In this chapter, focus is on the SOI MOSFETs and a comparison with conventional MOSFETs in terms of threshold voltage roll-off, kink effect, parasitic bipolar action, transconductance and output resistance is presented. The current status and future direction of SOI technology in terms of materials and devices, present status and future developments are discussed. After a brief presentation of the two SOI process families (fully depleted SOI and partially depleted SOI processes), attention will be drawn on the advantages of SOI devices and also which SOI specific effects must be accounted for by designers. Main mechanisms involved in the operation of fully and partially depleted SOI MOSFETs are then discussed.

Consideration is given to the design challenges faced with SOI technologies: the possibility of scaling scenarios that can replace the conventional scaling under the condition where gate insulator thickness is no longer scaled. It will be demonstrated that, on the basis of scalability and flexibility arguments, SOI is capable of further extending the limits and performance of bulk-Si technology.

5.1   Fabrication of SOI wafers

There are various methods to fabricate the SOI wafers and among them, separation by implanted oxygen (SIMOX) and UNIBOND (also known as Smart-cut) are considered most important for 300-mm SOI wafer fabrication. Several other methods including zone melting recrystallisation (ZMR), ELTRANS (epitaxial layer transfer), ELO (epitaxial lateral overgrowth) and SOS (silicon on sapphire) are used in research or specialised applications. Many of these techniques have been tried and tested for several years and almost all of them are present in a manufacturable technology. Fabrication of SOI substrates has been reviewed by S. Cristoloveanu [17].

Until the early 1990s, SOS was the most mature substrate technology for fabricating SOI devices. Several fabrication technologies exist for fabricating SOS devices [2]. SOS is fabricated by epitaxial growth of a Si film on sapphire (Al_2O_3). The electron mobility in SOS devices ($\sim$250–350 cm^2/Vs) is significantly lower compared to bulk silicon mobility ($\sim$1300–1400 cm^2/Vs). This is largely due to the high defect density and the compressive stress present in silicon on sapphire, which leads to a reduction in the hole effective mass. This stress is expected to produce an increase in the hole mobility. However, owing to the presence of surface defects, hole mobility shows a value nearly identical to that of the bulk mobility. As a result of these characteristics, SOS is not the ideal material for fabricating advanced SOI devices. The benefit of reduced parasitic capacitance is offset by the reduced carrier mobilities and the added cost of SOS technology.

Within the epitaxial growth category, SOI fabrication is accomplished using one of two different methods. The first technique is described as epitaxial lateral overgrowth (ELO). Here the fabrication process involves epitaxial growth of silicon from seeding windows in the SiO_2 layer. The silicon growth occurs initially in the oxide windows and then laterally over silicon dioxide islands or devices capped with an insulator. ELO process requires a post-epitaxy thinning of the Si film, which is achieved by using a patterned oxide. The silicon film in excess is removed, leaving an isolated Si island in the buried oxide.

The ELO technique has been used to fabricate three-dimensional and dual-gate device structures. The other technique to fabricate SOI devices epitaxially is known as lateral solid phase epitaxy (LSPE) technique. The LSPE technique is based on the lateral epitaxial growth of crystalline silicon through the controlled crystallisation of amorphous silicon. This technique is a good choice for fabrication processes in the low temperature range. In ZMR technique, a polySi layer is deposited on the BOX and is recrystallised by scanning a source of energy (laser) across the wafer. As for ELO, ZMR is limited by the lateral extension of defect-free and single crystal islands.

Full isolation by oxidised porous silicon (FIPOS) involves converting a silicon crystalline layer to porous silicon, followed by a subsequent oxidation of the porous silicon layer. The advantage of this approach is that porous silicon oxidises much more readily without affecting the crystalline silicon on top of it. The original FIPOS process produced SOI material of extremely high quality and the devices fabricated using this technique showed excellent characteristics. However, this process technology suffers from wafer warpage owing to the formation of porous silicon and then oxidising it.

In the early development stage, SOI technology was based on SIMOX, which refers to the deep implantation of oxygen ions into a Si wafer to create a buried insulator (BOX). The SIMOX technique (see Figure 5.1) involves formation of a buried layer of SiO_2 by implantation at high energy ($\sim$200 keV) oxygen ions beneath the surface of a silicon wafer. The quality of the active silicon layer is a function of the process and in particular the temperature of the anneal following implantation, where annealing at high temperature (1320 °C, for 6 h) restores the crystalline quality of the film. SIMOX wafers have good thickness uniformity, low defect density (except threading dislocations: 10^4–10^6 cm^{-2}), sharp Si–SiO_2 interface, robust BOX and high carrier mobility.

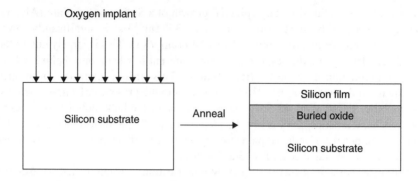

Figure 5.1   SIMOX technique

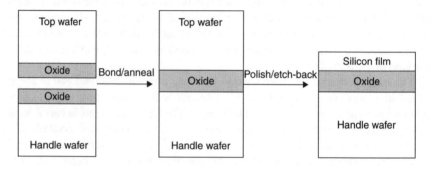

Figure 5.2   Wafer bonding

Experiments have shown that oxygen dose of the order of $10^{18}/\text{cm}^2$ must be used to get a box-like profile. SIMOX comes in several versions: (1) thin and thick Si films fabricated by adjusting the implant energy, (2) low-dose SIMOX ($\leq 4 \times 10^{17}$ O^+/cm^2) with a 100 nm or thin BOX, (3) standard SIMOX (oxygen dose 1.8×10^{18} O^+/cm^2 that typically has a 0.2 μm thick Si film and a 0.4 μm thick BOX, (4) double SIMOX, where the Si layer sandwiched between the two oxides serves for buried interconnects, wave guiding, additional gates or electric shielding and (5) interrupted oxides that can be viewed as SOI regions integrated into a bulk-Si wafer. Ultra-thin SOI layers in particular require not only epitaxial layer thickness control but also hydrogen annealing to smooth the SOI surface without reducing the film thickness. However, in SIMOX technology continuous improvements to the base fabrication technology are being implemented for device performance enhancement.

Wafer bonding (WB) and etch-back or BESOI has become an important alternative state-of-the-art SOI technology, in which an oxidised Si wafer is bonded to a second Si wafer (see Figure 5.2). After bonding, the wafer is thinned down to reach the target thickness of the silicon film. In principle, BESOI can provide defect-free Si overlayer and compete with SIMOX for thick-film SOI applications. Etch-stop layers are needed and they can be achieved by doping steps (p^+/p^-, P/N), epitaxy of a sacrificial layer (SiGe) or porous silicon. Presently, BESOI is a strong competitor to SIMOX for supply of quality production SOI wafers, although control of thickness of very thin layers remains to be resolved.

As the SOI wafer is a composite substrate with an active top Si layer decoupled from the support wafer, currently the Smart-Cut technique, which is a combination of ion implantation and bonding techniques, is being widely used. Smart-Cut-based manufacturing of Soitec's SOI wafers (trade name UNIBOND) started in the late 1990s [18, 19]. Differing from traditional layer-transfer techniques, which are based mainly on wafer bonding and etch-back or epitaxial lift-off, the Smart-Cut approach uses a thermal activation process as an 'atomic scalpel'. This process offers better control, and a single donor substrate can be reused many times for further layer transfers (see Figure 5.3).

The Smart-Cut technique employs the deep implantation of hydrogen to generate microcavities. In order to solve the control problem of thin film thickness in BESOI, a starting wafer is implanted with hydrogen ions at a low depth below a previously

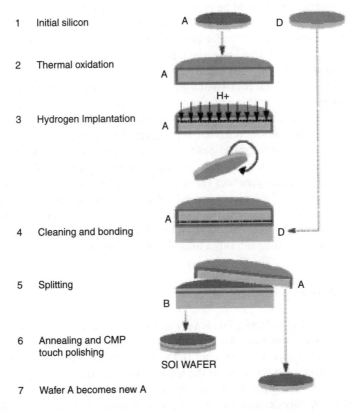

Figure 5.3 Typical process steps involved in Smart-Cut technique [Source: SOITEC.]

grown oxide. In the second step, the hydrogen-implanted wafer is bonded to an oxidised wafer. After bonding and annealing, the assembly can be accurately split at the level of the hydrogen implant, since hydrogen ions weaken the bonding of Si atoms. Smart-Cut features many advantages with regard to SIMOX and BESOI. Smart-Cut is currently deployed commercially for SOI, SOQ (silicon-on-quartz or, more precisely, single crystalline Si on a fused silica substrate) and strained-Si on insulator. The advantages of of Smart-Cut technology are as follows:

- It is well-suited to high-volume manufacturing.
- The thicknesses of the top Si and BOX layers can be easily controlled.
- The process is scalable to any wafer diameter.
- It uses standard IC manufacturing equipment.

Recently, SOITEC has introduced strained-Si on insulator (strained-SOI) substrates for the next generation of 65 nm devices, especially for high-speed circuits, wireless and broadband communication applications [20,21]. Strained-SOI combines two high-performance technology solutions: strained-Si and ultra-thin SOI. Strained-SOI, however, is a generic term describing a wafer that integrates both strained-Si and SOI technologies, which could be either sSOI, or strained-Si on SiGe-on-insulator (SSGOI). The production of strained-SOI substrates requires growing epitaxial SiGe and Si layers, combined with wafer bonding and layer transfer to a handle wafer. Depending on the process and target application, different combinations of Si and SiGe layers on insulator can be obtained.

Strained-Si on insulator consists of a thin layer of strained-Si bonded directly onto an insulated handle wafer. It is better suited for a fully depleted device architecture, with strained-Si film thicknesses typically below 50 nm. Since sSOI has the same basic structure as the conventional SOI, it is expected that the actual film thicknesses for sSOI will follow similar guidelines to those that the industry's roadmap indicates for SOI. SSGOI consists of a bulk silicon support, topped by a layer of insulator, and a relaxed-SiGe layer that generates the strain for the silicon film above it. It can be tailored for both partially depleted and fully depleted applications, with strained-Si film thicknesses of 7.5–22.5 nm. The total strained-Si and SiGe bilayer is typically 40–80 nm thick. sSOI is likely to be used for high-performance logic applications for 45 nm and below requiring ultra-thin fully depleted structures.

5.2   SOI devices

The SOI MOSFET (see Figure 5.4) is very similar to a conventional MOSFET with identical modes of operation, the only difference being that the device is formed on a silicon layer above an underlying BOX. There are two different types of SOI MOSFET, based on the thickness of the silicon overlayer on which the devices are fabricated. These two types of device types are partially depleted (PD) and fully depleted (FD). A fully depleted SOI MOSFET is, as the name implies, fully depleted of majority carriers at threshold. This tends to occur in relatively thin lightly doped silicon layers. On the other hand, partially depleted structure has a thick enough

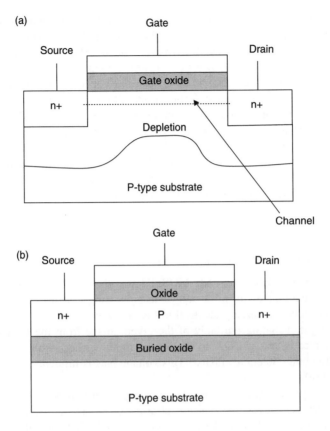

Figure 5.4 *(a)* *Cross-sectional view of conventional bulk n-MOSFET and* *(b) cross-sectional view of an SOI n-MOSFET*

silicon layer to have a neutral region above the buried oxide. This is known as a 'floating region', as its electric potential is not fixed.

Full depletion of the silicon film not only depends on silicon thickness but also on the gate length, the doping level of the silicon film, the gate oxide thickness and the flat-band voltage. For partially depleted SOI transistor, d.c. operation is governed by the floating body voltage, which is related to the body charge. The major parasitic effect in SOI-MOSFETs is related to the build-up of a charge in the silicon layer originated from impact ionisation at the drain junction. This charge cannot be removed, primarily because no contact with the Si film is available. There are various consequences of this charge, which are generically referred to as floating-body effects: kink effect, drain current overshoot, hysteresis and instabilities associated with parasitic bipolar transistor action and premature breakdown. When the body potential becomes high enough the barrier at the source junction is lowered and an injection of carriers due to lateral bipolar action occurs, injecting additional minority carriers from the source (emitter) into the body (base).

SOI technology is now utilized in many industrial and consumer products requiring high-performance logic in sub-90 nm technologies. The design issues resulting from the floating-body in partially depleted SOI device structure, such as parasitic bipolar effect and V_t variation, are now well understood and circuit/design techniques to mitigate these effects have been developed. SOI also provides new degrees of freedom to create many forms of multiple gate devices and offers advantages in many niche applications. For example, in microsensors the Si/BOX interface gives a perfect etch-stop mark, making it possible to fabricate very thin membranes. A device with the best possible electrostatic control, the gate-all-around (GAA) transistor was first proposed by Colinge *et al.* [22], which can be fabricated by etching a cavity into the BOX. A free-standing membrane, which will form the body of the transistor, is obtained. This membrane is subsequently oxidised and wrapped with a polySi gate [23].

5.2.1   Partially depleted SOI transistors

The partially depleted floating-body MOSFET was the first SOI transistor adopted for volume production, owing principally to its processing similarities to bulk CMOS. The PD–SOI device is largely identical to the bulk device, except for the addition of a BOX layer, isolating the body of the given device from the bodies of other devices. Isolated below by BOX, to the left and right by trench isolation or diffused junctions, the body potential 'floats'. This voltage bias is influenced by both static and dynamic effects.

In a partially depleted MOSFET, the gate depletion region does not extend all the way to the BOX layer, and thus a neutral region exists. As there is no external contact, the holes stored in the body can modulate the body potential and hence the threshold voltage. Reduction in the threshold voltage gives rise to an abrupt increase in the drain current. This is known as the 'kink' effect as a kink is observed in the I–V curves (see Figure 5.5), which is due to majority carriers, generated by impact ionisation influencing the body potential. In weak inversion and at high drain bias, a positive feedback is responsible for negative resistance regions, hysteresis in log $I_d - V_g$ curves and eventually transistor latch.

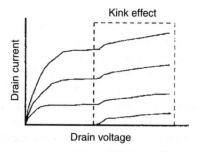

Figure 5.5   Kink effect in an SOI MOSFET

The floating body may also cause undesirable transient effects [24]. Under conditions of fast gate ramps, the body potential tends to rise with the gate potential, which reduces the threshold voltage and increases the drain current, giving rise to a phenomenon called drain current overshoot. Majority carriers are expelled from the depletion region and collected in the neutral body. The drain current decreases gradually with time during electron-hole recombination. In addition to the usual kink seen in the drain current characteristics due to the presence of a floating body, one can observe an additional kink immediately after the first kink. The occurrence of the second kink is due to the presence of a parasitic npn bipolar transistor that is inherent in the floating-body structure. The current overshoot is useful, because an extra drive current is available, which benefits fast switching logic circuits. A reciprocal under-shoot occurs when the gate is switched from strong to weak inversion: the current now increases with time as the majority carrier generation allows the depletion depth to shrink gradually.

At high drain voltage, impact ionisation generate more electron-hole pairs, and thus modify the shape of the transient. Speed improvement in SOI can occur not only from the reduction of drain and source capacitances but also from the static and dynamic threshold voltage lowering due to the floating-body. In both fully and partially depleted MOSFETs with submicron gate length, problems with the lateral bipolar transistor that existed for longer gate length can be alleviated by using reduced drain voltage. However, care must be taken using TCAD to accurately predict the onset of breakdown, which is governed by film thickness and doping.

The self-heating effect, induced by the poor thermal conductivity of the BOX, is responsible for the mobility degradation, threshold voltage lowering and negative differential conductance. The threshold voltage is reduced by charge sharing effect and drain-induced barrier lowering (DIBL) in the film. Improved DIBL is due to the penetration of the electric field, into the BOX and underlying Si substrate. Such fringing fields are responsible for an increase in the potential at the lower interface, as if the back gate (substrate) was driven from depletion to weak inversion. In other words, the drain acts as a virtual back gate. This drain-induced virtual substrate biasing is critical in sub-0.1 μm fully depleted MOSFETs, where the front-channel threshold voltage is lowered and the subthreshold swing is degraded.

Unfortunately, thin SOI MOSFETs are prone to another short channel mechanism: drain-induced virtual substrate biasing (DIVSB) [25]. Increasing the drain bias allows the fringing fields to penetrate into the buried dielectric and underlying substrate (see Figure 5.6). A depletion region can form underneath the buried insulator, which modifies its apparent capacitance. A more serious problem is the modification of the potential distribution in the BOX with a net increase in the potential at the film-BOX interface, much as if a virtual positive bias was applied to the back gate (substrate).

The fringing fields into the buried insulator and substrate depletion region stand as a key limiting factor for SOI MOSFET with channel length scaled beyond 0.1 μm. The thin, fully depleted SOI transistors are particularly sensitive to DIVSB because the fringing fields drive the back interface from depletion to weak inversion [26]. Owing to interface coupling, the front-channel threshold voltage is lowered and the subthreshold swing is degraded, which leads to an overall degradation of the transistor performance.

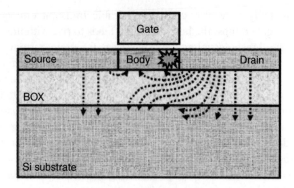

Figure 5.6   Schematic representation of fringing fields and hot-spot in an SOI transistor

5.2.2   Fully depleted SOI transistors

For the fully depleted MOSFET, the depletion regions throughout the silicon reaches the BOX. This reduces all the capacitances associated with the silicon layer. In order to maintain balance of electric fields and capacitance within the SOI device and reduce short channel effects, the active silicon thickness must be reduced with scaling. In this fully depleted FET, the threshold of the device is now defined by the amount of charge obtainable within the isolated body.

Fully depleted SOI has been shown to be an attractive candidate for deep sub-micron CMOS low-power and high-speed applications. The subthreshold slope of a fully depleted SOI MOSFET is near the ideal value of 60 mV/decade. This is because the gate depletion width is very large and the associated depletion capacitance in series with the gate capacitance becomes negligibly small. The steeper subthreshold slope for the fully depleted SOI MOSFET permits a lower threshold voltage for the same I_{off} current, which in turn allows the devices to be used at lower supply voltages in low power applications. Appropriate scaling of the ratio of film thickness to gate length can reduce short-channel effects and enhance output currents due to higher mobility in lightly doped thin films.

Aggressive CMOS scaling involves reducing the thickness of the thin silicon film. The superior short-channel behaviour of appropriately scaled fully depleted SOI devices can ultimately allow for use of undoped silicon layers, even at deep-submicron gate lengths, yielding higher mobility and current drive. Fully depleted devices require stringent control on the thickness of the active silicon film and hence pose tighter controls on the fabrication process. For example, silicidation of ultra-thin films (<50 nm) may consume the entire silicon film in the S/D areas.

The key parameters in SOI are gate length, film thickness and the nature and thickness of the BOX [27]. Increasing the doping reduces the threshold voltage roll-off $V_t(L)$ resulting from both CSE and DIBL. There is no influence of the film thickness when the device is partially depleted. However, Figure 5.7 shows the desirability of

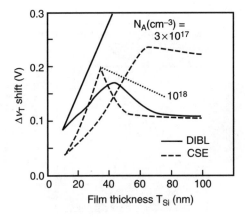

Figure 5.7   Threshold voltage shift in sub-100 nm SOI MOSFETs: influence of film thickness and doping on the CSE and DIBL effects [After S. Cristoloveanu, Solid-State Electron., Vol. 45, 2001(1403–1411).]

using ultra-thin, undoped SOI even for very short gate lengths, as long as the film thickness is scaled to the gate length.

In SOI MOSFETs, inversion channels can be activated at both the front Si–SiO$_2$ interface and the back Si–BOX interface (back-gate biasing). A strong coupling develops between the gate bias and the inversion charge, leading to enhanced drain current. The front- and back-surface potentials become coupled too and the electrical characteristics of one channel vary with the bias applied to the opposite gate [28]. This interface coupling causes the front-gate measurements to depend on the back-gate bias and quality of the BOX and interface. Hence different I_d-V_g equations apply to fully depleted SOI MOSFETs, compared to bulk MOSFETs.

5.3   Double-gate MOSFET

As MOSFETs scale into the sub-100 nm regime, the gate oxide thickness required for adequate short channel effect (SCE) suppression reaches below 2.0 nm. The direct tunnelling current through such a thin gate oxide starts to adversely affect the stand-by power consumption. In highly doped layers random dopant fluctuations makes it very difficult to control short channel effects. The improved electrostatic control offered by an alternative device structure, the double-gate (DG) MOSFET illustrated in Figure 5.8, will enable MOSFET scaling below 45-nm node on ITRS roadmap.

Distinct features of this novel device include application of a second gate and use of a thin, fully depleted undoped silicon layer. The dual-gate configuration offers multiple options about how to bias the two gates. Bias configurations may be categorised as symmetric, asymmetric or ground-plane (GP). A symmetric DG MOSFET results when two gates have the same workfunction and a single input voltage is applied to both gates. An asymmetric DG MOSFET either has synchronised but

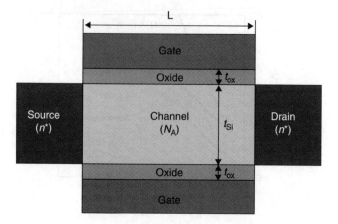

Figure 5.8   Schematic of the cross-section of a DG MOSFET with fully depleted channel

different input voltages to two identical gates or has the same input voltage to two gates that have distinct workfunctions. The names symmetric or asymmetric essentially reflect the presence or absence of symmetry of the vertical electric field inside the silicon. A ground-plane MOSFET (GP MOSFET) has one of its gates biased to a constant voltage, while the other gate voltage is variable.

The operation of DG SOI MOSFETs is based on the concept of volume inversion: the formation of front and back inversion channels causes, by continuity, the spreading of minority carriers in the volume of a thin SOI film. It was predicted at an early stage that the DG MOSFET will represent the ultimate MOS transistor [29, 30], as the scalability is improved by the double-gate electrostatic control [29, 31]. Planar DG MOSFETs with symmetrical configuration have been fabricated with GAA [2], Delta [32] and lateral epitaxial growth technologies [33].

When biased in volume inversion, a 3-nm-thick DG MOSFET shows an increase in transconductance as compared to SG operation [see Figure 5.9(a)] [34]. This difference is explained using the Poisson and Schrodinger equations, solved self-consistently. Both the vertical potential profile and spatial distribution of charge are symmetrical, but carriers are no longer confined to the oxide interfaces [Figure 5.9(b)] [34, 35].

As the film thickness decreases below 20 nm, the behaviour of the electron mobility becomes very complex [34–36]. There are several competing mechanisms:

1.   Carrier confinement increases, leading to enhanced phonon scattering.
2.   The size-induced quantisation is beneficial for the mobility.
3.   The surface roughness and Coulomb scattering increase and involve the two Si–SiO$_2$ interfaces.

Monte Carlo simulations show that the carrier mobility may increase as the film thickness is reduced below 10 nm [37]. The electric field is negligible in the middle of

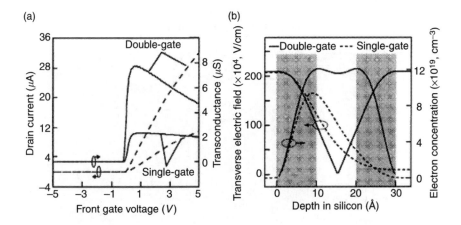

Figure 5.9 *(a) Drain current and transconductance measured in a 3-nm SOI MOSFET operated in SG and DG modes. (b) Quantum distributions of minority carriers and electric field ($V_g - V_t = 2\ V$); in the grey regions, the carrier mobility is presumably degraded by surface roughness [After S. Cristoloveanu, Solid-State Electron., Vol. 45, 2001(1403–1411).]*

the DG MOSFET where most of the inversion charge is located. As electron–phonon scattering strongly depends on the field, the mobility is presumably enhanced in the centre of the film, but it is also reasonable to assume that surface roughness degrades the mobility near each interface but where carrier concentration is now much lower [grey areas in Figure 5.9(b)].

5.4 TCAD simulation of SOI devices

Silvaco provides an integrated simulation software for studying various aspects of SOI technology. The ATHENA/SUPREM4 and ATLAS/S-PISCES simulators allow device engineers to study deep sub-micron physical effects. TCAD simulation includes models for simulation of both fully and partially depleted SOI transistors based on accurate process geometry and doping profiles. In the following, some results of ATHENA/ATLAS simulation will be presented that illustrate a few common effects found in short-channel SOI MOSFETs.

A short-channel SOI MOSFET created using ATHENA is shown in Figure 5.10. The structure of the device is very similar to that of a standard MOSFET device, but the presence of a thick layer of insulating material under the depletion region changes some of the device characteristics. Major requirements for SOI process simulation are

- definition of regions;
- deposition/oxidation of silicon;

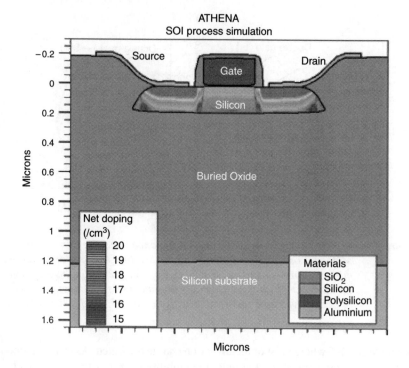

Figure 5.10   Short-channel SOI MOSFET created using ATHENA. Local oxidation is used to isolate the active layer [Source: Silvaco International.]

- local oxidation to form silicon islands;
- shallow implant and RTA diffusion process steps;
- CMP module silicon island isolation techniques;
- LOCOS and Trench;
- Silicides.

SOI MOSFETs are much more difficult to simulate owing to the presence of significant floating body effects, carrier heating and non-localised transport. The presence of a floating potential region in an SOI device results in an ill-conditioned matrix, and the solution of SOI structures may become a numerically unstable problem. However, these problems can be resolved as long as small steps in voltage are used. ATLAS allows the implementation of a hierarchy of basic transport models and advanced models accounting for non-localised and lattice heating effects. In solving the semiconductor equations in SOI structures, Drift-Diffusion or Energy Balance formulation may be used. However, Energy Balance is necessary for very short-channel devices to account for velocity overshoot. A non-localised approach allows accurate simulation of impact ionisation. Basic models for SOI simulation include a choice of different mobility models. Inclusion of a model for impact ionisation is essential to model kink effect, snapback and hysteresis.

As SOI devices exhibit self-heating effects, a substantial elevation of temperature within the SOI device occurs, which can modify the output I–V characteristics of the device. SOI transistors exhibit significant heating due to the low thermal conductivity of the buried insulator. Thicker BOX leads to heat dissipation problems, because the heat is trapped in the upper silicon layer by the insulator. Thermal and electrical effects can be coupled through self-consistent calculations. Models for this effect include heat generation, heat flow, and effects of local temperature on physical constants.

Difference between drain I–V characteristics of long channel and short channel device is often observed. Also a transition takes place from pronounced kink effect at low gate voltage, when power dissipation is low but impact ionisation at the drain is high, to negative resistance at high gate voltage when self-heating predominates, owing to higher power dissipation. As the drive current is increased the silicon lattice temperature increases, and significantly degrades the carrier mobility so that, despite increasing the drain voltage, the drive current decreases.

If the film thickness is reduced so that the film becomes fully depleted, the kink is eliminated. The increased recombination around the source junction of fully depleted devices also helps to eliminate drain current overshoot in ultra-thin film transistors, and a lowering of the drain breakdown voltage occurs. [38, 39]. The reduced breakdown voltage at zero gate voltage is attributed to the shape of the drain region, with the reduced drain depth of ultra-thin SOI devices giving rise to a significant increase in lateral drain field, and a consequent increase in impact ionisation. Holes generated by impact ionisation serve as a source of base current for the lateral bipolar transistor, and in the subthreshold region, bipolar snapback effects have been observed experimentally [40, 41]. To understand the lateral bipolar effect fully in thin SOI MOSFETs and to establish design criteria for submicrometer n-channel transistors, a two-dimensional finite difference simulation has been employed that incorporates the following important features: (1) two carrier modelling of current flow, (2) parasitic bipolar effects including snapback, (3) breakdown voltage, (4) punchthrough due to back-channel conduction and (5) negative differential output resistance caused by thermal effects.

For accurate SOI device modelling with ultra-thin films quantum simulations are often indispensable. Quantum confinement of carriers in semiconductor devices can be incorporated in ATLAS through a self-consistent Schrodinger–Poisson solver that allows calculation of bound state energies and associated carrier wave function. A density gradient model allows simulation of confinement effects on carrier transport. The spatial distribution of the electron concentration in the semiconductor thin film is calculated classically and quantum mechanically (Figure 5.11) for a double-gate transistor. The fundamental difference between the classical and quantum calculation is that the classical electron concentration is maximum at the semiconductor interfaces, while the quantum electron concentration is maximum away from the semiconductor interfaces. If the film is thin enough the maximum electron concentration is located in the middle of the film giving volume inversion (see Figure 5.12).

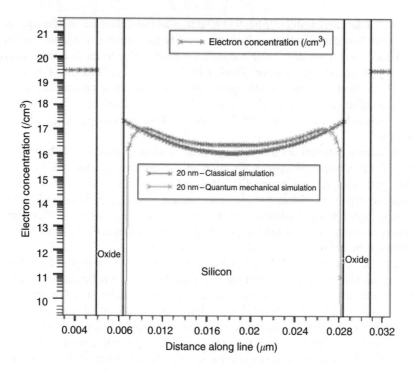

Figure 5.11   Special distribution of electron concentration using 'classical' or 'Schrodinger' calculation [Source: Silvaco International.]

5.5   Simulation using MASTAR

MASTAR [42] is a computing tool developed for the analytical calculation of the electrical characteristics of advanced CMOS devices such as planar bulk transistors, DG or SOI devices. The calculation is based on analytical drift diffusion equations, which depend directly on the major technological parameters, such as gate length, channel doping, oxide thickness, and so forth. This application that has been used to generate data for the ITRS roadmap allows rapid evaluation of key parameter such as threshold voltage, 'on' current, 'off' current and gate delay. Moreover, the influence of 'physical' secondary parameters such as mobility, poly depletion and quantum effect can be visualised. Owing to a possible limited validity of the drift diffusion models with respect to very short devices, the MASTAR predictions should be considered as 'worst case'.

Fenouillet-Beranger *et al.* [43] have used MASTAR to compare various devices architectures (bulk, SOI and SON) and process modules (metal gate and strained-Si channel) in a consistent way. This program makes some empirical projections as to likely enhancement in performance of CMOS bulk devices, through use of SOI or strained-Si material. It is shown that the use of MASTAR helps projection of a device

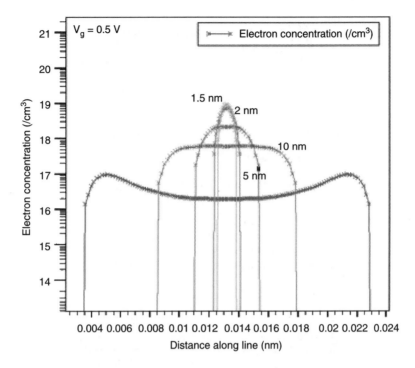

Figure 5.12   Influence of silicon film thickness on electron concentration distribution [Source: Silvaco International.]

architecture roadmap and shows precisely which architectures, modules and materials will be needed at a given CMOS node. Specifically MASTAR is capable of predicting the degradation in DIBL, subthreshold slope and I_{off} vs. I_{on}. Figure 5.13 illustrates the impact of the SOI BOX thickness on electrical parameters such as DIBL [see Figure 5.13(a)] and subthreshold slope [see Figure 5.13(b)] for various T_{Si} (5, 7.5, 10 nm) and a gate oxide of 1 nm, for a gate length of 40 nm and low channel doping. This study reveals alternative ways in which acceptable levels of short-channel effect and DIBL (<100 mV/V) can be achieved. The first is to reduce silicon film thickness to values down to 7.5 and 5 nm. The second is the use of a thin BOX with a thickness lower than 40 nm. For higher thicknesses no SCE effect improvement is observed. Another important aspect that is observed in Figure 5.13(a) and (b) is that the thicker the silicon film is, the more efficient the thin BOX. In summary, ultra-thin Si-film and BOX permit both subthreshold slope and DIBL below 100 mV/V in the shortest devices (see Figure 5.13), especially for metal gate (MG) with non-doped channel, where SCE cannot be reduced by increase in film doping. Thin BOX is thus considered a beneficial feature, in spite of the fact that, in long-channel devices, it leads to a slight degradation in the subthreshold slope, owing to higher coupling between the device body and the bulk. The latter is, however, considered a low price to pay for gaining a much better electrostatic integrity of very short devices.

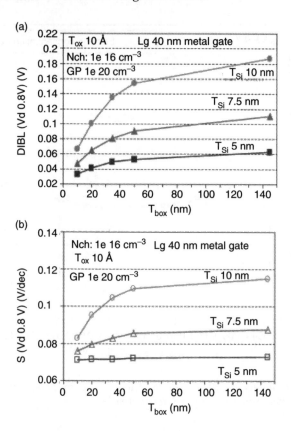

Figure 5.13   *Simulated FD-SOI device with $T_{Si} = 5$, 7.5, and 10 nm, L_g-40 nm,*
$T_{ox_{eq}} = 10$ Å, metal gate, non-doped channel, with ground plane (GP).
Curves (a) DIBL and (b) vs. BOX thickness. The ultra-thin Si-film
(between 7.5 and 5 nm) and relaxed BOX permit to maintain the sub-
threshold slope below 100 mV/dec and DIBL ~100 mV. The relaxed-Si
film thickness (10 nm) requires a thin BOX to achieve the same target.
Thin BOX thickness is efficient for BOX thickness under 40 nm [After
T. Skotnicki et al. Simulation results using MASTAR.]

The use of a metal gate is a very interesting way to reduce DIBL. However, the metal gate workfunction has to be correctly chosen to maintain a low threshold voltage on long- and short-channel MOSFET. Figure 5.14 presents the threshold voltage behaviour of a midgap gate (TiN) with undoped channel at $V_{dd} = 0.1$ V and $V_{dd} = 1$ V for simulated devices with T_{Si} 5 nm, thin and thick BOX and a low channel doping. In the case of a thin BOX the threshold voltage of the long-channel gate length ($L_g = 0.1 \mu$m) is around 0.55 and 0.45 V for the thick BOX. At high V_{dd} the threshold voltage of the device with the thick BOX decreases owing to larger short-channel effects down to 0.1 V, which is the target value. For 10 nm BOX, owing

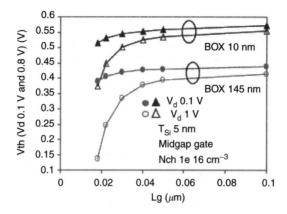

Figure 5.14   *Threshold voltage evolution of simulated devices with midgap gate vs.*
L_g at V_d = 0.1 and 0.8 V for two BOX thicknesses 10 and 145 nm, T_{Si} =
5 nm and the channel is undoped [After T. Skotnicki et al. Simulation
results using MASTAR.]

to reduced short-channel effects the threshold voltage is maintained to a higher value
around 0.4 V. These results indicate that with a midgap gate it is very difficult to keep
a low threshold voltage value. Two solutions can be envisaged for decreasing this
value of 0.4 V; one is to reduce the workfunction of the gate by around 200 mV for
NMOS transistors and the other is to counter dope the channel.

A comparison of metal gate and polySi gate is shown in Figures 5.15, 5.16
and 5.17. The simulated transistor has the following features: L_g = 22 nm, gate oxide
thickness: 13 Å, BOX T_{BOX} 10 nm, and a silicon thickness of 5 nm. In Figure 5.15, the
counter doping has to be relatively high (5×10^{18} cm^{-3}) to bring down threshold voltage to the same value obtained with polySi gate and doped channel. In Figures 5.16
and 5.17, an increase of DIBL and subthreshold slope is observed for devices with
counter doped channel. Thus, a transistor with a midgap gate and a counter doped
channel compared with polySi gate degrades SCE. The gain in I_{on} performance
induced by the suppression of polySi depletion for metal gate is compensated by the
loss of mobility, owing to the high counter doping necessary to reduce the threshold
voltage.

5.6   Strained-Si MOSFETs on SOI

Compared to bulk-Si, enhancement of electron mobilities and hole mobilities on
SOI are found to be >50 per cent and 15–20 per cent, respectively [44]. However
although theoretical studies have predicted higher carrier mobility in ultra-thin
single-gate SOI samples compared to standard Si devices, this mobility increase
(estimated in the best case at 10 per cent) is obtained only for very low silicon
layer thicknesses (<5 nm) and for a high inversion charge concentration. In other

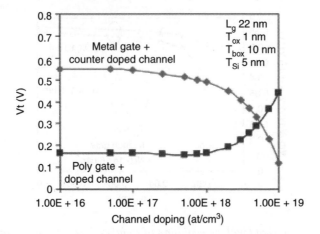

Figure 5.15   Threshold voltage evolution of simulated devices with midgap gate vs. channel counter doping, compared with polySi gate vs. channel doping, at $V_d = 0.1\,V$ [After T. Skotnicki et al. Simulation results using MASTAR.]

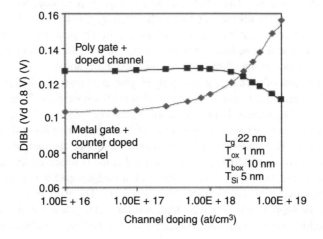

Figure 5.16   DIBL evolution of simulated devices with midgap gate vs. channel counter doping, compared with polySi gate vs. channel doping, at $V_d = 0.1\,V$ [After T. Skotnicki et al. Simulation results using MASTAR.]

cases, that is, for thicker silicon thicknesses, a mobility degradation is experimentally observed [45]. One may wonder whether it would be possible to combine the two structures (strained-Si inversion layer and SOI inversion layers) to enjoy the advantages of each and at the same time overcome the deficiencies they present separately.

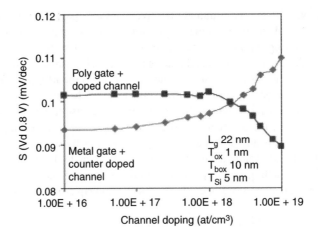

Figure 5.17   *Subthreshold slope evolution of simulated devices with midgap gate vs. channel counter doping, compared with polySi gate vs. channel doping, at $V_d = 0.1\,V$. BOX thickness is $10\,nm$, $T_{ox} = 1\,nm$, $T_{Si} = 5\,nm$ for $L_g = 22\,nm$ [After T. Skotnicki et al. Simulation results using MASTAR.]*

5.6.1   Simulation of SSGOI MOSFETs

SOI MOSFETs with strained-Si channel on SiGe-on-insulator (SSGOI) substrates have been fabricated by combining the SIMOX technology with regrowth of strained-Si films [46]. It has been demonstrated that strained-SOI n- and p-channel MOSFETs have mobilities typically 1.6 and 1.3 times higher than conventional Si MOSFETs, respectively. SSGOI structures provide a good control of short channel effects, have a lower parasitic capacitance and higher radiation tolerance and, moreover, present mobility values that are much higher than those found in conventional SOI MOSFETs. However, an accurate knowledge of high- and low-field transport in strained-Si for evaluating its true potential for submicron CMOS devices under the impact of non-equilibrium transport is still lacking.

In this section, design and simulation of p-channel strained-Si MOSFETs on SOI using ATLAS is discussed. The design considerations are illustrated by comparing with the experimental data obtained from n-polySi gate strained-Si channel SOI MOSFETs. The dependence of the carrier mobility in strained-Si/SiGe-OI inversion layers on the germanium mole fraction and the strained-Si thickness, T_{Si}, is presented.

Both theoretical and experimental studies have shown both the electron and hole mobility enhancements when silicon is grown pseudomorphically on relaxed $Si_{1-x}Ge_x$ [47]. In the conduction band, tensile strain splits the six-fold degeneracy and lowers the two-fold degenerate perpendicular Δ-valleys with respect to the four-fold in-plane Δ-valleys in energy space. The strain causes the six-fold degenerate valleys of the silicon conduction band minimum to split into two groups: two lowered valleys with the longitudinal effective mass axis perpendicular to the interface and four raised valleys with the longitudinal mass axis parallel to the interface. Such energy splitting suppresses inter-valley carrier scattering between the two-fold and four-fold degenerate valleys and causes preferential occupation of the two-fold valleys where the in-plane conduction mass is lower. Similarly, the effect of tensile strain on the valence band needs to be considered. The combination of a lower effective mass and reduced inter-valley scattering gives rise to higher electron/hole mobility. Moreover, the lower inter-valley scattering rates make energy-relaxation times higher, which results in a significant electron velocity overshoot [48]. Therefore, several material parameters for strained-Si and strained/relaxed SiGe are needed for the device modelling, namely, the band offsets encountered in strained layers, bandgap narrowing due to strain, doping-induced bandgap narrowing and the carrier mobility.

Silicon strained in biaxial tension results in a type II band offset and this allows tailoring of the band structure to confine both holes and electrons. To obtain accurate results for p-MOSFET simulations, it is necessary to account for effects associated with inversion layers. This is done by modelling mobility as having a dependence on both transverse and longitudinal fields. It has been shown that the value of $v_{sat} = 1.1 \times 10^7$ cm/s ($T = 300$ K) is the same regardless of the germanium mole fraction of the relaxed-SiGe alloy underneath the strained-Si layer. To take into account the enhanced mobility in strained-Si with increasing strain (i.e., with Ge mole fraction, x, in the substrate) and the effects associated with inversion layers, a modified CVT mobility model has been used in the simulation [49].

If the trap density at the interfaces with the SiO_2 is kept low, Coulomb scattering is very weak, and its effects could be ignored. Therefore, only phonon scattering and surface-roughness scattering are taken into account in the simulation. However, alloy scattering was ignored. Electrons/holes are also scattered by the random nature of the SiGe alloy. This is a fundamental limitation that cannot be removed (unless the alloy can be grown in an ordered form), but it is expected to have a weak effect in Si/SiGe system because only the tail of the electron distribution penetrates the SiGe layer [50].

The schematic structure of simulated p-MOSFETs is shown in Figure 5.18. This device structure is chosen to compare quantitatively the simulation with the experimental results, as there is reliable experimental data available in Reference 51. This device has an n^+-polySi gate, channel length (L_g) of 0.85 μm, channel width (W) of 5 μm, 8.5 nm oxide thickness and a 20-nm thick strained-Si layer, 290-nm-thick germanium in $Si_{1-x}Ge_x$ ($x = 0.1$) layer and 85-nm BOX thickness. A uniform n-type doping level of 5×10^{14} and 5×10^{15} cm^{-3} is assumed in strained-Si and control-SOI MOSFETs. The source and drain regions are doped to an impurity concentration of 10^{20} cm^{-3} and have a junction depth of 0.3 μm. The fixed oxide charge density is assumed to be 1×10^{10} cm^{-2} and the interface between the strained-Si and SiGe is assumed to be defect free.

The simulated I_d–V_{ds} characteristics and, for comparison, reported experimental results (data reproduced from Figure 10 in Reference 51) have been shown in Figure 5.19(a) and (b), respectively. Qualitative comparison from

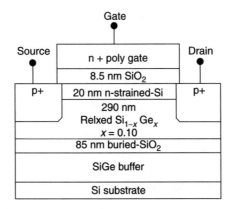

Figure 5.18   Schematic diagram of an SSGOI p-MOSFET [After S. K. Samanta, 'Ultrathin oxide and oxynitride gate dielectric films for silicon hetero-FETs', PhD Thesis, Indian Institute of Technology, Kharagpur, 2003.]

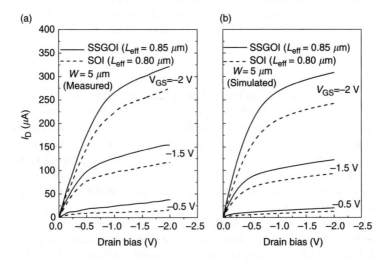

Figure 5.19   Drain characteristics comparing sSOI and SSGOI: (a) measured and (b) simulated [After S. K. Samanta, 'Ultrathin oxide and oxynitride gate dielectric films for silicon hetero-FETs', PhD Thesis, Indian Institute of Technology, Kharagpur, 2003.]

Figure 5.19(a) and (b) shows a close similarity between the experimental and simulation results at room temperature. The drain current at saturation region is enhanced by about 18 per cent, compared to control-SOI p-MOSFETs in both cases (measured and simulated).

The simulated subthreshold characteristics of strained-SOI devices with 10 per cent Ge concentration at the same channel is shown in Figure 5.20(a). The simulated subthreshold slope (S) of SSGOI MOSFETs is found to be 115 mV/decade.

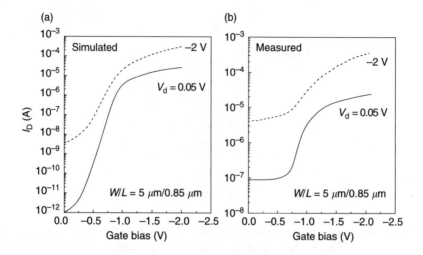

Figure 5.20   (a) Simulated and (b) measured subthreshold characteristics [After S. K. Samanta, 'Ultrathin oxide and oxynitride gate dielectric films for silicon hetero-FETs', PhD Thesis, Indian Institute of Technology, Kharagpur, 2003.]

However, large amount (10^{-7}) of drain leakage currents at low gate bias is observed from reported experimental results [51], which is shown in Figure 5.20(b). This high drain leakage current is due to the misfit dislocation in the SiGe layer and the back-side leakage current of BOX layer is due to its poor quality. Leakage current can be reduced by optimising the SIMOX process. Defect-free SiGe layer has, however, been considered in the simulation. The punchthrough effect is also observed at high drain bias for both measured and simulated data, because of thicker strained-Si/SiGe layers upon the BOX layer.

The simulated output characteristics of SSGOI MOSFETs at room temperature are plotted for different germanium mole fractions and gate-source voltages in Figure 5.21. As the drain-source voltage is increased the transverse field at the drain edge decreases. This reduction and increasing carrier confinement with increasing valence band offset produces a separation of the output curves with increasing mole fraction of Ge at high drain-source voltages.

Figure 5.22(a) shows the significant enhancement of drain current of SSGOI MOSFETs compared to SOI MOSFETs for all the drain-source voltages at room temperature. A plot of the ratio of the drain current values for ($x = 0.3$) and ($x = 0$) [Figure 5.22(b)] shows the largest enhancement in drain current arising from the increase in mobility. The channel conductance is plotted in Figure 5.23 vs. drain-source voltage for different x. Greatest enhancement in current in comparison with standard SOI MOSFETs is obtained at low V_{ds}, in the linear operation region, and it is associated with increased mobility. The saturation region conductance is similar in all cases since the saturation velocity is not dependent on Ge factor x.

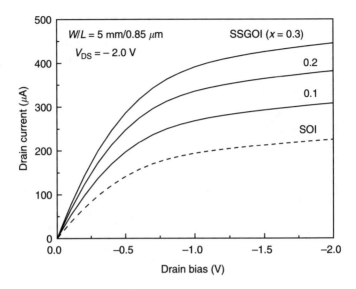

Figure 5.21   *Output characteristics with different Ge mole fraction [After S. K. Samanta, 'Ultrathin oxide and oxynitride gate dielectric films for silicon hetero-FETs', PhD Thesis, Indian Institute of Technology, Kharagpur, 2003.]*

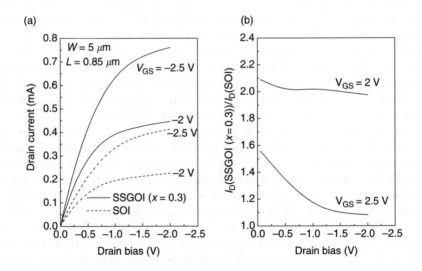

Figure 5.22   *Ratio between the drain current corresponding to (x = 0) and (x = 0.3) vs. V_{gs} [After S. K. Samanta, 'Ultrathin oxide and oxynitride gate dielectric films for silicon hetero-FETs', PhD Thesis, Indian Institute of Technology, Kharagpur, 2003.]*

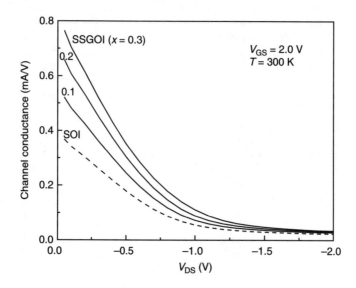

Figure 5.23   Channel conductance vs. drain-source voltage [After S. K. Samanta, 'Ultrathin oxide and oxynitride gate dielectric films for silicon hetero-FETs', PhD Thesis, Indian Institute of Technology, Kharagpur, 2003]

The effect of Ge content x on the linear transconductance of SSGOI p-MOSFETs is shown in Figure 5.24 where the transconductance has been plotted as a function of gate voltage for different Ge-content (in the relaxed-$Si_{1-x}Ge_x$: $x = 0.1, 0.2$ and 0.3) devices. The germanium mole fraction leads to a greater confinement of the holes in strained-Si/SiGe layers and increases the phonon scattering rate. However, the increase in the germanium mole fraction produces an increase in the strain in the Si layer and, as a consequence, a decrease in the effective mass and a reduction in the inter-valley scattering rate and the enhancement mobility behaviour.

As seen from Figure 5.24, when compared to a device with $x = 0.1$, the devices with $x = 0.2$ and $x = 0.3$ exhibit a hole mobility (and hence the transconductance) enhancement factor of 1.2 and 1.4, respectively. Therefore, the decrease in effective mass of holes and inter-valley scattering rate are the important factors to enhance mobility. At low V_{gs}, the output current is small and remains nearly constant up to about -0.5 V. The gate voltage at which peak transconductance occurs depends on the value of V_{ds}. Two peaks are seen at about -1.0 and -1.5 V. The peak at -1.0 V corresponds to hole confinement at the strained-Si/SiGe-buffer interface. However, at high gate voltage, the holes at the SiO_2/strained-Si interface dominate the channel conduction and the device becomes a surface channel device.

The transconductance and mobility of these devices is shown in Figure 5.25(a) and (b), respectively, for different values of strained-Si thickness (T_{Si}). It is observed that for $T_{Si} = 10$ nm, only one prominent peak of g_m is present in Figure 5.25(a) at gate voltage of -1.0 V. The relative position and size of the peak decreases with

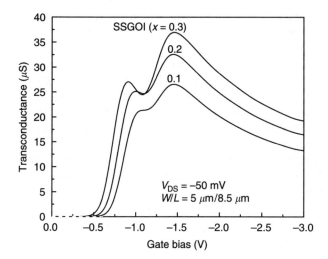

Figure 5.24   *Simulated linear transconductance characteristics: V_{ds} for an n^+-gate SSGOI p-MOSFET with Ge content (x = 0.1, 0.2 and 0.3) [After S. K. Samanta, 'Ultrathin oxide and oxynitride gate dielectric films for silicon hetero-FETs', PhD Thesis, Indian Institute of Technology, Kharagpur, 2003.]*

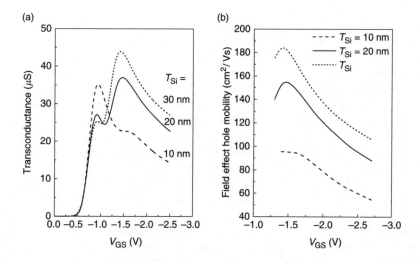

Figure 5.25   *(a) Transconductance and (b) effective mobility vs. gate voltage for different values of T_{Si} [After S. K. Samanta, 'Ultrathin oxide and oxynitride gate dielectric films for silicon hetero-FETs', PhD Thesis, Indian Institute of Technology, Kharagpur, 2003.]*

increasing thickness of strained-Si layer. The holes at the strained-Si/SiGe-buffer interface dominate the channel conduction for $T_{Si} < 2$ nm while the device becomes a surface channel device for $T_{Si} > 2$ nm. If the germanium mole fraction is maintained at $x = 0.3$, below the critical thickness, the strain remains unchanged, and fundamentally the same splitting occurs between the heavy hole (HH) and light hole (LH) bands of the silicon valence band minimum. Therefore, the hole effective mass is essentially independent of T_{Si}. However, the greater confinement of holes in the strained-Si/SiGe inversion layers, with a thinner silicon layer, leads to a greater phonon scattering rate. As the thickness of the strained-Si layer is reduced, phonon scattering rate increases, while the effective mass remains essentially the same, and therefore a degradation in hole mobility is expected, which is shown in Figure 5.25(b).

5.6.2   MC simulation of sSOI MOSFETs

In this section, the capability of the Monte Carlo device simulator SPARTA [52–55] to simulate devices under biaxial tensile strain will be discussed. The example considered here is an advanced 25-nm strained-Si directly on SOI (sSOI) n-MOSFET, which corresponds approximately to a scaled version of the realisation of the sSOI MOSFETs [56]. Quasi-ballistic transport has been observed from extracted channel velocity, which is far above the saturation velocity, and also from strong anisotropy between the crystallographic ⟨100⟩ and ⟨110⟩ directions.

Quasi-ballistic transport occurs when electrons near equilibrium are suddenly exposed to a high field and is strongly enhanced by strain, as can be seen from the overshoot peaks in Figure 5.26. With their initially low energy, they do not suffer much scattering over a short distance (in time or space), so that the gain in velocity during the free-flight leads to overshoot effects that are very sensitive to the band curvature. The anisotropy of the full band structure in the quasi-ballistic transport regime is analogously present in nanoscale devices, which is the reason for high strain-induced performance enhancement as demonstrated by measurements [57] and Monte Carlo simulation [53]. This increased contribution of quasi-ballistic transport, which is not captured by classical device simulation (discussed above), makes Monte Carlo simulation necessary for simulation of nanoscale devices.

In the DESSIS and SPARTA [55] command files, the level of the pseudomorphic strain in the strained regions is specified by the germanium mole fraction in the virtual SiGe substrate. In addition, the Ge content needs to be specified in the Monte Carlo section of the SPARTA command file, so that the corresponding band structure table is used. In the SPARTA simulation, it is also possible to switch between the crystallographic ⟨110⟩ and ⟨100⟩ channel orientations. The resulting comparison of the characteristics of the sSOI n-MOSFET output is shown in Figure 5.27.

From strain dependence of the Monte Carlo simulations alone, Figure 5.28 shows the velocity profiles 0.13 nm below the gate oxide, corresponding to the on-state of the output characteristics in Figure 5.27. It can be seen that along almost the whole 25-nm channel the electrons travel with a velocity much higher than the saturation value of 10^7 cm/s.

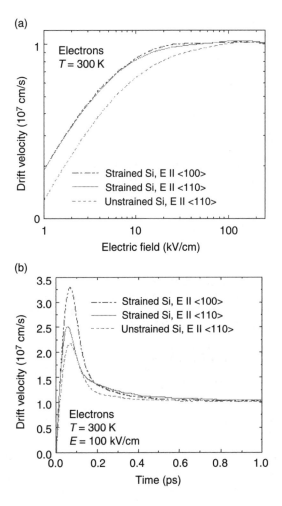

Figure 5.26 *(a) In-plane velocity vs. field characteristics of electrons in unstrained silicon along ⟨100⟩ and ⟨110⟩ directions in strained-Si; (b) transient in-plane velocity overshoot of electrons after a sudden application of a field of 100 kV/cm [Source: Application Notes, Integrated Systems Engineering AG, Switzerland.]*

The anisotropy of the on-current (Figure 5.27) stemming from the anisotropic velocities (Figure 5.28) demonstrates that the on-state is governed by quasi-ballistic transport, which is analogous to the anisotropic velocity overshoots in bulk strained-Si in Figure 5.26. Figure 5.29 shows I_d–V_d characteristics as simulated with SPARTA, the drift-diffusion (DD) model and the hydrodynamic (HD) model. The DD simulations have been performed without considering avalanche generation. It is seen that while DD underestimates the drain current, HD overestimates it. Consequently, the results of SPARTA could be used as a reference for fine-tuning the energy relaxation time. Nevertheless, the relaxation time so calibrated is valid only for the given

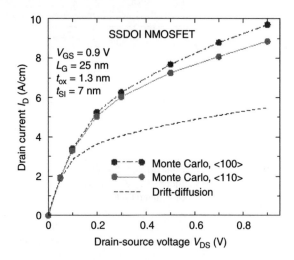

Figure 5.27   Monte Carlo output characteristics of the 25-nm n-type sSOI MOSFET for strained-Si with ⟨100⟩ and ⟨110⟩ channel orientations in comparison to the drift-diffusion result [Source: Application Notes, Integrated Systems Engineering AG, Switzerland.]

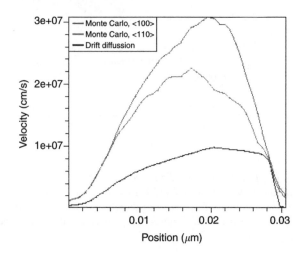

Figure 5.28   Velocity profiles along the channel, 0.13 nm below the gate oxide, in unstrained-Si and strained-Si n-type sSOI MOSFETs [Source: Application Notes, Integrated Systems Engineering AG, Switzerland.]

gate length and channel orientation, that is, ⟨110⟩. The results of such a calibration are shown in Figure 5.30 where the energy relaxation time in strained-Si was reduced from 0.35 to 0.11 ps. A detailed discussion on comparisons of DD, HD and Monte Carlo is available in Reference 58.

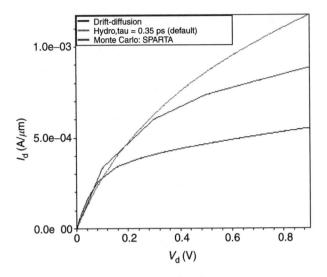

Figure 5.29 I_d–V_d *characteristics as obtained from drift-diffusion, hydrodynamic and Monte Carlo for the channel oriented along* ⟨110⟩; *the energy relaxation time for hydrodynamic uses the default value [Source: Application Notes, Integrated Systems Engineering AG, Switzerland.]*

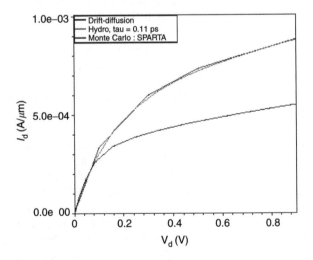

Figure 5.30 I_d–V_d *characteristics as obtained from drift-diffusion, hydrodynamic and Monte Carlo simulations; the hydrodynamic simulation uses a calibrated energy relaxation time, which is valid only for a given gate length (0.25 μm) and channel orientation (⟨110⟩) [Source: Application Notes, Integrated Systems Engineering AG, Switzerland.]*

5.7   Modelling of multi-gate SOI devices

In implementation of multiple gate SOI devices, device designers have sought alternate design concepts to achieve the desired current drive capability and electrostatic control. Recent investigations have resulted in several novel device structures including the FinFET [59–61], the tri-gate, the quantum wire MOSFET, the Ω-Gate or the Π-Gate MOSFET and the Quadruple-Gate or the GAA MOSFET [62]. These multi-gate devices are 3D variants of the conventional, planar MOSFET offering improved current drive, transconductance and subthreshold characteristics. In addition, they have the added feature that their fabrication technology is compatible with the already well-understood, low-cost and reliable silicon IC technology.

The primary problem in the mass fabrication of these novel devices is the difficulty in accurate patterning of thin, narrow device dimensions for the channel body or fin. As device dimensions continue to shrink, fabrication processes tend to become more complex. As a result, modelling of multi-gate devices using state-of-the-art commercial device simulators now requires 3D as opposed to 2D simulation.

TCAD has an important role in the introduction of FinFETs, and must address new complexities in the range of device parameters that must be defined. There is also a need for simulation studies that compare the behaviour of multi-gate devices [63].

On the basis of 3D device simulations, Kranti and Armstrong [64] have reported the possible device performance enhancements for triple-gate (TG) and double-gate (DG) FinFETs for high performance (HP), low operating power (LOP) and low standby power (LSTP) logic technologies according to ITRS 65-nm node specifications. The authors have studied in detail the impact of spacer width (s), lateral source/drain doping gradient (d), aspect ratio (AR), fin thickness (T_{fin}) and height (H_{fin}) along with gate workfunction (ϕ_{m}) on the device performance. The authors have presented a design guideline on SCEs, intrinsic delay ($\tau = C_{\text{gg}}V_{\text{dd}}/I_{\text{on}}$), off-current ($I_{\text{off}}$) and $I_{\text{on}}/I_{\text{off}}$ to meet ITRS requirements for TG and DG FinFETs.

3D device simulations using ATLAS [49] have been carried out to explore the design space for triple-gate (Figure 5.31a and b) and double-gate FinFETs (Figure 5.31c). A BOX (t_{BOX}) of 100 nm, doping (N_{a}) of the p-type SOI layer of 10^{15} cm^{-3} and gate workfunction of 4.52–4.72 eV were chosen for the devices. H_{fin} was varied from $L_{\text{g}}/2$ to $2L_{\text{g}}$, whereas T_{fin} was varied from $L_{\text{g}}/2$ to L_{g}, where L_{g} is the physical gate length of the devices. The spacer width (s) was varied from $(0.25)L_{\text{g}}$ to $(1.0)L_{\text{g}}$ for all the devices and S/D doping profile defined by its gradient at the gate edges [65–67] was varied from 2 to 8 nm/decade, as shown in Figure 5.31(d). The equivalent oxide thickness (EOT) in inversion, including finite inversion layer capacitance, gate depletion and physical oxide thickness to account for quantum effects was defined by ITRS guidelines. The sidewall spacer of SiO$_2$ and gate electrode thickness of 25 nm was assumed in all the simulations. The double-gate FinFET was simulated by electrically isolating the top gate electrode using a thicker top gate oxide thickness of 50 nm, as shown in Figure 5.31(c).

Simulations were performed with DD models using a modified expression of saturation velocity (v_{sat}) and the empirical parameter beta ($\beta = 1$) proposed by Granzner *et al.* [68, 69] to account for electron transport for nanoscale gate length

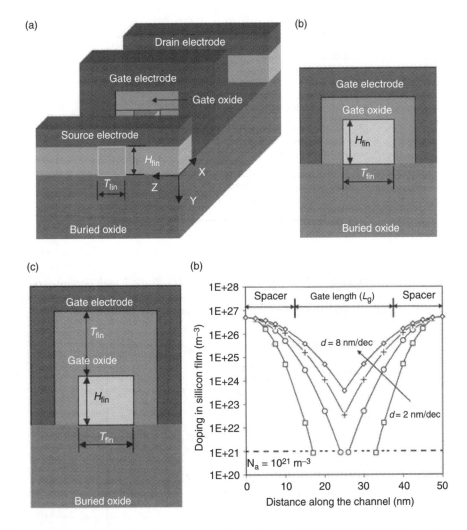

Figure 5.31 (a) Schematic diagram of a triple-gate (TG) FinFET. (b) 2D cut-plane (y-z plane) at x = L_g/2 of a triple-gate FinFET. (c) 2D cut-plane (y-z plane) at x = L_g/2 of a double-gate FinFET. (d) Variation of source/drain doping gradient along the channel at a fixed spacer of (0.5)L_g [After A. Kranti and G. A. Armstrong, Semicond. Sci. Technol., Vol. 21, 2006 (409–421).]

devices. The gate-length-dependent saturation velocity (in units of 10^7 cm/s) [69] was defined by the expression $v_{sat}(L_g) = 2.0 + (19.2/L_g^{1.43})$. Simple modification in the saturation velocity of carriers provides a reasonable estimation of on current when compared with Monte Carlo simulation for DG MOSFETs over a wide range of gate lengths (10–100 nm) [69].

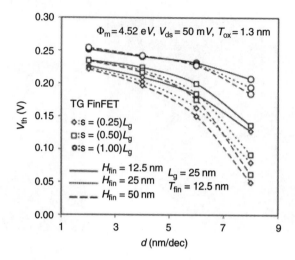

Figure 5.32   Variation of threshold voltage of the TG FinFET with doping gradient for various values of spacer and fin height [After A. Kranti and G. A. Armstrong, Semicond. Sci. Technol., Vol. 21, 2006(409–421).]

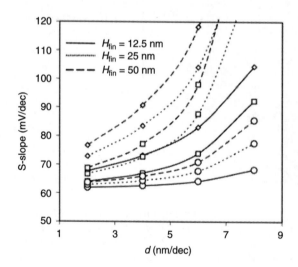

Figure 5.33   Variation of S-slope of the TG FinFET as a function of doping gradient for various values of spacer and fin height [After A. Kranti and G. A. Armstrong, Semicond. Sci. Technol., Vol. 21, 2006 (409–421).]

Figures 5.32–5.34 show the dependence of threshold voltage (V_{th}), DIBL and subthreshold slope (S) on H_{fin}, s and d. V_{th} is defined as the gate bias when the drain current reaches $500\,nA \times (W_g/L_g)$ and DIBL as V_{th} (at $V_{ds} = 1.1\,V$)-V_{th} (at $V_{ds} = 0.05\,V$). With an increase in H_{fin}, the top gate loses control over the

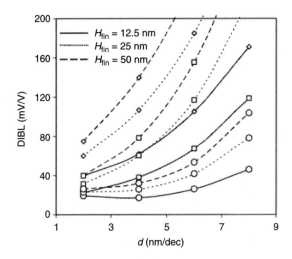

Figure 5.34   *Variation of DIBL of the TG FinFET with doping gradient for various values of spacer and fin height [After A. Kranti and G. A. Armstrong, Semicond. Sci. Technol., Vol. 21, 2006(409–421).]*

channel and the two side gates mainly govern current conduction. This results in a degradation of SCEs – lowering of V_{th} and increase in DIBL and S-slope. Therefore, from the viewpoint of controlling SCEs, TG devices should be designed with lower aspect ratios. Also, as shown in the figure, lateral source/drain doping gradient along with spacer width offers another degree of freedom apart from the important device parameters such as H_{fin}, T_{fin}, AR and T_{ox} to minimise SCEs. Larger s values along with lower AR result in V_{th}, S-slope and DIBL values that are nearly independent of d. It is, however, better to design TG devices with smaller values of d along with larger s values as this combination results in a shorter effective channel length $((L_{eff})_{WI})$ in the weak inversion region, as shown in Figure 5.35. $(L_{eff})_{WI}$ was extracted by comparing I_{off} (I_{ds} at $V_{gs} = 0$ V, $V_{ds} = 1.1$ V) for a device having finite d and s with an ideal device with $s = d \rightarrow 0$ [67].

Clearly, adjusting s and d is a more viable option for controlling SCEs and thus achieving lower I_{off} in the ultra-short gate length regime than to vary the gate length and/or fin height. It should be noted that many possible combinations of s and d can be obtained to achieve nearly the same $(L_{eff})_{WI}$ that is, $s = (0.25)L_g$ with $d = 3$ nm/decade, $s = (0.50)L_g$ with $d = 4.5$, $s = (0.75)L_g$ with $d = 6$ and $s = (1.0)L_g$ with $d = 8$ all achieve $(L_{eff})_{WI}$ of $\sim$30 nm. Therefore, optimisation of s and d for short-channel immunity must be considered together with the on–off current values as a high doping gradient that is, higher values of d as the gate edge not only reduces $(L_{eff})_{WI}$, (that is, increases SCEs), but also draws more carriers into the spacer region near the gate edges, thus increasing I_{on} (I_{ds} at $V_{gs} = V_{ds} = 1.1$ V). Thus, lateral S/D doping gradient along with the spacer width defines a design trade-off between acceptable SCEs and parasitic series resistance and thus requires careful optimisation of the overall performance (I_{on}, I_{off} and I_{on}/I_{off}) of the device.

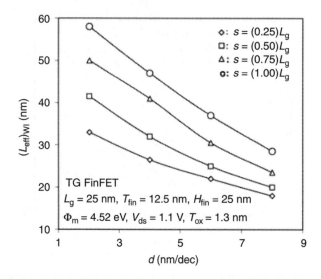

Figure 5.35   *Dependence of extracted effective channel length ($L_{eff\,WI}$) in the sub-threshold region with spacer width and doping gradient with AR = 2 [After A. Kranti and G. A. Armstrong, Semicond. Sci. Technol., Vol. 21, 2006(409–421).]*

Modelling of tri-gate SOI device has been carried out by INTEL using DESSIS [70]. The goal of this study was to understand the device performance under aggressive scaling conditions. The baseline structure for this study has a gate length, body height and width of 30 nm. The simulations show that the device shows a near-ideal, fully depleted performance with good DIBL and steep subthreshold gradient. The simulations also help understand the role of the corner transistors in influencing device operation. Their results suggest that the tri-gate device could eventually replace conventional planar MOS transistors in the near future.

G. Pei *et al.* [71] have reported the subthreshold characteristics for multiple dimension FinFET transistors using a 3D numerical device simulator, FIELDAY. In this report, design considerations for FinFET devices have been explored using 3D numerical simulations and analytical modelling. Analytical expressions have been derived using solutions of the 3D Laplace equation for the subthreshold behaviour in these nanoscale devices. Simulation results reported here compare well with experimental measurements on fabricated device structures.

5.8   Summary

The use of TCAD to investigate design and applications of devices based on SOI technology has been reviewed. SOI technology allows integration of high-performance innovative devices that can push forward the present frontiers of the CMOS downscaling. As one moves from 0.1 μm generation and below, SOI offers a number of advantages in low power, communication circuits and system-on-chip and may ultimately replace bulk CMOS technology. SOI technology improves performance over bulk CMOS technology by 25–35 per cent, equivalent to two years of bulk CMOS advances and offers the low power advantages. Indeed some perceived SOI disadvantages (self-heating, hot carriers and early breakdown) are no longer such a significant problem for operation at low voltage. The importance of using strained-Si alongside SOI technology to yield significant improvements in mobility has been highlighted.

As devices such as the FinFET are essential to a 3D structure, the importance of adapting the TCAD tools to investigate various design issues has been described. On the basis of 3D device simulations, predictions for device performance enhancements for TG and DG FinFETs for high performance (HP), low operating power (LOP) and low standby power (LSTP) logic technologies according to ITRS 65-nm node specifications have been made.

References

1 T. Yaur, and T. H. Ning, *Fundamentals of Modern VLSI Devices*. Cambridge University Press, Cambridge, 1998.

2 J.-P. Colinge, *SOI Technology: Materials to VLSI, 2nd Ed.* Kluwer, Boston, 1997.

3 G. G. Shahidi, A. Ajmera, F. Assaderaghi, R. J. Bolam, E. Leobandung, W. Rausch, *et al.*, 'Partially-depleted SOI technology for digital logic', in *IEEE ISSCC Tech. Dig.*, pp. 426–427, 1999.

4 K. Ueda, K. Nii, Y. Wada, I. Takimoto, S. Maeda, T. Iwamatsu, *et al.*, 'A CAD-compatible SOI/CMOS gate array having body-fixed partially-depleted transistors', in *IEEE ISSCC Tech. Dig.*, pp. 288–289, 1997.

5 Y. Ohtomo, S. Yasuda, M. Nogawa, J. Inoue, K. Yamakoshi, H. Sawada, *et al.*, 'A 40Gb/s 8×8 ATM switch LSI using 0.25 μm CMOS/SIMOX', in *IEEE ISSCC Tech. Dig.*, pp. 154–155, 1997.

6 M. Harada, T. Tsukahara, and J. Yamada, '0.5–1V 2GHz RF front-end circuits in CMOS/SIMOX', in *IEEE ISSCC Tech. Dig.*, pp. 378–379, 2000.

7 M. Kumar, Y. Tan, J. Sin, L. Shi, and J. Lau, 'A 900MHz SOI fully-integrated RF power amplifier for wireless transceivers', in *IEEE ISSCC Tech. Dig.*, pp. 382–383, 2000.

8 T. Eimori, T. Oashi, H. Kimura, Y. Yamaguchi, T. Iwamatsu, T. Tsuruda, *et al.*, 'ULSI DRAM/SIMOX with stacked capacitor cells for low-voltage operation', in *IEEE IEDM Tech. Dig.*, pp. 45–48, 1993.

9 Y. Koh, M. Oh, J. Lee, J. Yang, W. Lee, C. Park, *et al.*, '1 Giga bit SOI DRAM with fully bulk compatible process and body-contacted SOI MOSFET structure', in *IEEE IEDM Tech. Dig.*, pp. 579–582, 1997.

10 J.-W. Park, Y.-G. Kim, I.-K. Kim, K.-C. Park, H. Yoon, and K.-C. Lee, 'Performance characteristics of SOI DRAM for low-power application', in *IEEE ISSCC Tech. Dig.*, pp. 434–435, 1999.

11 T. Fuse, Y. Oowaki, T. Yamada, M. Kamoshida, M. Ohta, and T. Shino, 'A 0.5 V 200 MHz 1-Stage 32b ALU using a body bias controlled SOI pass-gate logic', in *IEEE ISSCC Tech. Dig.*, pp. 286–287, 1997.

12 A. Ebina, T. Kadowaki, Y. Sato, and M. Yamaguchi, 'Ultra low-power CMOS IC using partially-depleted SOI technology', in *IEEE CICC Tech. Dig.*, pp. 57–60, 2000.

13 Y. Hirano, T. Matsumoto, S. Maeda, T. Iwamatsu, T. Kunikiyo, K. Nii, *et al.*, 'Impact of 0.10 μm SOI CMOS with body-tied hybrid trench isolation structure to break through the scaling crisis of silicon technology', in *IEEE IEDM Tech. Dig.*, pp. 467–470, 2000.

14 G. K. Celler, and S. Cristoloveanu, 'Frontiers of silicon-on-insulator', *J. Appl. Phys.*, vol. 93, pp. 4955–4978, 2003.

15 S. Cristoloveanu, and G. Reichert, 'Recent advances in SOI materials and device technologies for high temperature', in *Proc. Devices and Sensors Conference, High-Temperature Electronic Materials*, pp. 86–93, 1998.

16 S. Cristoloveanu, and V. Ferlet-Cavrois, 'Introduction to SOI MOSFETs: context, radiation effects, and future trends', *Int'l. J. High Speed Electron. Syst.*, vol. 14, pp. 465–487, 2004.

17 S. Cristoloveanu, 'Silicon on insulator technologies and devices: from present to future', *Solid State Electron.*, vol. 45, pp. 1403–1411, 2001.

18 M. Bruel, B. Aspar, B. Charlet, C. Maleville, T. Poumeyrol, A. Souble, *et al.*, 'Smart Cut: A promising new SOI material technology', in *IEEE Intl SOI Conf. Proc.*, pp. 178–179, 1995.

19 M. Bruel, 'Silicon on insulator material technology', *Electron. Lett.*, vol. 31, pp. 1201–1202, 1995.

20 B. Ghyselen, J.-M. Hartmann, T. Ernst, C. Aulnette, B. Osternaud, Y. Bongumilowicz, *et al.*, 'Engineering strained silicon on insulator wafers with the SmartCut^{TM} technology', *Solid State Electron.*, vol. 48, pp. 1285–1296, 2004.

21 G. Celler, and M. Wolf, 'Strained silicon on insulator: A quick guide to the technology, the processes, and the products, SOITEC', July 2003.

22 J. P. Colinge, M. H. Gao, A. Romano-Rodriguez, H. Maes, and C. Claeys, 'Silicon-on-insulator "gate-all-around device"', in *IEEE IEDM Tech. Dig.*, pp. 595–598, 1990.

23 R. A. Johnson, P. R. de la Houssey, C. E. Chang, P.-.F. Chen, M. E. Wood, G. A. Garcia, *et al.*, 'Advanced thin-film silicon-on-sapphire technology: microwave circuit applications', *IEEE Trans. Electron Devices*, vol. 45, pp. 1047–1054, 1998.

24 D. Munteanu, D. Weiser, S. Cristoloveanu, O. Faynot, J.-.L. Pelloie, and J. G. Fossum, 'Generation-recombination transient effects in SOI transistors: experiments and simulations', *IEEE Trans. Electron Devices*, vol. 45, pp. 1678–1683, 1998.

25 T. Ernst, C. Tinella, C. Raynaud, and S. Cristoloveanu, 'Fringing fields in sub-0.1 μm fully depleted SOI MOSFETs: Optimization of the device architecture', *Solid State Electron.*, vol. 46, pp. 373–378, 2002.

26 K. Oshima, S. Cristoloveanu, B. Guillaumot, S. Deleonibus, and H. Iwai, 'SOI MOSFETs with buried alumina: Thermal and electrical aspects', *J. Electrochem. Soc.*, vol. 151, pp. G257–G261, 2004.

27 Y. Ohmura, S. Nakashima, K. Izumi, and T. Ishii, '0.1-μm-gate, ultra-thin film CMOS device using SIMOX substrate with 80-nm-thick buried oxide layer', in *IEEE IEDM Tech. Dig.*, pp. 675–678, 1991.

28 H.-K. Lim, and J. G. Fossum, 'Threshold voltage of thin-film silicon on insulator (SOI) MOSFETs', *IEEE Trans. Electron Devices*, vol. 30, pp. 1244–1251, 1983.

29 D. Franck, S. Laux, and M. Fischetti, 'Monte Carlo simulations of a 30 nm dual gate MOSFET: How short can Si go?', in *IEEE IEDM Tech. Dig.*, pp. 553–556, 1992.

30 E. Rauly, O. Potavin, F. Balestra, and C. Raynaud, 'On the subthreshold swing and short channel effects in single and double gate deep submicron SOI MOSFET's', *Solid State Electron.*, vol. 43, pp. 2033–2037, 1999.

31 R. H. Yan, A. Ourmazd, and K. F. Lee, 'Scaling the Si MOSFET – from bulk to SOI to bulk', *IEEE Trans. Electron Devices*, vol. 39, pp. 1704–1710, 1992.

32 D. Hisamoto, T. Kaga, and E. Takeda, 'Impact of the vertical SOI "DELTA" structure on planar device technology', *IEEE Trans. Electron Devices*, vol. 38, pp. 1419–1424, 1991.

33 H. P. Wong, K. K. Chan, and Y. Taur, 'Self-aligned (top and bottom) double-gate MOSFET with a 25nm thick silicon channel', in *IEEE IEDM Tech. Dig.*, pp. 427–430, 1997.

34 T. Ernst, D. Munteanu, S. Cristoloveanu, T. Ouisse, S. Horiguchi, Y. Ono, *et al.*, 'Investigation of SOI MOSFETs with ultimate thickness', *Microelectronics Engineering*, vol. 48, pp. 339–342, 1999.

35 B. Majkusiak, T. Janik, and J. Walczak, 'Semiconductor thickness effects in the double-gate SOI MOSFET', *IEEE Trans. Electron Devices*, vol. 45, pp. 1127–1134, 1998.

36 F. Gamiz, J. B. Roldan, P. Cartujo-Cassinello, J. A. Lopez-Villanueva, and S. Rodriguez, 'Electron mobility in extremely thin single-gate silicon-on-insulator inversion layers', *J. Appl. Phys.*, vol. 86, pp. 62–69, 1999.

37 C. Fiegna, A. Abramo, and E. Sangiorgi, 'Single- and double-gate SOI MOS structures for future ULSI: a simulation study', in *Future Trends in Microelectronics*, Wiley, New York, 1999.

38 G. A. Armstrong, J. R. Davis, and A. Doyle, 'Characterization of bipolar snapback and breakdown voltage in thin film SOI transistors by two-dimensional simulation', *IEEE Trans. Electron Devices*, vol. 38, pp. 328–336, 1991.

39 G. A. Armstrong, and W. D. French, 'Simulation of ultra thin film SOI transistors using a non-local model for impact ionisation', *Solid State Electron.*, vol. 35, pp. 1761–1770, 1992.

40 G. A. Armstrong, and W. D. French, 'Improved physical modelling of bipolar effects in SOI transistors', in *Proc. 5th Int. Symposium on silicon-on-insulator technology*, pp. 104–112, 1992.

41 G. A. Armstrong, and W. D. French, 'Suppression of parasitic bipolar effects in thin film SOI transistors', *IEEE Electron Device Lett.*, vol. EDL-13, pp. 198–200, 1992.

42 T. Skotnicki, 'MASTAR Manual, ver. 4', 2005.

43 C. Fenouillet-Beranger, T. Skotnicki, S. Monfray, N. Carriere, and F. Boeuf, 'Requirements for ultra-thin-film devices and new materials for the CMOS roadmap', *Solid State Electron.*, vol. 48, pp. 961–967, 2004.

44 D. Adams, M. Austin, R. Rai-Choudhury, and J. Hwang, 'A performance comparison of advanced SOI technologies', *SOS/SOI Technology Workshop*, p. 79, 1988.

45 J. H. Choi, Y. J. Park, and H. S. Min, 'Electron mobility behavior in extremely thin SOI MOSFET's', *IEEE Electron Device Lett.*, vol. 16, pp. 527–529, 1995.

46 S. Takagi, N. Sugiyama, T. Mizuno, T. Tezuka, and A. Kurobe, 'Device structure and electrical characteristics of strained-Si-on-insulator (strained-SOI) MOSFETs', *Mat. Sci. Eng. B*, vol. 89, pp. 426–434, 2002.

47 M. V. Fischetti, and S. E. Laux, 'Band structure, deformation potentials, and carrier mobility in strained Si, Ge, and SiGe alloys', *J. Appl. Phys.*, vol. 88, pp. 2234–2252, 1996.

48 J. B. Roldan, F. Gamiz, J. A. Lopez-Villanueva, and J. E. Carceller, 'A Monte Carlo study on the electron-transport properties of highperformance strained-Si on relaxed $Si_{1-x}Ge_x$ channel MOSFETs', *J. Appl. Phys.*, vol. 80, pp. 5121–5128, 1996.

49 Silvaco International, *Silvaco-ATLAS User's Manual*, 2004.

50 J. H. Davies, *The Physics of Low-Dimensional Semiconductors: An Introduction*, Cambridge University Press, New York, 1998.

51 T. Mizuno, N. Sugiyama, A. Kurobe, and S. Takagi, 'Advanced SOI p-MOSFETs with strained-Si channel on SiGe-on-insulator substrate fabricated by SIMOX technology', *IEEE Trans. Electron Devices*, vol. 48, pp. 1612–1618, 2001.

52 F. M. Bufler, Y. Asahi, H. Yoshimura, C. Zechner, A. Schenk, and W. Fichtner, 'Monte Carlo simulation and measurement of nanoscale n-MOSFETs', *IEEE Trans. Electron Devices*, vol. 50, pp. 418–424, 2003.

53 F. M. Bufler, and W. Fichtner, 'Scaling of strained-Si n-MOSFETs into the ballistic regime and associated anisotropic effects', *IEEE Trans. Electron Devices*, vol. 50, pp. 278–284, 2003.

54 F. M. Bufler, and W. Fichtner, 'Scaling and strain dependence of nanoscale strained-Si p-MOSFET performance', *IEEE Trans. Electron Devices*, vol. 50, pp. 2461–2466, 2003.

55 ISE Integrated Systems Engineering AG., Zurich, Switzerland, *ISE-TCAD Manual*, 2004.

56 K. Rim, K. Chan, L. Shi, D. Boyd, J. Ott, N. Klymko, *et al.*, 'Fabrication and mobility characteristics of ultrathin strained-Si directly on insulator (SSDOI) MOSFETs', in *IEEE IEDM Tech. Dig.*, pp. 49–52, 2003.

57 K. Rim, S. Koester, M. Hargrove, J. Chu, P. M. Mooney, J. Ott, *et al.*, 'Strained Si NMOSFETs for high performance CMOS technology', in *IEEE VLSI Tech. Symp. Dig.*, pp. 59–60, 2001.

58 F. M. Bufler, A. Schenk, and W. Fichtner, 'Monte Carlo, hydrodynamic and drift-diffusion simulation of scaled double-gate MOSFETs', *J. Comput. Electron.*, vol. 2, pp. 81–84, 2003.

59 X. Huang, W. Lee, C. Kuo, D. Hisamoto, L. Chang, J. Kedzierski, *et al.*, 'Sub 50-nm FinFET: PMOS', in *IEEE IEDM Tech. Dig.*, pp. 67–70, 1999.

60 E. Liu, C. Lin, and X. Liu, 'Simulation of 100 nm SOI MOSFET with FinFET structure', in *Proc. Conf. Solid-state and Integrated-Circuit Technology*, pp. 883–886, 2001.

61 J. G. Fossum, M. M. Chowdhury, V. P. Trivedi, T.-J. King, Y.-K. Choi, J. An, *et al.*, 'Physical insights on design and modeling of nanoscale FinFETs', in *IEEE IEDM Tech. Dig.*, pp. 679–682, 2003.

62 J.-P. Colinge, 'The evolution of the silicon-on-insulator MOS transistor', in *Proc. ISDRS*, pp. 354–355, 2003.

63 J.-T. Park, and J. P. Colinge, 'Multiple-gate SOI MOSFETs: Device design guidelines', *IEEE Trans. Electron Devices*, vol. 49, pp. 2222–2229, 2002.

64 A. Kranti, and G. A. Armstrong, 'Performance assessment of nanoscale double- and triple-gate FinFETs', *Semicond. Sci. Technol.*, vol. 21, pp. 409–421, 2006.

65 R. J. Luyken, T. Schultz, J. Hartwich, L. Dreeskornfeld, M. Stadele, and L. Risch, *et al.*, 'Design considerations for fully depleted SOI transistors in the 25–50 nm gate length regime', *Solid State Electron.*, vol. 47, pp. 1199–1203, 2003.

66 R. J. Luyken, M. Stadele, W. Rosner, T. Schultz, J. Hartwich, L. Dreeskornfeld, *et al.*, 'Perspectives of fully-depleted SOI transistors down to 20 nm gate length', in *IEEE Int. SOI Conf.*, pp. 137–139, 2002.

67 T. C. Lim and G. A. Armstrong, 'Parameter sensitivity for optimal design of 65 nm node double gate SOI transistors', *Solid State Electron.*, vol. 49, pp. 1034–1043, 2005.

68 R. Granzner, V. M. Polyakov, F. Schwierz, M. Kittler, and T. Doll, 'On the suitability of DD and HD models for the simulation of nanometer double-gate MOSFETs', *Physica E*, vol. 19, pp. 33–38, 2003.

69 R. Granzner, V. M. Polyakov, F. Schwierz, M. Kittler, R. J. Luyken, W. Rosner, *et al.*, 'Simulation of nanoscale MOSFETs using modified drift-diffusion and hydrodynamic models and comparison with Monte Carlo results', *Microelectronics Engineering*, vol. 83, pp. 241–246, 2006.

70 B. S. Doyle, S. Datta, M. Doczy, S. Hareland, B. Jin, J. Kavalieros, *et al.*, 'High performance fully-depleted Tri-Gate CMOS transistors', *IEEE Electron Devices Lett.*, vol. 24, pp. 263–265, 2003.

71 G. Pei, J. Kedzierski, P. Oldiges, M. Ieong, and E. C.-C. Kan, 'FinFET design considerations based on 3-D simulation and analytical modeling', *IEEE Trans. Electron Devices*, vol. 49, pp. 1411–1419, 2002.

Chapter 6

Heterostructure bipolar transistors

Early bipolar junction transistors were too slow for practical applications in telecommunications. One approach to speed up the flow of the minority carriers from the emitter to the collector by incorporating an 'electric field' into the base region, the so-called '*drift transistor*,' was proposed by Herbert Kroemer in 1953 [1]. The drift transistor used the concept of a doping-engineered electric field in the base to reduce the electron base transit time. An eight-fold increase in the theoretical frequency was predicted as compared to Shockley's 'diffusion' bipolar transistors. This could be achieved by using not a uniform doping in the base but one that decreased exponentially from the emitter end to the collector end. While working out the detail, Herbert Kroemer realised that

a drift field may also be generated through a variation of the energy gap itself, by making the base region from a nonstoichiometric mixed crystal of different semiconductors with different energy gaps (for example, SiGe), with a composition that varies continuously through the base (translation from Reference 2.)

This was not yet the full general design principle, but it constituted the original conception (see Figure 6.1) of what has become known as the heterostructure bipolar transistor, and ultimately of the heterostructure device field in general [2]. Heterostructures, as it is known today, may be defined as heterogeneous semiconductor structures built from two or more different semiconductors, in such a way that the transition region or interface between the different materials plays an essential role in any device action [3]. Kroemer has pioneered the understanding of heterojunctions and heterointerfaces. His research on the theoretical prediction of band line-ups and the problems associated with connecting electronic wave functions across heterointerfaces provide many valuable insights into heterostructures [4].

Commonly used elements in present heterostructure technology are the elements from the central portion of the periodic table belonging to columns II through VI. They are Zn, Cd, Hg (from column II); Al, Ga, In (from column III); Si, Ge, C (from column IV); N, P, As, Sb (from column V); and S, Se, Te (from column VI). Interestingly,

Sonderdruck

aus

ARCHIV DER ELEKTRISCHEN ÜBERTRAGUNG

Zur Theorie des Diffusions- und des Drifttransistors
III. Dimensionierungsfragen

Von HERBERT KRÖMER*

Mitteilung aus dem Fernmeldetechnischen Zentralamt Darmstadt

(A. E. Ü. 8 [1954], 499—504; eingegangen am 24. Juli 1954)

....

Außer durch inhomogene Dotierung eines homogenen Halbleiters läßt sich ein Driftfeld auch dadurch erzeugen, daß man die Breite des verbotenen Bandes selbst ändert, indem man die Basiszone
$-V)\Big]^{-1/2}$.(6a)   aus einem nichtstöchiometrischen Mischkristall verschiedener Halbleiter mit verschiedenen Bandabständen (z.B. Ge-Si) herstellt, dessen Zusammensetzung sich innerhalb der Basis stetig ändert. Bei nicht zu hoher Dotierung bleiben dann die Emitterkapazitäten klein, obwohl selbst dann, wenn diese Dotierung konstant ist, ein Driftpotential von

$$\Delta V \approx E_{B,E} - E_{B,C} \qquad (1\,b)$$

erzeugbar wäre. Mit Ge-Si gäbe das etwa 0,4 eV = 16 kT.   *This is the start of Si-Ge*
Eine Variante dieses Verfahrens besteht darin, zwar in der Basiszone den homogenen Halbleiter mit inhomogener Dotierung beizubehalten, für die Emitterzone jedoch einen Halbleiter mit wesentlich größerem Bandabstand zu wählen[2]. Dann ist es nämlich möglich, die Störstellenkonzentration P_e im Emitter weit unter N_a zu senken, ohne daß der Wirkungsgrad des Emitters abnimmt. Dadurch nehmen aber gemäß Gl. (6a) auch die echten Kapazitäten ab, und unter Umständen kann ΔV noch über die

---

[2] Den Hinweis hierauf verdanke ich Herrn A. HÄHNLEIN; siehe hierzu auch LEHOVEC [3].

Figure 6.1   *The start of heterostructures: portion of a page from Reference 2*
[Source: C. K. Maiti]

silicon, the backbone of modern electronics is at the centre. Below Si is germanium, which is, however, rarely used, but Ge-Si alloys with composition-dependent position play an increasingly important role in today's heterostructure technology. In fact, historically Ge-Si was the first heterostructure device system proposed, although it was also the system that took longest to bring to practical maturity, largely because of the 4 per cent mismatch between the lattice constants of Si and Ge.

In principle, every element in column III may be combined with every element in column V to form a so-called III–V compound. Two or more discrete compounds may be used to form alloys. A common example is aluminium-gallium arsenide, $Al_xGa_{1-x}As$, where x is the fraction of column III sites in the crystal occupied by Al atoms and $1 - x$ is occupied by Ga atoms. Hence, one may have a continuous range of materials and it becomes possible to make compositionally graded heterostructures, in which the composition varies continuously rather than abruptly throughout the device structure. Similar to the III–V compounds, every element from column VI may be used together with every element in column II to create II–VI compounds.

The fundamental design principle of heterostructures is that their variable energy gaps can be used to control the distribution and flow of electrons and holes separately and independently of each other as shown in Figure 6.2. In a heterostructure, where the energy gap becomes position dependent, and the two band-edge slopes are no longer equal, the two forces are no longer equal in magnitude. It would, for example, be possible to have a force acting only upon one kind of the carriers [see Figure 6.2(b)] or to have forces that act in the same direction for both types of carriers [see Figure 6.2(c)]. Only electrical forces in homogeneous crystals can never do this. This is why Kroemer called these forces 'quasi-electric' and commented as *they present a new degree of freedom for the device designer to enable him to obtain effects that are very difficult to obtain using only 'real' electric fields* [3].

The underlying design principle of heterostructure devices was reported for the first time in a 1957 paper by Kroemer [5], although the wide-gap emitter idea, which appears to have been presented principally to cover alternative design possibilities, a procedure typical in patents, was offered by W. Shockley in his patent application in 1948 [6]. On the basis of the concept of graded energy gap, the band diagram for a pnp transistor having a base region with a graded gap, to speed up minority carrier flow from emitter to collector, is shown in Figure 6.3. Note that Figure 6.3 shows a flat conduction band, as would be the case for a heavy uniform doping; the band diagram of Figure 6.2(b) represents essentially the base region of graded energy gap concept while the Figure 6.2(c) illustrates the generality of the design principle.

For the proposed graded-gap base structure, however, no technology was available in the 1950s. The other possibility envisaged was a design in which the emitter was made from a wider-gap semiconductor than the base, with a quasi-abrupt transition at the interface between the two, leading to a band diagram as shown in Figure 6.4. It was also realised that a wide-gap emitter has advantages of its own [7,8]. Kroemer pointed out that the advantage of the wide bandgap emitter over a conventional bipolar transistor in current gain was due to an exponential factor proportional to the difference in the base and emitter energy gaps. For example, one of the problems

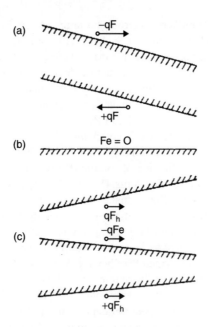

Figure 6.2   *Quasi-electric fields. Energy band diagrams showing how Kroemer used quasi-electric fields to engineer the force experienced by electrons and holes: (a) shows the tilted energy bands expected in the presence of a true electric field; (b) shows the bands with a quasi-electric field with no force on electrons but a force on the holes; and (c) 'shows quasi-electric fields forcing electrons and holes in the same direction. Nobel Lecture [After H. Kroemer, Rev. Mod. Phys., Vol. 73, 2001(783–793)]*

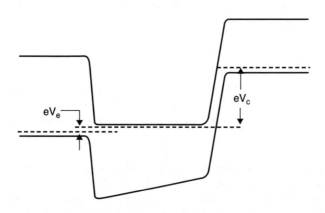

Figure 6.3   *pnp transistor with a base region with a graded gap, to speed up minority carrier flow from emitter to collector. pnp transistors were the preferred design for the Ge-based transistors of the mid-1950s. Nobel Lecture [After H. Kroemer, Rev. Mod. Phys., Vol. 73, 2001(783–793)]*

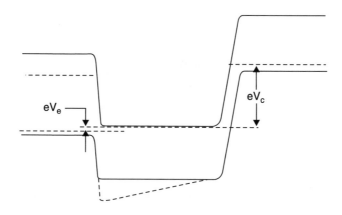

Figure 6.4 *Wide-gap emitter. The energy-gap variation has been compressed into a quasi-abrupt transition at the emitter-to-base interface. The base region still has a uniform energy gap without the transport-aiding quasi-field, but there is now a potential barrier for the escape of electrons from the base into the emitter that is larger than the barrier for holes entering the base from the emitter. Nobel Lecture [After H. Kroemer, Rev. Mod. Phys., Vol. 73, 2001(783–793)]*

with bipolar transistors is minimising the back-injection of majority carriers from the base (electrons in a pnp transistor) into the emitter. In a homojunction transistor, this requirement limits the base doping, which has other undesirable consequences, like a large base resistance. A wide-gap emitter greatly suppresses this back-injection current, roughly by a factor $\exp(-\Delta E_g/kT)$, where ΔE_g is the difference in energy gaps. In his classic 1982 paper on HBTs and ICs [8], Kroemer advanced the concept of the 'inverted' transistor, more commonly referred to now as the collector-up structure. The basic advantage of the 'inverted' transistor is a lower collector-base capacitance and consequently increased RF gain and higher speed.

6.1 The first SiGe transistor

While at RCA in 1957, Kroemer made an unsuccessful attempt to build a Ge transistor with a Ge-Si alloy emitter [9]. The idea was to utilize the fact that the Au-Si phase diagram exhibits a low-melting (370 °C) eutectic. Small grains of the eutectic powder was put onto a Ge chip and alloyed at a temperature somewhere around 500–600 °C. The Au-Si alloy would melt and penetrate into the Ge chip, dissolving some Ge. Upon cooling, a Ge-Si alloy emitter would recrystallise (see Figure 6.5). Although one or two working transistors were obtained, the large thermal strains generated during the solidification of the eutectic caused the Ge chip to crack. The work was followed up by Diedrich and Jotten [10], but the technology was unpromising and SiGe HBTs had to wait several decades for their practical realisation.

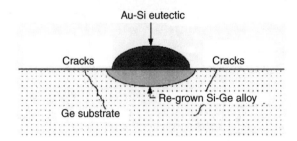

Figure 6.5   Attempt to realise a Ge transistor with a Ge-Si alloy emitter. A piece of Au-Si eutectic was alloyed into a Ge base, forming a SiGe alloy emitter upon cooling. Nobel Lecture [After H. Kroemer, Rev. Mod. Phys., Vol. 73, 2001(783–793)]

6.2   Issues related to heterostructure

When two semiconductor materials with significantly different lattice constants are grown upon each other, whether graded or not, huge strains rapidly build up with increasing thickness, and eventually misfit dislocations form, making the film useless. As a result, the need for lattice matching is obvious. Historically, the importance of lattice matching was recognised almost from the beginning, especially for bipolar applications, and a lattice mismatch below 0.1 nm (0.2 per cent) was suggested as the most promising one, indicating a recognition of the stringency of the lattice-matching needs. With the emergence of quantum wells, superlattices and other heterostructures calling for very thin layers, the issue of strain induced by lattice mismatch has lost some of its relevance [3]. In sufficiently thin structures, a large strain can be accommodated without dislocation formation, to the point that the modification of the energy band structure of a heterostructure by deliberate introduction of strain has become an important device design principle. The evolution of successful SiGe HBTs is perhaps the most striking example [11, 12].

The main reason for the continued dominance of the (Al,Ga)As alloy system in heterostructure studies is due to the fact that AlAs and GaAs have essentially the same lattice parameter. A display of the energy gaps of III–V semiconductors vs. their lattice constants is shown in Figure 6.6, with interconnect lines representing binary alloys. Bulk GaAs is obtained as high-quality single crystals with low dislocation densities, especially in semi-insulating form, making it an ideal substrate readily available for the growth of different heterostructures. The only bad aspect to the (Al,Ga)As system is the chemical affinity of aluminium to oxygen leading to many residual defects in (Al,Ga)As. A second natural substrate is InP, widely used for both optoelectronic and high-speed device applications that call for energy gaps less than that of GaAs. There is no binary III–V compound lattice matched to InP, but InP is widely used in devices, combined with a wide variety of alloys ranging from (Ga,In)As to Al(As,Sb). In 1983, Kroemer proposed the use of (Ga,In)P, which is lattice matched to GaAs. (Ga,In)P has drawn some attention as an alternative to

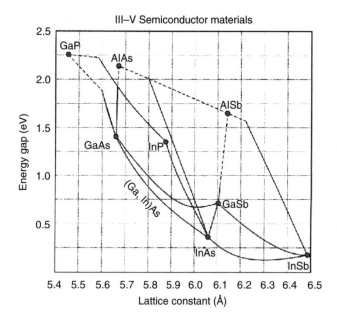

Figure 6.6   *Energy gap of various III–V compounds vs. lattice constant. Nobel Lecture [After H. Kroemer, Rev. Mod. Phys., Vol. 73, 2001(783–793)]*

(Al,Ga)As, especially in HBTs, for which the band alignment at the (Ga,In)p-GaAs interface is more favourable.

In the early 1960s, the most successful heterojunctions demonstrated were based on the Ge-on-GaAs system [13]. If lattice matching were the only constraint, then the Ge-GaAs system would be the ideal heterosystem, which, however, has been found to be otherwise. This has been explained in the literature due to 'valence mismatching,' meaning that the number of electrons provided by the atoms is not equal to the canonical number of exactly two electrons per covalent bond [3]. Covalent bonds between Ge on the one hand and Ga or As on the other are readily formed, but they are not electrically neutral, as was first pointed out by Harrison *et al.* [14]. Although the authors discuss only the GaAs-Ge interface, their argument applies to other interfaces combining semiconductors from different columns of the periodic table. In the specific case of compound semiconductor growth on a column IV elemental semiconductor, the additional problem of antiphase domains on the compound side arises [15].

The above discussion pertains to the most widely used (001)-oriented interface. The interface charge at a valence-mismatched interface actually depends on the crystallographic orientation. It has been shown by Wright *et al.* [16] that an ideal (112) interface exhibits neither an interface charge nor antiphase domains, and it was in fact possible to demonstrate GaP-on-Si interfaces that had a sufficiently low defect density that they operated as emitters in a GaP-on-Si HBT [16, 17].

Performance improvements in Si bipolar have so far been achieved through the vertical scaling of the transistor, such as an increase in the collector doping concentration to reduce the base-collector junction depletion width and to increase the current level. As a result, speed improvement is generally accompanied by an increase in the operating current density. In high-speed digital as well as in analog/mixed-signal applications, high speed, high transconductance, low noise and low power performance are of crucial importance. Also, for the implementation of broadband systems extensive efforts have been made to improve the speed performance of modern Si bipolar transistors, and the bipolar transistor has re-established itself as the leading contender for these applications.

The concept of combining silicon and germanium into an alloy for use in high-speed transistor engineering is an old one, and was envisioned by Kroemer in his early research on drift transistors discussed above [2, 5, 18]. However, because of non-availability of appropriate technique for growing lattice-matched SiGe alloy on Si, this concept could be made to practical reality only in the past 20 years. The incorporation of silicon-germanium (SiGe) in the mainstream Si technology during the past ten years has changed the scenario completely. Graded epitaxial-base SiGe heterojunction bipolar transistors have several advantages over conventional implanted-base silicon bipolar junction transistors (BJTs) and are competing the III–V technologies [19].

The silicon-germanium heterojunction bipolar transistor is the first practical bandgap-engineered device to be realised in silicon. As a result of the bandgap engineering made possible by including a small amount of germanium in the base, one achieves

1. improved emitter injection efficiency;
2. reduced emitter charger storage;
3. reduced base transit time;
4. reduced output conductance;
5. improved current gain at low temperature.

In addition to directly improving the intrinsic device performances via bandgap engineering, the inclusion of an epitaxial SiGe-base in a HBT enhances the device design window in the following ways:

1. reduced basewidth;
2. improved profile control;
3. lower electric field at the emitter-base junction.

Many of the advantages of the SiGe HBTs have opened up several new device design windows, which is usually quite rigid for high-performance BJTs. Compared to III–V technologies, SiGe-technology offers the following advantages:

1. low cost through compatibility with silicon CMOS;
2. uniformity and high yield across large wafers;
3. high thermal conductivity;
4. lower operating voltage.

The SiGe HBT offers several advantages over a Si BJT [19]:

1. Owing to the reduced base bandgap, a much higher base doping concentration can be used, decreasing the base resistance.
2. A reduction in the base transit time results in higher f_T and f_{max}.
3. An increase in collector current density will allow high current gain with low base resistance.
4. Early voltage is higher at a given cut-off frequency.

The presence of Ge in the base of a bipolar transistor enhances the collector current at fixed base-emitter voltage over a comparably constructed Si BJT. The J_c enhancement depends exponentially on the emitter-base boundary value of Ge-induced band offset, and linearly on the Ge grading across the base. This dependence offers a trade-off for the Ge profile optimisation. A box Ge profile is better for current gain enhancement than a triangular Ge profile. The Ge-induced J_c enhancement is exponentially dependent on reciprocal temperature, and thus cooling will produce a strong magnification of the enhancement in gain. Thus, SiGe HBTs exhibit superior cryogenic performance, without the heavy doping and freeze-out effects generally seen in Si BJTs at reduced temperatures [20]. This is because most of the key parameters of SiGe HBTs are thermally activated functions of the amount of bandgap reduction in the base region [21,22]. The bandgap reduction is proportional to the Ge content, and hence small amounts of Ge are sufficient to attain the desired cryogenic performance in SiGe HBTs. For an extensive discussion of the SiGe HBT design issues, the reader may refer to several texts [19,23,24].

The silicon-germanium heterojunction bipolar transistor is the first practical bandgap-engineered device to be realised in silicon. SiGe HBT technology combines transistor performance competitive with III–V technologies with the processing maturity, integration levels, yield, and hence, cost commonly associated with conventional Si fabrication. Progress in SiGe HBTs has been very rapid, from the first laboratory demonstration to the installation of a complete IC technology in a 200-mm CMOS/DRAM manufacturing for commercial RF and microwave applications [25,26]. Several companies have put first-generation SiGe HBTs into production with values of f_T and f_{max} around 50 GHz. Third-generation technology produced transistors with values of both and f_{max} around 120 GHz, while a 2003 state-of-the-art device has f_T of 375 GHz [27]. SiGe HBTs are well suited for wireless and optical telecommunications market in the 1–20 GHz range. Technical progress in bringing SiGe HBT to reality has been exceptionally rapid and IBM has been the major player. The first functional SiGe HBT was announced in 1987 [25]. Worldwide attention was directed toward the technology in 1990 with the demonstration of a non-self-aligned SiGe HBT with a cut-off frequency (f_T) of 75 GHz [28]. Later in 1990, the first emitter-coupled logic (ECL) ring oscillators using self-aligned SiGe HBTs were produced [29]. The first SiGe BiCMOS technology was reported in 1992 [30], and the first large-scale integrated (LSI) circuit (a 1.2 GSample/s digital-to-analog converter) in 1993 [31]. SiGe HBTs with frequency response greater than 100 GHz were demonstrated in 1994 [32–34]. Since 1995 to present, many circuits, too numerous to list individually, targeting digital, RF and microwave applications have been demonstrated. In 2001, a 40 Gb/s communication subsystem SiGe BiCMOS technology exhibited $f_T = 120$ GHz [35, 36].

6.3   SiGe materials

In this section, a review on the materials used in SiGe HBTs and their principle of operation are given. The key to the evolution of SiGe heterojunction bipolar technology has been developments in epitaxy systems that have allowed the growth of pseudomorphic (strained) $Si_{1-x}Ge_x$ to be achieved at temperatures low enough for integration in CMOS technology. When bandgap engineering is used to construct a transistor, a perfect crystalline structure is essential, since any defects created can degrade circuit yields. Si and Ge are completely miscible over the entire compositional range, giving rise to alloys with a diamond crystal structure. There is a 4.17 per cent mismatch in the lattice constant of Si and Ge at room temperature, and the mismatch increases slightly with temperature. Hence, when a SiGe layer is deposited on a Si substrate, the lattice mismatch can lead to the growth of either a strained or a relaxed layer. The lattice constant at room temperature is given by Vegard's rule as

$$a_{Si_{1-x}Ge_x} = a_{Si} + x(a_{Ge} - a_{Si}) \tag{6.1}$$

for low atomic concentrations, x, of Ge [37]. When growth conditions favour pseudomorphic (or strained) growth, the deposited SiGe films are under biaxial compressive stress and the SiGe film adopts the lattice constant of the underlying Si. Relaxed or unstrained growth occurs when misfit dislocations are generated at the interface. As germanium is not lattice-matched to silicon, growth of SiGe on silicon introduces compressive strain. For a given Ge concentration, there is a limit as to how thick the SiGe layer can be and remain stable; thicker layers will tend to incorporate dislocations. The theoretical stability criterion developed by Matthews and Blakeslee [38] is equivalent to a certain maximum integrated Ge content across the film.

SiGe epitaxial growth studies conducted by Bean [39] show that the optimal temperature for a good-quality epitaxial growth is around 550 °C for moderate Ge concentrations (up to 15 per cent). At lower Ge concentrations (<10 per cent), however, optimal growth temperatures can be somewhat higher, increasing with decreasing Ge content. SiGe alloys, because of their larger lattice constant, show a smaller fundamental bandgap compared to Si. Strain in the film further reduces the bandgap and also results in a further modification of the band structure. For example, the splitting of degenerate valence and conduction bands can result from strain. The band alignment for strained-SiGe films on Si substrate is type-I in nature (i.e., the entire bandgap of SiGe layer is contained within the Si bandgap), with most of the difference occurring in the valence band. On the other hand, a strained-Si film grown on an unstrained SiGe substrate acquires a type-II band alignment, with the conduction and valence bands of Si shifted upwards in energy compared to the substrate.

The thickness of the deposited SiGe layer is an important device design consideration. Van der Merwe [40] introduced the concept of 'critical thickness' based on equilibrium theory, defining 'critical thickness' as the film thickness below which, it was energetically favourable to contain the misfit by elastic energy stored in the distorted crystal and above which it was favourable to store part of the energy in misfit dislocations at the heteroepitaxial interface. Another definition for the critical thickness was introduced by Matthews and Blakeslee [38]. They defined critical thickness in terms of the mechanical equilibrium of a pre-existing threading dislocation, as the thickness at which the force of the threading dislocation residing at the heteroepitaxial interface is equal to that component of force per unit length acting on the threading component of the dislocation in the plane of the epitaxial layer. Figure 6.7 shows the stability curve for strained-SiGe epitaxial films.

The Matthews and Blakeslee curve, shown on the left in the solid line, represents the theoretically expected critical thickness, whereas Stiffler's curve [41], shown on the right in the dashed line, represents the empirical critical thickness for ultra-high vacuum/chemical vapour deposition grown SiGe films on a blanket Si wafer. The difference in the two curves is not unexpected because the critical thickness depends on the material system, substrate type, the amount of surface defects, temperature, growth process and various other parameters. Deposited SiGe films that lie below the stability curve are thermodynamically stable and can be processed using conventional furnace or rapid thermal annealing, ion implantation without generating defects. Films

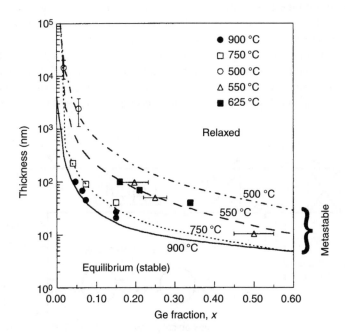

Figure 6.7   The critical thickness as a function of effective strain for SiGe films grown on Si

lying above the stability curve are thermodynamically 'metastable' and will relax to their natural lattice constant if exposed to temperatures above the original growth temperature, or if stressed by fabrication steps such as ion implantation or rapid thermal annealing.

6.3.1   UHVCVD growth of SiGe films

The first successful growth of SiGe epitaxial films was demonstrated using molecular beam epitaxy (MBE) in 1987 [42]. Later, a number of other techniques such as rapid-thermal chemical vapour deposition (RTCVD) [43], reduced-pressure (or atmospheric-pressure) chemical vapour deposition (RPCVD/APCVD) [44], and ultra-high vacuum chemical vapour deposition [45] have also been used to successfully grow the SiGe films. Of all the above methods, UHVCVD has shown the ability for high volume production through large-scale integration and batch processing. This section gives a brief description of IBM's UHVCVD SiGe technology.

Figure 6.8 shows the cross-section of a UHVCVD reactor [45]. UHVCVD relies on hydrogen passivation for preparing the silicon surface for epitaxy rather than the conventional high-temperature oxide desorption. When Si is etched in hydrofluoric acid (HF), an adlayer of hydrogen is formed, which reduces the reactivity of the silicon surface with respect to oxidants such as water and oxygen by about 13 orders of magnitude. The HF passivated wafers are loaded into the load-lock of the UHVCVD apparatus. After pump down below 10^{-6} Torr, wafers are transferred under flowing hydrogen into the UHVCVD section of the apparatus, and growth commences immediately. The gaseous sources employed for alloying or doping are silane, germane, diborane and phosphine. Films are deposited at temperatures in the range of 400–500 °C. Film growth rates may be varied from 0.01 to 10 nm/min as a function of temperature and film germanium content, and typical rates employed are between 0.4 and 4.0 nm/min. These limits are used to ensure precise dimensional control, of the order of 1–2 atomic layers. This level of precision is required if one is to compete

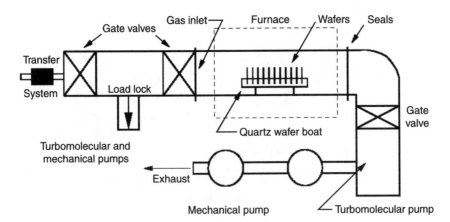

Figure 6.8   Schematic cross-section of a UHVCVD reactor

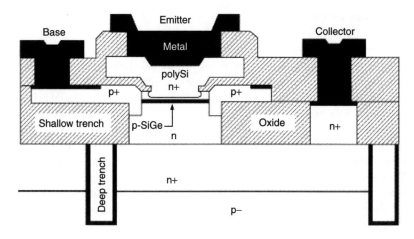

Figure 6.9   Typical cross-section of a SiGe HBT [After J. Tang et al., IEEE Trans. Microwave Th. and Tech., Vol. 50, 2002(2467–2473). ©2002 IEEE]

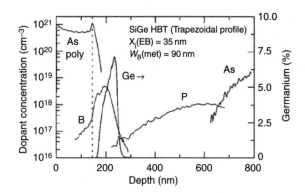

Figure 6.10   Typical secondary ion mass spectroscopy (SIMS) doping profile of a SiGe HBT [After S. Zhang et al., IEEE Trans. Electron Devices, Vol. 49, 2002(429–435). ©2002 IEEE]

effectively with the control of ion implantation. The deposition of compositionally graded germanium profiles within the base of the SiGe heterojunction bipolar transistors is routinely practised because it enhances the device performance for relatively low-average Ge content. The typical cross-section of the UHVCVD SiGe HBT is shown in Figure 6.9.

Figure 6.10 shows a typical secondary ion mass spectroscopy (SIMS) depth profile of an advanced SiGe HBT, which has been optimised for room temperature operation. The in situ boron-doped graded SiGe base is deposited across the entire wafer using the UHVCVD technique. The Ge profile is graded across the neutral base and has a trapezoidal shape with 2 per cent Ge at the e-b depletion edge and a 5 per cent grading across the base. The metallurgical basewidth is about 90 nm.

6.4   Physics of SiGe HBTs

Pseudomorphic $Si_{1-x}Ge_x$ has a bandgap smaller than Si and hence makes possible bandgap engineering concepts that were previously only achievable in III/V or II/VI technologies. When incorporated into the base of a bipolar transistor $Si_{1-x}Ge_x$ gives a reduction in the potential barrier to electrons in the emitter. The essential differences between an Si BJT and an SiGe HBT are best illustrated using the schematic energy band diagram shown in Figure 6.11. Also the SiGe HBT and Si BJT are taken to be of identical geometry, and it is assumed that the emitter, base and collector-doping profiles of the two devices are identical, apart from the Ge in the base of the SiGe HBT. In this case, an ideal graded-base SiGe HBT with constant doping in the emitter, base and collector regions is considered. In such a device construction, the Ge content is linearly graded from 0 per cent near the metallurgical emitter-base (e-b) junction to some maximum value of Ge content near the metallurgical collector-base (c-b) junction and then rapidly ramped back down to 0 per cent Ge. However, a wide variety of practical SiGe profile designs, ranging from constant (box) Ge profiles to triangular (linearly graded) Ge profiles, and including the intermediate case of the Ge trapezoid (a combination of box and linearly graded profiles) are possible and the analyses have been made [19, 23, 24].

The addition of Ge in the base of the SiGe HBT causes a lowering of the conduction band, thus resulting in a bandgap reduction. The band offset at the emitter-base junction is $\Delta E_{g,Ge}(x = 0)$ and that at the collector-base junction is $\Delta E_{g,Ge}(x = W_b)$. The compositional grading of Ge across the base region results in a position dependence of the band offset, which can be conveniently expressed as a bandgap grading term $[\Delta E_{g,Ge}(grade) = \Delta E_{g,Ge}(W_b) - \Delta E_{g,Ge}(0)]$. This position-dependent band offset induces a quasi-drift field in the neutral base, which aids the transport of

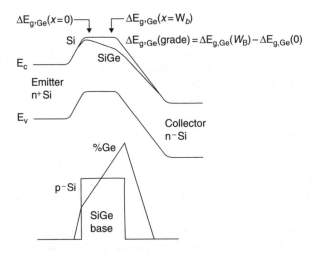

Figure 6.11   Energy band diagram of a graded-base SiGe HBT compared to an identically constructed Si BJT

minority carriers from emitter to collector, thus improving the frequency response. This also results in an enhanced collector current and hence enhanced gain. This enhanced gain can be traded for increased base doping and decreased basewidth, and hence improved high frequency performance.

To understand the operation of a SiGe HBT, one generally models enhancement in current gain, β, which is given by

$$\beta = \frac{\delta I_c}{\delta I_b} \tag{6.2}$$

where I_c is the collector current and I_b is the base current. Assuming a narrow base region and negligible base recombination, β is related to the physical parameters of the device by the well-known equation

$$\beta = \frac{D_{nb} n_{iB}^2 W_E N_E}{D_{ne} n_{iE}^2 W_B N_B} \tag{6.3}$$

where D_{nb} is the diffusion constant for electrons (in the case of an npn device) in the base, D_{ne} is that for holes in the emitter, W_B is the width of the base, W_E is that of the emitter, N_B is the doping concentration in the base, N_E is that in the emitter, n_{iB} is the intrinsic carrier concentration in the base and n_{iE}, that in the emitter.

In order to achieve a high β, a large N_E, a small value of N_B, and a small W_B are desirable. However, the doping and the width of the base must be large enough to withstand the depletion occurring in the junctions, so that a punchthrough of the base does not take place. The theoretical consequences of these bandgap changes on the collector current density can be expressed using the generalised Mall–Ross relation as

$$J = \frac{q(e^{qV_{be}/kT} - 1)}{\int_0^{W_b} \dfrac{N_b(x)dx}{D_{nb}(x)n_{ib}^2(x)}} \tag{6.4}$$

where W_b is the quasi-neutral basewidth at bias V_{be}, and D_{nb} is the minority electron diffusivity in the base. The position-dependent effective intrinsic carrier density in the base is given by:

$$n_{ib}^2(x) = \gamma n_{io}^2 e^{\Delta E_{gb}^{app}/kT} e^{\Delta E_{g,Ge(grade)}(x/W_b kT)} e^{\Delta E_{g,Ge}(0)/kT} \tag{6.5}$$

where $\Delta E_g^{app}/kT$ is the heavy doping-induced apparent bandgap narrowing of the base. γ is the ratio of density-of-states product between Si and SiGe $(N_c N_v(\text{SiGe})/N_c N_v(\text{Si}))$ and accounts for the reduction in the effective density-of-states with increasing Ge content. In the present analysis, the SiGe HBT and the Si BJT are assumed to be of identical geometry, and the emitter, base and collector profiles of the two devices are assumed to be the same, apart from the Ge in the base of the SiGe HBT. For simplicity, a Ge profile, which is linearly graded from the e-b to the c-b junctions, is considered. The same procedure can be employed in deriving the solutions for more general Ge profiles. The analysis also assumes low-injection conditions and Boltzmann statistics.

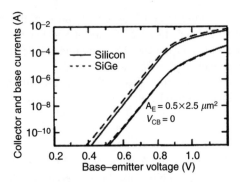

*Figure 6.12   Typical Gummel characteristics of an SiGe HBT as compared to an Si
                 BIT of comparable construction*

In ideal Si BJTs and SiGe HBTs, the emitter profiles are identical and thus base current will be the same. Hence the β-ratio is equal to the J_c ratio. Using (6.4) and (6.5), the ratio of the collector current density and hence current gain ratio can be expressed as

$$\frac{J_{c,SiGe}}{J_{c,Si}} = \frac{\beta_{SiGe}}{\beta_{Si}} = \gamma \eta \left(\frac{\Delta E_{g,Ge\,(grade)}/kT)e^{\Delta_{g,Ge}(0)/kT}}{1 - e^{-\Delta E_{g,Ge\,(grade)}/kT}} \right) \tag{6.6}$$

where $\eta = D_{nb}(SiGe)/D_{nb}(Si)$ accounts for the differences in the electron mobilities in the base. The measured Gummel characteristic of a SiGe HBT and an identically constructed Si BJT with identical base current can be seen in Figure 6.12. This plot confirms the enhancement in collector current in an SiGe transistor, predicted in (6.6). With reference to the band diagram in Figure 6.11, J_c in a SiGe HBT is exponentially dependent on the e-b boundary value of the Ge-induced band offset and linearly proportional to the Ge bandgap grading factor. The use of a graded Ge profile also leads to an exponential increase in Early voltage V_A and hence increased output conductance, as given in (6.7). The Early voltage is a measure of how much the neutral base profile can be depleted with reverse bias on the c-b junction.

$$\frac{V_{A,SiGe}}{V_{A,Si}} = e^{\Delta E_{g,Ge\,(grade)}/kT)} \left[\frac{1 - e^{-\Delta E_{g,Ge\,(grade)}/kT}}{\Delta_{g,Ge\,(grade)}/kT} \right] \tag{6.7}$$

An important figure of merit for analog applications such as high-speed data converters and precision current sources is the 'current gain–Early voltage product' (βV_A product). The βV_A product of a SiGe HBT as compared to that of an identically constructed Si BJT is given by the simple ratio

$$\frac{\beta V_{A,SiGe}}{\beta V_{A,Si}} = \gamma \eta e^{\Delta E_{g,Ge}(0)/kT} e^{\Delta E_{g,Ge\,(grade)}/kT} \tag{6.8}$$

Equation (6.8) shows that the βV_A product of a SiGe HBT depends exponentially on both the emitter-base band offset and the germanium grading in the base, and hence can be well controlled in applications that require it. For example, a triangular

Ge profile is better for Early voltage enhancement than a box Ge profile. As the Ge-induced V_A enhancement is exponentially dependent on reciprocal temperature, cooling will produce a strong magnification of the enhancement. In a well-designed SiGe HBT, the Ge profile design that balances the optimisation of β, V_A and βV_A would be a Ge trapezoid, with a small (e.g., 5 per cent) Ge content at the e-b junction, and a larger (e.g., 15 per cent to 20 per cent) Ge content.

The unity gain cut-off frequency f_T is an important figure of merit in RF and microwave circuit applications. Base transit time τ_b typically limits the maximum f_T in conventional Si BJTs. The built-in electric field induced by the Ge grading across the neutral-base effectively decreases τ_b since the carriers are now more rapidly accelerated across the base. The ratio of base transit time is given by:

$$\frac{\tau_{b,SiGe}}{\tau_{b,Si}} = \frac{2}{\eta}\left(\frac{kT}{\Delta E_{g,Ge\,(grade)}}\right)\left[1 - \frac{1 - e^{-\Delta E_{g,Ge\,(grade)}/kT}}{\Delta E_{g,Ge\,(grade)}/kT}\right] \tag{6.9}$$

Emitter transit time τ_e is typically not a limiting parameter in state-of-the-art polySi emitter devices. However, it may become more important as the vertical profile is scaled to thinner dimensions with further technology evolution. τ_e is inversely proportional to the a.c. current gain, and hence is lower in SiGe HBTs. This combination of lower τ_e and τ_b in SiGe HBTs results in a significant improvement of frequency response compared to similarly constructed Si BJTs.

6.5 Figures of merit of SiGe HBTs

To assess the capabilities and the performance of electronic devices, several figures of merit (FOM) are often used. RF figure of merit of a transistor are Unity gain frequency, f_T, maximum oscillation frequency, f_{max}, base resistance r_b, and the minimum noise figure NF_{min}. RF transistors are unconditionally stable at any operating frequency above a critical frequency f_k. At operating frequencies below f_k, however, the transistor is conditionally stable and certain termination conditions can cause oscillations. The stability behaviour of a transistor can be described by the stability factor k as introduced by Rollett [46]:

$$k = \frac{2\text{Real}(y_{11})\text{Real}(y_{22}) - \text{Real}(y_{12}y_{21})}{|y_{12}y_{21}|} \tag{6.10}$$

where y_{11}, y_{12}, y_{21} and y_{22} are the frequency-dependent y-parameters of the transistor. If $k > 1$, the transistor is unconditionally stable. Usually, f_{max} is more significant than f_T and is a good measure of transistor performance not only for power gain in small-signal and large-signal amplifiers, but also for wideband analogue amplifiers and even for non-saturating logic gates [47].

Cut-off frequency, f_T, can be determined by converting the measured s-parameters to h-parameters and plotting $|h_{21}|$ in decibels vs. log frequency. In general, for a well-behaved device, $|h_{21}|$ will decrease at 20 dB/dec of frequency, and therefore, f_T (the frequency value at 0 dB) can be extrapolated from the measured data [48]. Physically,

f_T can be related to the device profile via the total emitter to the collector delay. From the small-signal equivalent circuit, the equations for f_T can be derived as

$$f_T = \frac{1}{2\pi}\left(\frac{kT}{qI_c}(C_{eb} + C_{cb}) + \tau_e + \tau_{eb} + \tau_b + \tau_c + \tau_{bc}\right)^{-1} \qquad (6.11)$$

where C_{eb} and C_{cb} are the emitter-base and the collector-base depletion capacitances; τ_e, τ_b and τ_c are the base, emitter and collector transit times, respectively. τ_{eb} and τ_{bc} are the emitter-base and collector-base space charge region delay time, respectively. Thus, one can observe that for fixed bias current, improvements in τ_b and τ_e due to the presence of SiGe will directly translate into an enhanced f_T. In terms of transistor power gain, one can approximate the maximum oscillation frequency (f_{max}) figure of merit by

$$f_{max} = \sqrt{\frac{f_T}{8\pi C_{bc}R_b}} \qquad (6.12)$$

where R_b is the small-signal base resistance, and C_{bc} is the total collector-base capacitance. A derivation of f_T and f_{max}, together with relevant assumptions and discussion, can be found in References 23 and 19.

Thus, on the basis of the above discussions, several conclusions regarding the effects of Ge on the frequency response of a SiGe HBT can be made [19]:

1. For a fixed bias current, the presence of Ge in the base region affects the transistor's frequency response through the base and emitter transit times.
2. The f_T enhancement for a SiGe HBT over a Si BJT depends on Ge grading across the base.
3. A triangular Ge profile is better for cut-off frequency enhancement than a box Ge profile is.
4. The Ge-induced f_T enhancement depends strongly on temperature in contrast to a Si BJT.
5. Improved f_T in the entire useful range of I_c offers important opportunities for various communication applications.

For precision analog and RF circuits, parameter stability over both temperature and bias must be ensured. Thus, the specific impact on actual SiGe HBT devices and circuits is both profile design and application dependent, and thus they need careful design considerations. Various second order effects, neglected in this section have to be taken into account through TCAD. For example, high injection effects at high current density often results in barrier effects associated with the collector-base heterojunction can strongly degrade both d.c. and a.c. performance. Other examples include the impact of neutral base recombination (NBR) on SiGe HBT operation. Discussion on the implications of these and other second-order effects have been covered in several texts [19, 23, 24]. Use of TCAD to resolve some of these issues is covered in the following section.

6.6 Simulation of SiGe HBTs

TCAD device simulation involves an optimisation relating to the trade-offs involved in output parameters such as speed, leakage, noise, breakdown voltage and power consumption as a function of input design parameters such as transistor geometry, doping profiles, materials and material compositions. While d.c. simulation is sufficient for optimisation of breakdown voltages, turn-on voltages or leakage currents, a.c. simulation is required for speed, noise and power issues.

In the following, as an example of TCAD optimisation of a SiGe bipolar transistor, we have chosen the work of Palankovski and Selberherr [49], who have investigated polySi emitter SiGe HBTs epitaxially grown by chemical vapour deposition process for both the 0.8 μm and 0.35 μm technology nodes with an implanted CMOS n-well. The SiGe base has a triangular Ge profile. The base–emitter junction was formed by rapid thermal processing, which causes out-diffusion of As from the polySi emitter layer into the crystalline Si.

The process simulation was performed with DIOS [50] from the blank wafer to the final device with implant profiles and annealing steps calibrated to one-dimensional SIMS profiles. To study the influence of the selectively implanted collector (SIC) doping on device performance, four SiGe HBT structures with emitter areas of $6 \times 0.8 \ \mu$m^2 have been investigated both experimentally and by means of process simulations, followed by two-dimensional device simulation. The only process step in which the four HBTs (referred to as Dev.1, Dev.2, Dev.3 and Dev.4) differ is the combination of energy and dose used for the SIC implants, as summarised in Table 6.1. Both DESSIS [51] and MINIMOS-NT [52] were used for two-dimensional device simulation. The only fitting parameters used in the simulation were the contribution of bandgap narrowing to the conduction band (here about 80 per cent and 20 per cent for donor and acceptor doping, respectively), and the concentration of traps in the Shockley–Read–Hall recombination model (10^{13} cm^{-3}).

The only fitting parameters used in the simulation were the contribution of bandgap narrowing to the conduction band (here about 80 per cent and 20 per cent

Table 6.1   Summary of key process and device parameters [After V. Palankovski and S. Selberherr, Proc. Symp. on Diagnostics and Yield: Advanced Silicon Devices and Technologies for ULSI Era, 2003(1–11)]

Device	Energy (keV)	Dose (cm^{-2})	f_T (GHz)	BV_{ceo} (V)	$f_T \cdot BV_{ceo}$ (GHz $\cdot$ V)
Dev.1	480	7×10^{12}	32	4.0	128
Dev.2	480	3×10^{13}	40	3.7	148
Dev.3	300	7×10^{12}	33	3.1	102
Dev.4	300	3×10^{13}	42	2.3	97

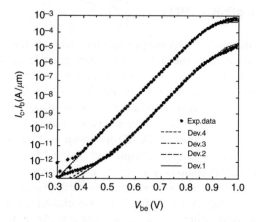

Figure 6.13   Forward Gummel plots at $V_{cb} = 0\,V$. Comparison between measurement and simulation [After V. Palankovski and S. Selberherr, Proc. Symp. on Diagnostics and Yield: Advanced Silicon Devices and Technologies for ULSI Era, 2003(1–11)]

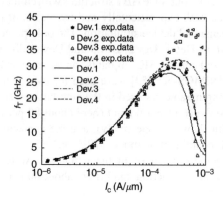

Figure 6.14 f_T vs. I_c at $V_{ce} = 1.5\,V$. Comparison between measurement and drift-diffusion simulation with DESSIS [After V. Palankovski and S. Selberherr, Proc. Symp. on Diagnostics and Yield: Advanced Silicon Devices and Technologies for ULSI Era, 2003(1–11)]

for donor and acceptor doping, respectively) and the concentration of traps in the Shockley–Read–Hall recombination model (10^{13} cm^{-3}). Some simulation results with measured device data are presented below. Both device simulators correctly reproduce the measured forward Gummel plot at 300 K (see Figure 6.13) with default models. The slight increase of collector current I_c with dose and energy at high bias is due to the differences in the base push-out effect.

It is observed from Figures 6.14 and 6.15 that both DESSIS and MINIMOS-NT failed to explain the experimentally observed similarity in peak f_T for some devices

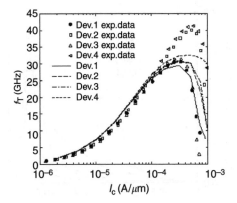

Figure 6.15 f_T *vs.* I_c *at* $V_{ce} = 1.5\,V$. *Comparison between measurement and drift-diffusion simulation with MINIMOS-NT [After V. Palankovski and S. Selberherr, Proc. Symp. on Diagnostics and Yield: Advanced Silicon Devices and Technologies for ULSI Era, 2003(1–11)]*

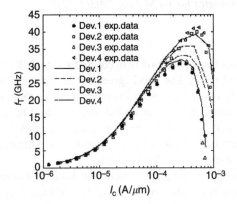

Figure 6.16 f_T *vs.* I_c *at* $V_{ce} = 1.5\,V$. *Comparison between measurement and hydrodynamic simulation with MINIMOS-NT [After V. Palankovski and S. Selberherr, Proc. Symp. on Diagnostics and Yield: Advanced Silicon Devices and Technologies for ULSI Era, 2003(1–11)]*

(Dev.3 and Dev.4) while the presence of about 50 per cent more phosphorus in the collector of the two low-dose devices (Dev.1 and Dev.3) provided an acceptable qualitative agreement. It is known that with shrinking device dimensions non-local effects, such as velocity overshoot, become more pronounced. Neglecting these effects can be reason for underestimating f_T [53].

Simulations with the hydrodynamic transport model improved the results quantitatively (see Figure 6.16). Figure 6.17 shows the velocity overshoot over the greater part of the base region, which is about twice the saturation velocity limit in the drift-diffusion case (10^7 cm/s). This correlates to the higher electron energy in the

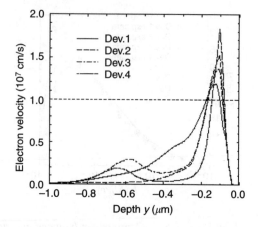

Figure 6.17   *Electron velocity overshoot in the base–collector space charge region at $V_{ce} = V_{be} = 0.88\,V$ [After V. Palankovski and S. Selberherr, Proc. Symp. on Diagnostics and Yield: Advanced Silicon Devices and Technologies for ULSI Era, 2003(1–11)]*

collector and explains the increase of f_T in comparison to drift-diffusion simulations (see Figures 6.14 and 6.15). The good agreement at low currents is very important since HBTs typically operate at much lower frequencies than at the maximum f_T. Simulations prove that in this range optimisation of the SIC implant do not have an influence on f_T, that is, the base–emitter capacitance and not the base–collector capacitance is dominating. The maximum f_T was found to have stronger dependence on the dose than on the energy of the implants.

Furthermore, an important figure of merit, the product of open circuit base breakdown voltage times unity gain frequency $BV_{ceo} \cdot f_T$ (see Table 6.1) reaches a maximum for high SIC implant energies (deep implant) and high SIC doses. The higher f_T for high-dose/low-energy SIC implants is due to smaller basewidth and delayed onset of the base push-out effect due to the higher collector doping.

In the simulation, for SiGe HBTs from the 0.35 μm technology node, several important physical effects, such as surface recombination, impact ionisation (II) generation and self-heating (SH), were modelled and included in the simulation in order to get a good agreement with measured forward (see Figure 6.18) and output characteristics (see Figure 6.19). It is noted that the simulation results without SH effects cannot match the experimental data, especially at high power levels. The only fitting parameters used in the simulation were the contribution of BGN to the conduction band, the trap charge density in the Shockley–Read–Hall recombination model (10^{14} cm^{-3}), the velocity recombination for holes in the polySi contact model [54] used at the emitter contact and the substrate thermal resistance. Both I_c and I_b increase owing to self-heating at a given bias condition. As the change is relatively higher for I_b in order to maintain it at the same level, V_{be} and, therefore, I_c, decreases. A closer look

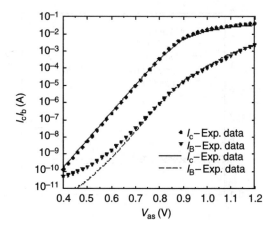

Figure 6.18   *Forward Gummel plots at $V_{cb} = 0\,V$. Comparison between measurement data and simulation at room temperature [After V. Palankovski and S. Selberherr, Proc. Symp. on Diagnostics and Yield: Advanced Silicon Devices and Technologies for ULSI Era, 2003(1–11)]*

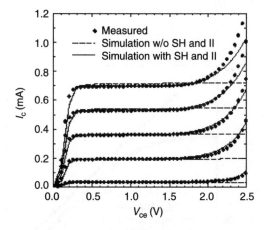

Figure 6.19   *Output characteristics: simulation with and without self-heating (SH) and impact ionisation compared to measurement data. I_b is stepped by 0.4 μA from 0.1 to 1.7 μA [After V. Palankovski and S. Selberherr, Proc. Symp. on Diagnostics and Yield: Advanced Silicon Devices and Technologies for ULSI Era, 2003(1–11)]*

at the increasing collector current I_c at high collector-to-emitter voltages V_{ce} and constant base current reveals the interplay between self-heating and impact ionisation (see Figure 6.20). While impact ionisation leads to strong increase of I_c, self-heating decreases it. Figure 6.21 shows excellent agreement between simulated and measured f_T.

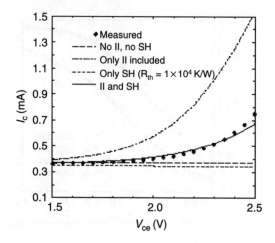

Figure 6.20   Output characteristics for $I_b = 0.9 \mu A$: I_c at high V_{ce} reveals the inter-play between self-heating (SH) effect and impact ionisation generation [After V. Palankovski and S. Selberherr, Proc. Symp. on Diagnostics and Yield: Advanced Silicon Devices and Technologies for ULSI Era, 2003(1–11)]

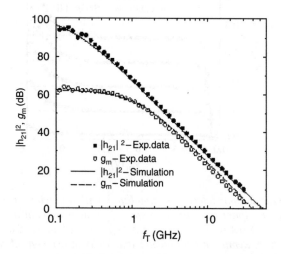

Figure 6.21   Short-circuit current gain h_{21} and matched gain g_m vs. frequency at $V_{ce} = 1 V$ and current density $J_C = 76\, kA/cm^2$ (measurements with circles) [After V. Palankovski and S. Selberherr, Proc. Symp. on Diagnostics and Yield: Advanced Silicon Devices and Technologies for ULSI Era, 2003(1–11)]

6.7   Transit time in SiGe HBTs

The drift-diffusion approximation can lead to inaccuracy in the prediction of device characteristics, particularly when the width of the base is reduced below 30 nm. In this instance it is necessary to perform a simulation involving energy balance (EB) [19]. Owing to the small basewidths of less than 30 nm in SiGe HBTs, the electron transport becomes more and more ballistic. This calls into question the validity of the standard simulation tools like the drift-diffusion (DD) or hydrodynamic (HD) model used to design high-frequency HBTs. As discussed earlier, the base transit time no longer dominates the cut-off frequency of high-speed HBTs, and the DD model overestimates the base transit time, bearing the possibility of wrong guidelines for transistor design optimisation. Also, DD model leads to an underestimation of the peak cut-off frequency by 10 per cent. Close to high injection, differences in the emitter transit times of the HD and MC model are observed that are mainly related to small differences in the Gummel plot. Transit times of SiGe HBTs have been investigated in the quasi-stationary limit by consistent drift-diffusion (DD), hydrodynamic (HD) and full-band Monte Carlo simulations [53]. The quasi-ballistic transport in the base and collector leading to a strong velocity overshoot is modelled using HD model, and corresponding transit times were found to be in good agreement with the MC results.

   The authors calculated the band structure with the non-local empirical pseudopotential method [55]. The conduction and valence band edge were modelled following Reference 56 and are the same for all three models. Apparent bandgap narrowing due to heavy doping, as described in Reference 57, was also included in simulation. All transport parameters of the DD and HD models were generated by MC bulk simulations to ensure consistency of the simulation models [58]. The heat flux of the HD model was reduced to 25 per cent. Using the three simulation models, transit times were evaluated and some simulation results are presented below.

   Base and collector transit times are shown in Figure 6.22. A good agreement is obtained between the HD and MC models for the base transit time resulting from classical and particle simulations. Since the DD model yields a velocity that is considerably smaller than the one of the other two models, the base transit time of the DD model is the largest one. Consequently, the DD model indicates that base transit time is dominant for the collector current levels where the HBT shows its peak performance, whereas this is not the case for the HD and MC models.

   In the case of the emitter transit time (see Figure 6.23) the DD and MC models agree well, whereas the HD model yields a transit time being lower than the DD result for a collector current larger than 0.7 mA/μm^2. Close to high injection, the HD emitter transit time deviates from MC results, which is mainly due to the small differences in the Gummel characteristics of both models. In contrast to the base and collector regions for which the qualitative behaviour of the transit times of the different simulation models is generally valid, no generally valid conclusions can be drawn for the emitter region.

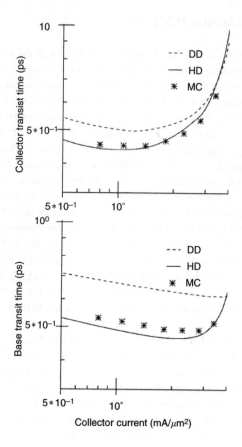

Figure 6.22   Base and collector transit times for $V_{ce} = 0.8$ V [After C. Jungemann et al., IEEE Trans. Electron Devices, Vol. 48, 2001(2216–2220). ©2001 IEEE]

The cut-off frequency is shown in Figure 6.24. As expected, the DD model yields the smallest cut-off frequency of about 95 GHz owing to the higher transit times in the base and in the Ge overlap region of the collector. Since the other transit times are of a similar magnitude, the impact of this error on the cut-off frequency is only about 10 per cent. However, all other models show the peak at the same collector current of about 2.0 mA/μm². The MC model peaks at 106 GHz and the HD model at 113 GHz. The larger cut-off frequency of the HD model is mainly due to its lower emitter transit time (see Figure 6.23).

6.8   SiGe HBTs at low temperature

While SiGe HBT technology is being primarily exploited for room temperature applications, it has long been recognised that bandgap engineering using SiGe

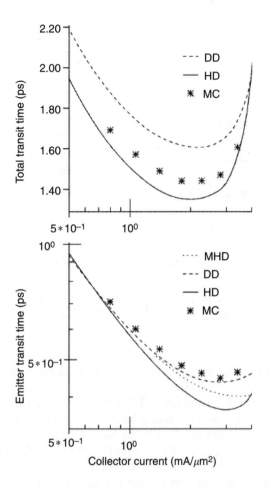

Figure 6.23   *Total and emitter transit times for $V_{ce} = 0.8\,V$ [After C. Jungemann et al., IEEE Trans. Electron Devices, Vol. 48, 2001(2216–2220). ©2001 IEEE]*

can provide superior cryogenic performance [20]. Examples of existing cryogenic electronics applications include satellite systems, deep-space and planetary space missions, very high precision instrumentation and detector electronics, superconductor-semiconductor hybrid electronic systems and very-low-noise receivers for astronomy. The transit time components of SiGe HBTs have been calculated in the low-temperature regime through simulation. The effect of low temperature on each transit time component has been investigated and an increase in f_T by more than 1.8 times has been obtained when the simulation temperature is lowered from 300 K to 120 K [59].

Low-temperature investigations provide not only the potential behaviour of the devices for specific applications but also are a powerful tool for a better

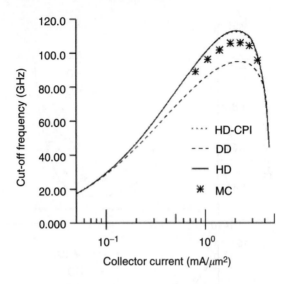

Figure 6.24   Cut-off frequency for $V_{ce} = 0.8\,V$ [After C. Jungemann et al., IEEE Trans. Electron Devices, Vol. 48, 2001(2216–2220). ©2001 IEEE]

understanding of the physics governing device performance at room temperature. Performance enhancements in HBTs, when operated at low temperature, have been reported in the literature. In contrast to BJTs, the Si/SiGe emitter/base heterojunction bandgap counter balances the doping level difference, resulting in enhancement of current gain by a factor of 10 or more from 300 to 77 K, and at the same time, to an increase of f_T by 30–60 per cent [20,60,61]. In the following, results of simulation of high-frequency figures of merit vs. temperature carried out on SiGe bipolar transistors with several base doping and activation annealing conditions are discussed. Cryogenic performance of a SiGe HBT technology is also presented.

The Gummel characteristics at 300 and 85 K are shown in Figure 6.25 for $V_{cb} = 0$ V. The base–emitter turn-on voltage increases with cooling, as expected, owing to the exponential decrease of the intrinsic carrier concentration with cooling. The base and emitter regions in this device are both doped above the Mott-transition and ensure that carrier freeze-out does not negatively impact the base or emitter resistance below 100 K. As can be seen in Figure 6.25, at 85 K this device is capable of very high current density operation (>25 mA/μm^2), and thus the high collector doping level effectively limits the impact of heterojunction barrier effects at low temperatures, which can be a key design issue for the cryogenic operation of SiGe HBTs [20]. The current gain increases monotonically with cooling, from 600 at 300 K, to 3800 at 85 K, as shown in Figure 6.26. Two mechanisms are responsible for this improvement with cooling: (1) the Ge-induced band offset in this device (exponentially) increases the current gain with cooling, and (2) the heavily doped base region partially offsets the

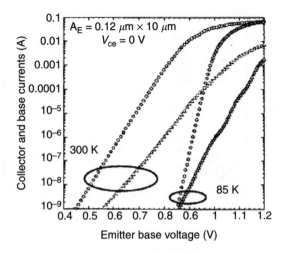

Figure 6.25   *Gummel characteristics at 300 and 85 K for a 0.12 × 10.0 μm SiGe HBT [After Banerjee et al., IEEE BCTM Proc., 2003(171–173). ©2003 IEEE]*

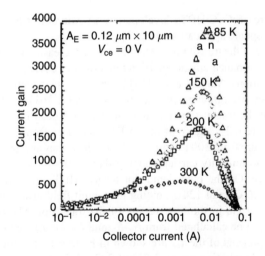

Figure 6.26   *Current gain as a function of bias current and temperature for a 0.12 × 10.0 μm SiGe HBT [After Banerjee et al., IEEE BCTM Proc., 2003(171–173). ©2003 IEEE]*

doping-induced bandgap narrowing associated with the emitter region. The strong decrease in the current gain above its peak value at 85 K is associated with the 'Ge-grading' effect, but the current gain remains above 2000 at 85 K at the current density at which peak f_T is reached, effectively minimising any emitter charge storage

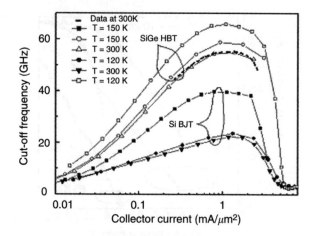

*Figure 6.27   Comparison of simulated and experimental (dashed line) cut-off fre-
quency vs. collector current at room temperature and performance
enhancement at low temperature [After S. Mandal et al., Proc. MIEL,
2004(315–318). ©2004 IEEE]*

at low temperatures. SiGeC HBTs developed by ST Microelectronics have also been
evaluated for cryogenic applications [62,63].

Understanding of the transit time behaviour with decreasing temperature is essential to get the performance advantage of heterostructure devices at low temperature
for analog applications. Mandal *et al.* [59] have shown using simulation, the possible high-frequency performance enhancement of a SiGe HBTs at low temperature.
The behaviour of the transit time components has also been investigated in detail.
As shown in Figure 6.27, the f_T of the Si-BJT increases from 21 to 38 GHz with
reducing temperature from 300 to 150 K and then decreases as temperature is further
decreased owing to minority carrier trapping that occurs in the neutral base region
at low temperature. On the other hand, for a SiGe HBT, the f_T increases from 40
GHz at 300 K to 66 GHz at 120 K, owing to in-built field inside the base resulting
from the grading of Ge in the base region of the transistor. Maximum oscillation
frequency, f_{max}, can be calculated from f_T, R_b and C_b for a given device geometry.
The various components of the τ_{ec} are shown in Figures 6.28 through 6.31. Transit
time components distribution for Si-BJT and SiGe HBT at different temperatures is
shown in Figures 6.32 and 6.33. It is observed that τ_b is the main component of the
total delay time in both Si and SiGe HBTs even at low temperature.

6.9   Summary

In this chapter, the design principle of heterostructure devices has been discussed.
Critical issues for numerical modelling of heterostructure devices have been

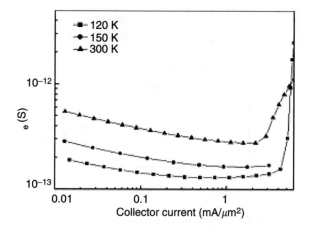

Figure 6.28   Simulated emitter transit time vs. collector current as a function of temperature [After S. Mandal et al., Proc. MIEL, 2004(315–318). ©2004 IEEE]

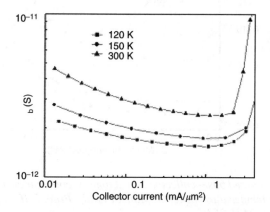

Figure 6.29   Simulated base transit time vs. collector current as a function of temperature [After S. Mandal et al., Proc. MIEL, 2004(315–318). ©2004 IEEE]

discussed. Several examples of simulation of devices employing SiGe HBTs have been considered. Attention has been given to simulation of various advanced technologies leading to high cut-off frequency and/or low transit time. Good agreement between simulation and measurement provides confidence in the use of device simulation for future device development. Transit time analysis of a SiGe HBT using drift-diffusion (DD), hydrodynamic (HD) and full-band Monte Carlo simulations has

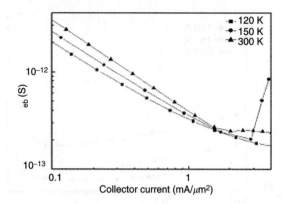

Figure 6.30   Simulated emitter–base transit time vs. collector current as a function of temperature [After S. Mandal et al., Proc. MIEL, 2004(315–318). ©2004 IEEE]

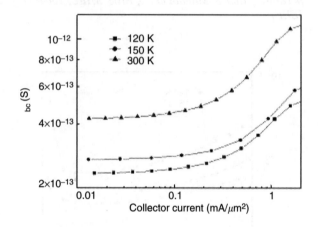

Figure 6.31   Simulated base–collector transit time vs. collector current as a function of temperature [After S. Mandal et al., Proc. MIEL, 2004(315–318). ©2004 IEEE]

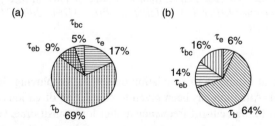

Figure 6.32   Transit time component distribution of (a) Si-BJT and (b) SiGe HBT at 150 K [After S. Mandal et al., Proc. MIEL, 2004(315–318). ©2004 IEEE]

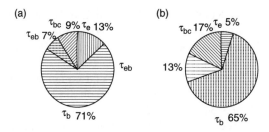

Figure 6.33 *Transit time component distribution of an SiGe HBT at tempera-ture (a) 300 K and (b) 120 K [After S. Mandal et al., Proc. MIEL, 2004(315–318). ©2004 IEEE]*

been discussed in detail. Detail transit time analysis at low temperature for a SiGe HBT has also been performed, which shows its applicability in low-temperature electronics.

References

1 H. Kroemer, 'Der Drifttransistor', *Naturwissensch.*, vol. 40, p. 5789, 1953.
2 H. Kroemer, 'Zur Theorie des Diffusions und des Drifttransistors: III Dimension-ierungsfragen', *Arch. Elektr. Ubertragung*, vol. 8, pp. 499–504, 1954.
3 H. Kroemer, 'Nobel Lecture: Quasielectric fields and band offsets: teaching electrons new tricks', *Rev. Mod. Phys.*, vol. 73, pp. 783–793, 2001.
4 W. R. Frensley, and H. Kroemer, 'Theory of energy-band lineups at an abrupt semiconductor heterojunction', *Phys. Rev. B*, vol. 16, pp. 2642–2652, 1977.
5 H. Kroemer, 'Quasielectric and quasimagnetic fields in nonuniform semicon-ductors. Republished from symposium on the role of solid state phenomena in electric circuits, Polytechnic Institute Brooklyn, NY, 1957, pp. 143–153', *RCA Rev.*, vol. 18, pp. 332–342, 1957.
6 W. Shockley. US Patent No. 2,569,347, 1951.
7 H. Kroemer, 'Theory of a wide-gap emitter for transistors', *Proc. IRE*, vol. 45, pp. 1535–1537, 1957.
8 H. Kroemer, 'Heterojunction Bipolar Transistors and Integrated Circuits', in *Proc. IEEE*, vol. 70, pp. 13–25, 1982.
9 H. Kroemer, 'Unpublished'.
10 H. Diedrich, and K. Jotten, in *Proc. of the Colloque International sur les Dispositifs a'Semiconducteurs, Paris (Editions Chiron, Paris)*, p. 330, 1961.
11 G. Abstreiter, 'Electronic properties of Si/SiGe/Ge heterostructures', *Physica Scripta*, vol. T-68, pp. 68–71, 1996.
12 U. Konig, 'Future applications of heterostructures', *Physica Scripta*, vol. T-68, pp. 90–101, 1996.
13 R. L. Anderson, 'Germanium-Gallium-Arsenide Heterojunctions', *IBM J. Res. Devices*, vol. 4, pp. 283–287, 1960.

14 W. A. Harrison, E. A. Kraut, J. R. Waldrop, and R. W. Grant, 'Polar heterojunction interfaces', *Phys. Rev. B*, vol. 18, pp. 4402–4410, 1978.

15 H. Kroemer, 'Polar-on-nonpolar epitaxy', *J. Cryst. Growth*, vol. 81, pp. 193–204, 1987.

16 S. L. Wright, M. Inada, and H. Kroemer, 'Polar-on-nonpolar epitaxy: sublattice ordering in the nucleation and growth of GaP on Si (211) surfaces', *J. Vac. Sci. Technol.*, vol. 21, pp. 534–539, 1982.

17 S. L. Wright, H. Kroemer, and M. Inada, 'Molecular-Beam epitaxial growth of GaP on Si', *J. Appl. Phys.*, vol. 55, pp. 2916–2927, 1984.

18 H. Kroemer, 'Heterostructure bipolar transistors: What should we build?', *J. Vac. Sci. Technol. B*, vol. 1, pp. 126–130, 1983.

19 C. K. Maiti, and G. A. Armstrong, *Applications of Silicon-Germanium Heterostructure Devices*. Inst. of Physics Pub., UK, 2001.

20 J. D. Cressler, J. H. Comfort, E. F. Crabbe, G. L. Patton, J. M. C. Strok, J. Y. C. Sun, *et al.*, 'On the profile design and optimization of epitaxial Si- and SiGe-base bipolar technology for 77 K applications-part I: Transistor d.c. design considerations', *IEEE Trans. Electron Devices*, vol. 40, pp. 525–541, 1993.

21 J. D. Cressler, E. F. Crabbe, J. H. Comfort, J. M. C. Strok, and J. Y. C. Sun, 'On the profile design and optimization of epitaxial Si- and SiGe-base bipolar technology for 77 K Applications-Part II: Circuit performance issues', *IEEE Trans. Electron Devices*, vol. 40, pp. 542–555, 1993.

22 J. D. Cressler, 'Silicon bipolar transistor: a viable candidate for high speed applications at liquid nitrogen temperature', *Cryogenics*, vol. 30, pp. 1036–1047, 1990.

23 J. D. Cressler, and G. Niu, *Silicon-Germanium Heterojunction Bipolar Transistors*. Artech House, London, 2003.

24 P. Ashburn, *SiGe Heterojunction Bipolar Transistors*. John Wiley & Sons, Ltd., Chichester, 2003.

25 S. S. Iyer, G. L. Patton, S. S. Delage, S. Tiwari, and J. M. C. Stork, 'Silicon-germanium base heterojunction bipolar transistors by molecular beam epitaxy', in *IEEE IEDM Tech. Dig.*, pp. 874–876, 1987.

26 D. J. Harame, J. H. Comfort, J. D. Cressler, E. F. Crabbe, J. Y. C. Sun, B. S. Meyerson, *et al.*, 'Si/SiGe epitaxial-base transistors-part I: Materials, physics, and circuits', *IEEE Trans. Electron Devices*, vol. 42, pp. 455–467, 1995.

27 J.-S. Rieh, B. Jagannathan, H. Chen, K. Schonenberg, S.-J. Jeng, M. Khater, *et al.*, 'Performance and design considerations for high speed SiGe HBTs of $f_T/f_{max} = 375\,GHz/210\,GHz$', in *Int. Conf. Indium Phosphide and Related Materials*, pp. 374–377, 2003.

28 G. L. Patton, J. H. Comfort, B. S. Meyerson, E. F. Crabbe, G. J. Scilla, E. D. Fresart, *et al.*, '75-GHz f_T SiGe-base heterojunction transistors', *IEEE Electron. Device Lett.*, vol. 11, pp. 171–173, 1990.

29 J. H. Comfort, G. L. Patton, J. D. Cressler, W. Lee, E. F. Crabbe, B. S. Meyerson, *et al.*, 'Profile leverage in self-aligned epitaxial Si or SiGe base bipolar technology', in *IEEE IEDM Tech. Dig.*, pp. 21–24, 1990.

30 D. L. Harame, E. F. Crabbe, J. D. Cressler, J. H. Comfort, J. Y.-C. Sun, S. R. Stiffler, *et al.*, 'A high-performance epitaxial SiGe-base ECL BiCMOS technology', in *IEEE IEDM Tech. Dig.*, pp. 19–22, 1992.

31 D. L. Harame, J. M. C. Stork, B. S. Meyerson, K. Y.-J. Hsu, J. Cotte, K. A. Jenkins, *et al.*, 'Optimization of SiGe HBT technology for high speed analog and mixed-signal applications', in *IEEE IEDM Tech. Dig.*, pp. 71–74, 1993.

32 E. Kasper, A. Gruhle, and H. Kibbel, 'High speed SiGe-HBT with very low sheet resistivity', in *IEEE IEDM Tech. Dig.*, pp. 79–82, 1993.

33 E. F. Crabbe, B. Meyerson, J. Stork, and D. Harame, 'Vertical profile optimization of very high frequency epitaxial Si- and SiGe-base bipolar transistors', in *IEEE IEDM Tech. Dig.*, pp. 83–86, 1993.

34 A. Schuppen, A. Gruhle, U. Erben, H. Kibbel, and U. Konig, 'Multi emitter finger SiGe-HBTs with F_{max} upto 120 GHz', in *IEEE IEDM Tech. Dig.*, pp. 377–380, 1994.

35 A. Joseph, D. Coolbaugh, M. Zierak, R. Wuthrich, P. Geiss, Z. He, X. Liu, *et al.*, 'A 0.18 μm BiCMOS technology featuring 120/100 GHz (f_T/f_{MAX}) HBT and ASIC-compatible CMOS using copper interconnect', in *IEEE BCTM Proc.*, pp. 143–146, 2001.

36 G. Freeman, Y. Kwark, M. Meghelli, S. Zier, A. Rylyakov, M. Sorna, *et al.*, '40 Gbit/sec circuits built from a 120 GHz f_T SiGe technology', in *IEEE 23rd GaAs IC Symp. Dig.*, pp. 89–92, 2001.

37 S. S. Iyer, G. L. Patton, J. M. C. Stork, B. S. Meyerson, and D. L. Harame, 'Heterojunction bipolar transistor using Si-Ge alloys', *IEEE Trans. Electron Devices*, vol. 36, pp. 2043–2063, 1989.

38 J. W. Matthews, and A. E. Blakeslee, 'Defects in epitaxial multilayers - I. Misfit dislocations', *J. Cryst. Growth*, vol. 27, pp. 118–125, 1974.

39 J. C. Bean, L. C. Feldman, A. T. Fiory, S. Nakahara, and I. K. Robinson, 'Ge_xSi_{1-x}/Si strained layer superlattice growth by molecular beam epitaxy', *J. Vac. Sci. Technol. A*, vol. 2, pp. 436–440, 1984.

40 J. H. van der Merwe, 'Crystal interfaces. Part I. Semi-infinite crystals (see also erratum p. 3420 (1963)', *J. Appl. Phys.*, vol. 34, pp. 117–122, 1963.

41 S. R. Stiffler, J. H. Comfort, C. L. Stanis, D. L. Harame, E. de Fresart, and B. S. Meyerson, 'The thermal stability of SiGe films deposited by ultrahigh-vacuum chemical vapor deposition', *J. Appl. Phys.*, vol. 70, pp. 1416–1420, 1991.

42 G. L. Patton, S. S. Iyer, S. L. Delage, S. Tiwari, and J. M. C. Stork, 'Silicon-germanium base heterojunction bipolar transistors by molecular beam epitaxy', *IEEE Electron Device Lett.*, vol. 9, pp. 165–167, 1988.

43 C. A. King, J. L. Hoyt, C. M. Gronet, J. F. Gibbons, M. P. Scott, and J. Turner, '$Si/Si_{1-x}Ge_x$ heterojunction bipolar transistors produced by limited reaction processing', *IEEE Electron Device Lett.*, vol. EDL-10, pp. 52–54, 1989.

44 M. Hong, E. de Fresart, J. Steele, A. Zlotnicka, C. Stein, G. Tam, *et al.*, 'High-performance SiGe epitaxial base bipolar transistors produced by a

reduced-pressure CVD reactor', *IEEE Electron Device Lett.*, vol. 14, pp. 450–452, 1993.

45 B. S. Meyerson, 'UHV/CVD growth of Si and Si-Ge alloys: Chemistry, physics and device applications', *Proc. IEEE*, vol. 80, pp. 1592–1608, 1992.

46 J. M. Rollett, 'Stability and power-gain invariants of linear twoports', *IRE Trans. Circ. Th.*, vol. CT-9, pp. 29–32, 1962.

47 D. J. Roulston, *Bipolar Semiconductor Devices*. McGraw Hill, Singapore, 1990.

48 G. Niu, S. Zhang, J. D. Cressler, A. J. Joseph, J. S. Fairbanks, L. E. Larson, *et al.*, 'Noise modeling and SiGe profile design tradeoffs for RF applications [HBTs]', *IEEE Trans. Electron Devices*, vol. 47, pp. 2037–2044, 2000.

49 V. Palankovski, and S. Selberherr, 'Critical modeling issues of SiGe semiconductor devices', in *Proc. Symp. on Diagnostics and Yield: Advanced Silicon Devices and Technologies for ULSI Era*, pp. 1–11, 2003.

50 ISE Integrated Systems Engineering AG., Zurich, Switzerland, *Tech. Rep. DIOS-ISE, Release 6.0*, 1999.

51 DESSIS-ISE, Release 8.5, 'Integrated Systems Engineering AG, Zurich, Switzerland', 2003.

52 T. Binder, K. Dragosits, T. Grasser, R. Klima, M. Knaipp, H. Kosina, *et al.*, '*MINIMOS-NT User's Guide*', in Institut fur Mikroelektronik, Technische Universitat, Wien, 1998.

53 C. Jungemann, B. Neinhus, and B. Meinerzhagen, 'Comparative study of electron transit times evaluated by DD, HD, and MC device simulation for SiGe HBT', *IEEE Trans. Electron Devices*, vol. 48, pp. 2216–2220, 2001.

54 Z. Yu, Ricco, and R. Dutton, 'A comprehensive analytical and numerical model of polysilicon emitter contacts in bipolar transistors', *IEEE Trans. Electron Devices*, vol. ED-31, pp. 773–784, 1984.

55 M. M. Rieger, and P. Vogl, 'Electronic-band parameters in strained $Si_{1-x}Ge_x$ alloys on $Si_{1-y}Ge_y$ substrates', *Phys. Rev. B*, vol. 48, pp. 14276–14287, 1993.

56 F. Schaffler, 'High-mobility Si and Ge structures', *Semicond. Sci. Technol.*, vol. 12, pp. 1515–1549, 1997.

57 R. J. E. Hueting, J. W. Slotboom, A. Pruijmboom, W. B. de Boer, E. C. Timmering, and N. E. B. Cowern, 'On the optimization of SiGe-base bipolar transistors', *IEEE Trans. Electron Devices*, vol. 43, pp. 1518–1524, 1996.

58 R. Thoma, A. Emunds, B. Meinerzhagen, H. J. Peifer, and W. L. Engl, 'Hydrodynamic equations for semiconductors with nonparabolic bandstructures', *IEEE Trans. Electron Devices*, vol. 38, pp. 1343–1352, 1991.

59 S. K. Mandal, G. K. Marskole, K. S. Chari, and C. K. Maiti, 'Transit time components of a SiGe-HBT at low temperature', in *Proc. MIEL*, pp. 315–318, 2004.

60 N. Zerounian, F. Aniel, R. Adde, and A. Gruhle, 'SiGe heterojunction bipolar transistor with 213 GHz f_T at 77 K', *Electron. Lett.*, vol. 36, pp. 1076–1078, 2000.

61 B. Banerjee, S. Venkataraman, Y. Lu, S. Nuttinck, H. Deukhyoun, Y.-J. E. Chen, *et al.*, 'Cryogenic performance of a 200 GHz SiGe HBT technology', in *IEEE BCTM Proc.*, pp. 171–173, 2003.

62 H. Baudry, B. Martinet, C. Fellous, O. Kermarrec, Y. Campidelli, M. Laurens, *et al.*, 'High performance 0.25 μm SiGe and SiGe:C HBTs using non selective epitaxy', in *IEEE BCTM Proc.*, pp. 52–55, 2001.

63 N. Zerounian, M. Rodriguez, M. Enciso, F. Aniel, P. Chevalier, B. Martinet, *et al.*, 'Transit times of SiGe:C HBTs using nonselective base epitaxy', *Solid State Electron.*, vol. 48, pp. 1993–1999, 2004.

Chapter 7

SiGe/SiGeC HBT technology

Before 1990, heterojunction bipolar transistors were available only in III–V compound semiconductor technologies. The addition of Ge in Si to form strained-SiGe and the incorporation of SiGe in conventional Si BiCMOS technology has created a revolution in the semiconductor industry [1,2]. The Ge added to form high-performance heterojunction bipolar transistors can operate at speed much higher than standard silicon bipolar transistors. State-of-the-art SiGe HBTs have values of f_T beyond 350 GHz and f_{max} 270 GHz [3] and operate at speed previously attainable only with GaAs technology. The performance improvement of SiGe HBTs is due to energy bandgap lowering resulting from adding Ge to the base of the device, and the principle of operation has been described earlier. SiGe HBTs differ from III–V technologies in that the breakdown voltage may be set by collector implants and not predetermined by layer growth. Therefore, the SiGe HBTs with multiple breakdown voltages can be fabricated in the same process by varying the collector implants.

In the high-speed arena, SiGe HBTs today are surpassing even the fastest III–V production devices. The key to this achievement is the superior parasitic control technology available to SiGe device designers, compared to what is available to III–V device designers. SiGe HBTs have the advantage of being built in the existing silicon fabrication using standard silicon production equipment. Research is under way on SiGe heterojunction MOSFETs, which in turn is expected to revolutionise the future of CMOS technology as well.

SiGe HBTs have also been integrated into standard CMOS logic technologies, resulting in the integration of high-performance analog and RF circuits with CMOS logic leading to SiGe-BiCMOS technology, although there is conflicting CMOS/HBT thermal budget requirements. The CMOS devices are used primarily for integrating digital logic functions with high-speed bipolar analog circuits. This allows fully integrated system-on-a-chip products, with the CMOS performing the low-frequency baseband signal processing.

Semiconductor chips used in wireless applications such as cellular phones, wireless networks and global positioning systems (GPS) must be inexpensive.

The performance and integration capability of SiGe technology has enabled the production of a wide range of new products for wireless and wired communications. Early SiGe products include chip sets for wireless cellular handsets and base stations [4,5], wireless local area network (WLAN), high-speed, high-capacity wired network applications and many other applications.

SiGe BiCMOS technology has become a popular choice for broadband applications because of the availability of the high-speed SiGe HBTs and the potential for highly integrated transceivers in robust silicon-based processes. However, the power consumption of SiGe high-speed building blocks [6] raises concerns about heat removal in a packaged environment, which may ultimately limit the achievable integration level for single-chip transceivers. Thus it is necessary to have a BiCMOS topology that allows operation from lower supply voltages without speed degradation.

Despite high cut-off frequencies in modern SiGe HBTs, the bandwidth of high-speed bipolar digital circuits is hindered by device parasitics such as the base resistance and the collector–base capacitance. For a given technology, the RC product cannot be reduced, since any increase in total emitter length to minimise base resistance results in a commensurate rise in base–collector capacitance.

BiCMOS technology development started in the early 1980s as a way of significantly enhancing digital performance. Analog BiCMOS processes were limited in applications, because of the high supply voltages prevalent in analog circuit design and the abundance of bipolar processes [7]. From a system point of view, analog BiCMOS implies a different focus than digital BiCMOS. Wireless applications are typically high-volume products and require simpler processes. Optimisation requires less emphasis on peak transistor speed and more emphasis on speed at low power, linearity, noise and passive elements. Current BiCMOS technology must satisfy the bipolar performance requirements as well as demonstrate full compatibility with existing libraries and methodologies [2]. For recent research activities in developing new silicon-based RF integrated circuit (RFIC) technologies including processes, active and passive devices, modelling and simulation, circuit applications and ESD protection, which are critical to realising true RF system-on-a-chip (SoCs), the reader may refer to the special issue on integrated circuits technologies for RF circuit applications [8].

Similar to Moore's law that predicts a two-fold increase in density every 18 months in CMOS logic and microprocessor technologies, there has been a similar performance trend in SiGe technologies where the SiGe HBT performance has approximately doubled every 2–3 years. As shown in Figure 7.1, cut-off frequency has increased from 50 GHz in 1997 to 350 GHz in 2003 [3, 9–13]. Figure 7.1 also shows the trend for SiGe BiCMOS technologies focused on applications requiring f_T in the range of 50–100 GHz [14–16]. These applications include wireless handsets, WLAN, GPS and storage. Today's state-of-the-art SiGe BiCMOS technology combines high-speed, low-noise SiGe HBTs, with aggressively scaled Si-CMOS, and offers on-chip passive components, to produce low-cost, high-yield, Si-based SoC solutions for applications in the 1–40 GHz range. Enhancements in passives enable system cost reductions by reducing the number of off-chip components and reducing chip size through passives scaling. Enhancements in passives that are more fully described in Chapter 11 include

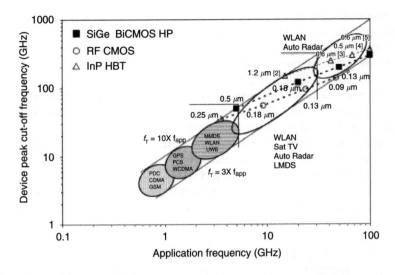

Figure 7.1   SiGe HBT f_T performance trends. RF application spectrum overlaid with technology performance. An acceptable range of f_T for each application is chosen as 3–10X [After A. J. Joseph et al., Proc. IEEE, Vol. 93, 2005(1539–1558). ©2005 IEEE]

increased MIM capacitance per area, improved inductor Q, improved varactors and tighter resistor tolerance. For a comprehensive review on SiGe BiCMOS technology for RF circuit applications, the reader may refer to the paper by Racanelli and Kempf [17], in which the authors have reviewed the state-of-the-art SiGe BiCMOS technology and future development trends for highly integrated RF platform solutions.

In this chapter, the challenges that were needed to overcome both to enable the growth of a SiGe epitaxial layer and to integrate the process with standard bipolar and CMOS processes are mentioned. This will lead to a discussion of the current state-of-the-art in SiGe processing technology along with a prediction of challenges anticipated with future, more advanced, SiGe technologies. The discussion will be based mostly on the information from IBM's 0.5, 0.24 and 0.18 μm SiGe BiCMOS technologies [18–20].

In the following, SiGe heterojunction bipolar technology, with particular emphasis on the influence of materials issues on the technology performance, is described. The discussion will be limited to only self-aligned, fully integrated, Si-technology-compatible SiGe HBT technologies that have been reported in the literature (see References 21 and 22 for a review on early SiGe technology). The history of development of SiGe HBTs at IBM has been reviewed by Harame and Meyerson [23], while the latest status of SiGe BiCMOS technology, modelling and design have been addressed in a special issue of the proceedings of the IEEE [24].

The first SiGe HBTs were mesa-type transistors fabricated using molecular beam epitaxy (MBE) and used a low-temperature processing. The collector, base and emitter layers were epitaxially grown by MBE without breaking vacuum. Mesa-defined transistor structures are fabricated using dry-etching techniques. A six-fold increase in the collector current was measured for a device with 12 per cent uniform germanium across a 100-nm basewidth, which corresponded to a total bandgap shrinkage of 45 meV [25]. Although the thermal cycle was kept relatively low (780 °C), the I–V characteristics (Gummel plot) were very non-ideal and essentially unusable owing to poor material quality and possible relaxation. However, eventual development of a manufacturable SiGe epitaxial (EPI) growth process was the fundamental breakthrough that underpinned SiGe technology evolution [26].

Conventional blanket epitaxial techniques use high temperature for surface cleaning and growth. A low-temperature epitaxial process is needed for integration of base epitaxial growth with SiGe processing. The first SiGe HBTs were fabricated using the conventional bipolar structures on patterned substrates and used an in situ doped graded SiGe base profile, readily accomplished with the CVD technique. The processing and structures employed the same isolation and polySi emitter processes found in conventional bipolar technology, setting a new and important direction for all SiGe epitaxial base work and fully compatible with the silicon technology.

A non-self-aligned transistor structure, in which the emitter opening was spaced away from extrinsic base or linkup regions, was used to simplify extracting vertical profile information from the resultant devices. Many process steps were borrowed from the conventional double-polySi self-aligned transistor technology. However, as the SiGe technology cannot satisfy all the requirements for specific applications, a series of derivative technologies that are based on the parent IBM's high-performance (HP) technology have been developed. These roughly fall into three groups,

1.   updating the older technology with improved passives and additional devices;
2.   low complexity, lower-cost (and lower performance) technologies for consumer markets such as wireless;
3.   high breakdown technologies for specialised applications. Process simplification can be achieved by trading performance for cost. This performance degradation, however, can be minimised through device scaling.

IBM's technology for SiGe is very mature since they were the first to commercialise SiGe technology, have published widely and offer several distinct versions of SiGe. IBM has also reported the strategy of having an SiGe BiCMOS roadmap with their mainstream HP and derivative components for specialised market segments. The demanding performance and integration requirements for both the digital and the interface optoelectronic components define the requirements for the HP SiGe BiCMOS roadmap.

Early IBM products demonstrated the performance and integration capability of SiGe BiCMOS leading to acceptance and large increase in the number of networking chips mostly at the 0.18 μm node. Advanced SiGe HBTs require a combination of vertical and lateral scaling [12, 27].

An advanced CMOS roadmap has been realised by incorporating base CMOS technology into the BiCMOS, which is facilitated by a base-after-gate process flow in IBM's 200 GHz f_T and 285 GHz f_{max} 0.13 μm SiGe BiCMOS technology, which is ideal for 40 Gb/s SOC implementation [12, 14]. SiGe technology from several manufacturers can be found in References 28–32. BiCMOS technologies combining the high RF performance of SiGe-based HBTs with the high level of CMOS integration are increasingly recognised as key enablers for single chip solutions for broadband and wireless communication systems. Status and future trends of 'C-free' SiGe BiCMOS technology have been reviewed in Reference 33. BiCMOS integration of C-doped SiGe (SiGeC) HBTs, which were initiated in the middle of the 1990s, is also addressed [34].

During the 1990s, two main approaches were used for the fabrication of SiGe HBTs [21, 22]. The first used the differential epitaxy that was put together to make the first epitaxial base transistor by modifying an existing bipolar structure and the second selective epitaxy [35, 36]. For the differential epitaxy process, the p^+SiGe base layer is grown after oxide isolation formation so that single-crystal material is formed where the silicon collector is exposed and polycrystalline material is formed over the oxide isolation. Several companies including IBM [37, 38] have pioneered the use of differential epitaxy in SiGe HBTs.

The differential epitaxy is carried out after the creation of shallow trench or LOCOS oxide isolation regions. Single crystal SiGe is produced where the n-Si collector is exposed and polycrystalline SiGe is grown over the field oxide. To achieve differential epitaxy, it is necessary to have good nucleation of polycrystalline SiGe on the isolation oxide around the perimeter of the base window. This is generally achieved by using silane as the silicon source gas during the SiGe growth. The Ge profile is often graded across the base to give a bandgap that decreases from emitter to collector. This gives a quasi-drift field, which aids carrier transport across the base, reduces the base transit time and enhances the value of f_T.

As an alternative to differential epitaxy, several companies have pioneered selective epitaxy [35, 36, 39]. In this approach, an overhanging p^+-polySi extrinsic base is created in an emitter window, prior to base epitaxy.

The hydrogen passivation preclean and UHVCVD deposition process, developed by IBM, meets most requirements of low-temperature defect-free SiGe epitaxial layers [40]. The low-temperature epitaxial (LTE) process produces very low defect epitaxial layer that is necessary for achieving high device yield. It also provides the capability for precise doping profile control that is necessary for HBT device performance [41–43]. However, the device yield has always been an issue with silicon bipolar and BiCMOS technologies. The yield is usually limited by dislocations in the silicon crystal that cause emitter to base, collector to base and/or collector to emitter shorts. The possibility of creating these dislocations is increased with SiGe HBTs, due to the potential for defect formation during EPI growth and defect formation due to the strain in the Ge layer.

As described above, the LTE process produces a relatively defect-free epitaxial layer and does not contribute to yield loss. Minimising the thermal cycles post epitaxial growth helps reduce the possibility of forming defects due to the strained layer. The primary cause for defect generation has been found to be related to stress from the edge of the shallow trench isolation (STI). The STI effect on HBT yield has been studied extensively with 0.5 μm generation technology. TEM analysis of low yielding samples showed dislocation defects propagating from either the top or bottom corners of the STI adjacent to the HBT. These results suggested that the HBT was not in stress equilibrium with the surrounding STI. It was found that the shape of the STI adjacent to the HBT affected how the transition from EPI silicon to polySi in the SiGe epitaxial film formed. The shape of this transition region has a big influence on the stress in the HBT. Optimisation of the STI shape reduced the stress, leading to fewer dislocations and significantly higher HBT yield.

7.1   SiGe-BiCMOS technology

BiCMOS technologies have the ability to integrate analog bipolar functions with CMOS logic on the same die. Coupled with the manufacturability of passive elements, BiCMOS silicon or SiGe technologies have potentials for designs that require a high level of RF and logic integration. From a system point of view, analog BiCMOS implies a different focus than digital BiCMOS. Logic gates are not constructed from Bipolar and CMOS devices, but rather the system is divided into separate analog Bipolar and CMOS digital logic sections. This has important methodology implications. The high-frequency sections require RF design skills and tools and tend to be relatively small in terms of area and total transistor count. The lower-frequency CMOS logic sections tend to be large in area and transistor count and require high-level design tools using hardware description language.

Early BiCMOS integration emphasised processes with only 3–4 additional masking steps to the CMOS processes and low-performance non-self-aligned npn transistors. Modern BiCMOS has high-performance self-aligned bipolar transistors that are incorporated in a modular fashion because modular process design is the only practical way to satisfy the CMOS, Bipolar and Interconnect requirements as well as the increased integration levels. The cost of modular design is significantly higher mask counts.

Another key issue for BiCMOS is lateral and vertical scaling of the bipolar transistor to meet the requirements of future applications. At low currents, device performance is parasitic capacitance limited, at moderate currents it is base transit time limited, and at high currents it is base pushout and resistance limited. Analog circuits have similar dependencies, and therefore it is crucial to reduce parasitics for low power operation. Vertical profile scaling for conventional ion-implanted base bipolar is much more problematic, because it largely consists of reducing the basewidth by decreasing the base implant energy. Because of dopant channelling during implantation, the implanted profile is difficult to control and very sensitive to implant energy. Therefore, it is important to use low-energy implantation to reduce basewidth. However, decreasing the energy of the implant results in higher surface concentrations, which results in heavily doped emitter–base junctions with high electric fields. These high electric fields are a reliability concern that is usually characterised by current gain degradation during reverse emitter–base bias stress [44]. Epitaxial base technology provides a way out of bipolar's vertical scaling dilemma by depositing the base dopant setback from the surface with a silicon spacer [43, 45].

Many BiCMOS technology applications require additional active devices beyond the npn transistors, p- and n-MOSFETs that are naturally part of the process. Circuit designers often require a pnp transistor for compatibility with circuits and circuit design techniques previously implemented in older bipolar technologies. Substrate pnp transistors are usually available but these are not useful for current mirrors. A bulk lateral pnp using the existing p-MOSFET wells and S/D implants is also available, but the transistor may have insufficient (low) current gain. It is difficult to simultaneously optimise a lateral pnp, p-MOSFET and npn bipolar with the same process. Lateral pnp optimisation requires deep junctions, thick epitaxy – which would degrade the npn bipolar and high surface well concentrations to prevent depletion of the neutral base. pFET optimisation requires shallow junctions, reduced well doping at the surface and moderate doping in the bulk to reduce short channel effects.

For smart power applications high-voltage LDMOS transistors have been added to existing BiCMOS processes [46, 47]. These devices are constructed by bringing the gate polySi over the LOCOS field oxide and dropping the voltage across a lightly doped drift region under the field oxide. The bipolar n^+ subcollector may be placed under the n-well in high-side driver LDMOS devices to prevent vertical punch-through and the heavily doped n^+ reach through is used for guarding to limit hole injection into the substrate [48].

In the conventional design methodology, communication systems applications involve a combination of GaAs, BiCMOS and CMOS chips. The requirements for monolithic wireless chips emphasise technology with high-Q passives (inductors, varactors and capacitors) in addition to the active devices, compact models with accurate noise figure and distortion analyses, signal integrity analysis and RLC parasitic extraction. Designers perform multiple design passes on each of the chips separately. For active discrete components, GaAs has so far been preferred over SiGe because it was a higher performance. However, SiGe BiCMOS technology allow now an entire multi-technology chip set [49–54]. Resistors are often fabricated by using existing implants in the process, for example, n-well, p-well, n^+ subcollector, n^+ reach through, p^+ extrinsic base polySi, n^+ emitter polySi, and so on. Circuit designers often prefer resistance values outside the range of existing layers in the process. A challenge for designers is to use the existing values in a BiCMOS process for circuit designs.

Approaches to generate high-Q inductors are as follows:

1. Reduce resistance of the coil.
2. Make the dielectric layer underneath the coil as thick as possible.
3. Reduce substrate loss.
4. If possible, connect the inductor differentially to reduce capacitance to ground.

Fabricating the inductor coils in a higher level of metallisation improves Q but it requires an additional metal layer and is not cost effective. Using a higher resistivity substrate can reduce substrate loss but excessively resistive substrate makes CMOS devices more prone to latch-up. Alternatively, deep trenches underneath the inductor can be used to reduce substrate losses. Another interesting technique to improve inductor Q is to connect the inductor differentially.

While there is a great diversity of HP BiCMOS process integration approaches, they all have similar objectives and problems and have used similar solutions. For a BiCMOS process flow design, one needs to consider the following: substrate selection, buried layers and epitaxial growth, collector isolation, wells, field oxide isolation, reach through, V_t and bipolar base implants, gate oxidation and polySi deposition as process modules common to all BiCMOS processes. After gate oxidation the decision about the sequence and bipolar structure used in the process is made. The process flows and approaches, however, may diverge. The structure, sequence, and thermal cycle must all be selected to satisfy the requirements for a particular application. A general process flow will be presented and used to highlight the most common approaches and process flows as well as significant issues for these approaches.

7.2   IBM SiGe-BiCMOS technology

The development of RF/analog mixed-signal methodology at IBM dates back to the support for the early bipolar technology used in bipolar-based mainframes and as an outgrowth of the analog BiCMOS processes. An excellent review on the evolution of IBM's SiGe performance and minimum lithographic feature size together with some derivative technologies may be found in references [55, 56]. The HBT cut-off frequency f_T has improved from 47 GHz in the 0.5 μm generation to 210 GHz in the 0.13 μm generation. The pace of development continues unabated, and there are no apparent barriers to scaling the SiGe HBTs beyond 350 GHz [57, 58].

IBM has four SiGe BiCMOS technologies and one CMOS RF technology qualified for high volume production [59–63]. The key technology characteristics for four generations of SiGe Bipolar/BiCMOS technology from IBM is shown in Table 7.1. The main purpose of the CMOS devices in a BiCMOS technology is to provide integrated logic functionality. SiGe HBT characteristics across four generations of BiCMOS technology is shown in Table 7.2. Incorporation of SiGe bipolar technology leads to many process differences, thus leading to significant device differences. However, the CMOS device characteristics in the BiCMOS technology closely match those of the base CMOS technology. Table 7.3 compares some of the key parameters of SiGe BiCMOS and RF CMOS technologies.

7.2.1   Base-during-gate process

The key issue when defining a SiGe BiCMOS process involves choosing an integration method that minimises the interaction of the HBT and CMOS processing steps. IBM's 0.5 μm SiGe BiCMOS technologies [20, 33, 55] utilized a 'base-during-gate' integration method. This was also referred to as base=gate because the extrinsic base of the bipolar was the same as the gate stack of the CMOS. With this approach, the growth and patterning of the npn SiGe epi-base also forms the FET polySi gates. Figure 7.2 shows the main process flow for the base CMOS process (known as the CMOS backbone) with bipolar and analog element steps that feed into the CMOS

Table 7.1 Key technology characteristics for four generations of SiGe Bipolar/ BiCMOS technology from IBM [After D. Harame et al., Proc. 4th IEEE Intl. Caracas Conf. on Devices, Circuits and Systems, 2002(D052-1- D052-17).]

	BiCMOS5HP	BiCMOS7HP	BiCMOS8HP
Lithography (μm)	0.5	0.18	0.18/0.13
Emitter width (μm)	0.42	0.2	0.12
npn(HP/HB)f_T (GHz)	47/27	120/30	210/—
npn(HP/HB)f_{max} (MAG, GHz)	65/55	100/—	~160/—
npn density	1.0X	2.0X	2.0X
CMOS supply (V)	3.3	1.8/2.5/3.3	NA
CMOS power (μW/MHz/gate)	0.3	0.03	NA
CMOS gate delay (ps)	180	33	NA
Rel. CMOS density (High-perf/I/O)	1X	7.5X/2X	NA
BEOL metal	Al	Cu	Cu
BEOL current – M1	1X	1.5X	1.5X
HBT structure	ETX	ETX	REB
Integration scheme	Gate during base	Base after gate	Base after gate

Table 7.2 HBT characteristics across generations of BiCMOS technology [After D. Harame, Lecture notes.]

	Units	5HP	5HPE	6HP	7HP	8HP
npn isolation		Deep trench	Shallow trench	Deep trench	Deep trench	Deep trench
Peak f_T	GHz	50	45	50	120	210
Peak f_{max}	GHz	65	50	65	100	185
f_T 30 μA	GHz	15	19	19	25	50
I_c at peak f_T	mA	0.6	0.5	0.5	1.0	1.0
BV_{ceo}	V	3.3	3.3	3.3	2.0	1.9
BV_{cbo}	V	10.5	10.5	10.5	6.5	6

backbone. In the 0.5 μm generation the base CMOS was used only for digital appli- cations and lacked analog modules. Therefore, additional processing modules for resistors and capacitors were added in the BiCMOS circuits. The integration method is key in ensuring that the addition of the SiGe HBT to the CMOS process has

Table 7.3   Comparison of thin-oxide CMOS parameters for SiGe BiCMOS and RF CMOS technologies [After D. Harame, Lecture notes.]

Parameter	Units	CMOS 6SFRF n-FET	CMOS 6SFRF p-FET	SiGe 7HP n-FET	SiGe 7HP p-FET
T_{ox} (thin)	nm	5.0	—	3.5	—
T_{ox} (thick)	nm	7.0/14.0	—	6.8	—
V_{supply}	V	2.5/3.3/6.5	—	1.8/3.3	—
L_{min}	μm	0.25	—	0.18	—
L_{eff}	μm	0.18	0.18	0.12	0.14
$V_{t,lin}$	mV	500	−500	355	−420
I_{dsat}	μA/μm	595	280	600	260
I_{off}	nA/μm	0.005	0.003	0.1	0.05
Body effect	—	0.23	0.27	0.19	0.25
$g_{m,sat}$	μS/μm	300	200	500	275
f_T	GHz	35	20	75	45
f_{max}	GHz	22	22	70	45

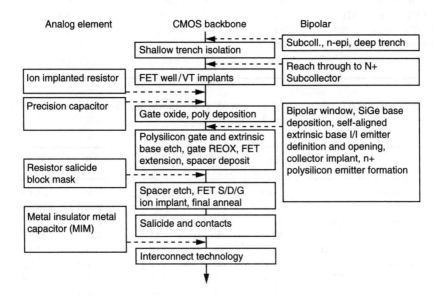

Figure 7.2   Schematic process flow diagram for the 0.5 μm Base-During-Gate (BDGate) production technology. The CMOS backbone represents the base CMOS process. Insertion points are shown for analog and bipolar process modules [After D. L. Harame et al., IEEE Trans. Electron Devices, Vol. 48, 2001(2575–2594). ©2001 IEEE]

no effect on the FET device characteristics. Similarly, it is important for bipolar device performance. The primary considerations are to minimise the impact of the CMOS processing steps on the thermal cycle interactions and the trade-off of overall process modularity vs. process sharing.

By integrating the various components of the bipolar transistor properly into the process flow (see Figure 7.2), the performance of both the CMOS and npn transistors is not degraded. The key is that the CMOS heat cycles occur before the bipolar device epitaxial base growth, so that the base dopant diffusion is kept to a minimum. To maintain CMOS ASIC compatibility, IBM qualified its commercial SiGe process at a low temperature for production in 1996, based on a method of depositing an epitaxial layer of SiGe alloy [19]. The alloy was co-deposited with the boron-doped intrinsic base using a novel form of LTE growth. The addition of the LTE process/tooling is the only requirement to convert an existing silicon BiCMOS fabrication to one that can also produce SiGe BiCMOS products. Sharing of layers and thermal cycles worked well by reducing the number of overall steps. However, changes in thermal cycles with subsequent generations of CMOS technologies created difficulties with this approach.

7.2.2 Base-after-gate process

To overcome thermal cycle incompatibility with IBM's 0.24 μm generation SiGe BiCMOS technology a 'base-after-gate' integration scheme was developed. In the base-during-gate SiGe BiCMOS technologies [9, 18, 19, 22], the npn base polySi is patterned at the same time as the FET gates; therefore, the npn base and emitter are subjected to the FET spacer and source/drain-activation thermal cycles. This approach could not be used for BiCMOS 6HP because the parent CMOS technology has large source/drain activation thermal cycles that would be detrimental to the HBT. In the 0.24 μm generation, the CMOS technology had high thermal cycles for the REOX (900 °C furnace oxidation) and arsenic-doped n-FET extension anneals, which prohibited the BDGate process from being used. Therefore, the BAGate integration approach for the 0.24 μm generation was developed to decouple the CMOS thermal cycle from the bipolar (see Figure 7.3).

In the 'base-after-gate' integration flow, major thermal cycles are performed prior to base deposition and a low thermal budget HBT module is used. BAGate technology has several advantages over BDGate technology. The CMOS is largely fabricated in one complete block before the SiGe base is deposited; therefore, the base CMOS technology steps can be completely copied without modification. Also, derivative technologies composed of high-voltage FETs in a similar or identical design-rule set can be easily developed from an existing production BiCMOS technology because the FET steps do not need to be modified. The bipolar is fabricated in a complete block after the CMOS without sharing any CMOS thermal cycles except for a final emitter drive-in RTA, making the BAGate an ideal process for the HBT. The concern with the BAGate process is that it is structurally a more complex process because the bipolar layers that are deposited over the CMOS topography must be completely removed.

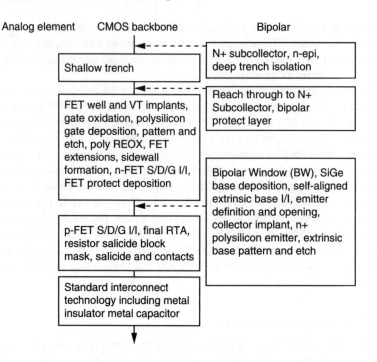

Figure 7.3   Schematic process flow diagram for the 0.25 µm and 0.18 µm Base-After-Gate (BAGate) production technology. The CMOS backbone represents the base CMOS process. Insertion points are shown for analog and bipolar process modules. Note the reduced number of analog and bipolar blocks compared to the BDGate process [After D. L. Harame et al., IEEE Trans. Electron Devices, Vol. 48, 2001(2575–2594). ©2001 IEEE]

This base-after-gate approach is not without its own challenges. All films deposited during the HBT module must be completely removed from over the FET gates without forming extraneous spacers. Advances in reactive ion etch (RIE) technology has made removal of these films possible. The RF and analog circuits that are built with SiGe BiCMOS technologies typically require additional components above and beyond the standard HBT and logic CMOS devices. These include analog FETs, precision resistors and capacitors, and high-Q inductors. The modularity of BAGate process supports rapidly developing derivatives and low-cost alternative BiCMOS technologies.

Examples of SiGe HBT technologies available from IBM with application specific focus are IBM's SiGe BiCMOS 5HPE, SiGe BiCMOS 7WL, and SiGe BiCMOS 8WL [15, 16]. These technologies are compared to their high performance counterparts SiGe 5HP, 7HP, and 8HP in Table 7.4 [9, 11, 13]. As seen in the Table 7.4 these technologies have lower npn f_T than their parent technologies.

Table 7.4   SiGe BiCMOS parameters by technology

	5HPE	7HP	7WL	8HP	8WL
Lithography (μm)	0.35	0.18	0.18	0.13	0.13
npn f_T/f_{max} (GHz)	44/80	120/100	60/110	200/280	100/200
npn BV_{ceo} (V)	3.3	2.0	3.3	1.77	2.5
CMOS L_g Drawn (μm)	0.36	0.18	0.18	0.13	0.13
CMOS V_{dd} (V)	3.3, 5.0	1.8, 2.5, 3.3	1.8, 2.5, 3.3	1.2, 2.5	1.2, 2.5, 3.3
Poly resistor R_s (Ω/sq.)	220	270	260	340	340
MIM cap. (fF/μm^2)	2.0	1.0	2.0	1.0	4.2
Peak inductor Q freq.	18	18	28	18	28
(2nH inductor)	(5 GHz)	(5 GHz)	(3 GHz)	(5 GHz)	(3 GHz)

7.3   SiGe HBTs on SOI

As described in Chapter 5, SOI technology improves performance over bulk CMOS technology by 25–35 per cent. SOI technology also brings power use advantages of 1.7 to 3 times – a key requirement for extending the battery life of small, hand-held devices. In the context of BiCMOS, the CMOS circuitry gains much advantage from SOI; namely simplified process flow and latch-up immunity together with enhanced MOSFET device performance. Bipolar transistors fabricated on SOI substrates offer the possibility of lower parasitic capacitances, but have a greater susceptibility to self-heating [64].

Silicon-on-insulator (SOI) technology also brings advantages of reduced collector capacitance and reduced cross-talk for mixed-signal circuits. In particular, the lower threshold voltage and low junction capacitance associated with SOI CMOS is particularly suited to low-voltage, low-power applications. BESOI (bond and etch back SOI) is preferred for HBT application due to the better Si and SiO$_2$ quality. A HBT on bonded wafer SOI has been reported by NEC for possible application in low-power, radio frequency electronics [65].

An SiGe technology for combining 0.2 μm self-aligned selective-epitaxial-growth (SEG) SiGe heterojunction bipolar transistors with CMOS transistors and high-quality passive elements has been developed for use in microwave wireless and optical communication systems [66]. The process sequence is shown in Figure 7.4. A standard bipolar process was carried out from the formation of the n$^+$ buried collector layer by ion implantation and diffusion of Sb to the trench isolations. Then, a 0.3-μm CMOS process module was inserted.

The technology has been applied to fabricate SiGe HBTs on SOI wafer based on a high-resistivity substrate. The fabrication process is almost completely compatible with the existing 0.2 μm BiCMOS process because of the essential similarity of the two processes. SiGe HBTs with STI and deep-trench isolations and Ti-salicide

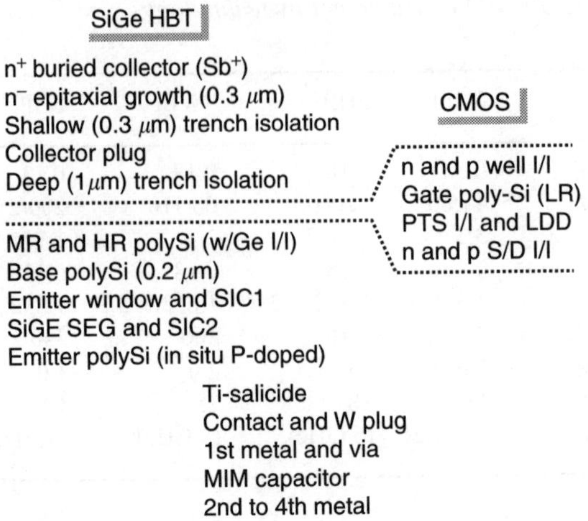

SiGe HBT

n⁺ buried collector (Sb⁺)
n⁻ epitaxial growth (0.3 μm)   CMOS
Shallow (0.3 μm) trench isolation
Collector plug   n and p well I/I
Deep (1μm) trench isolation   Gate poly-Si (LR)
....
                                                PTS I/I and LDD
MR and HR polySi (w/Ge I/I)   n and p S/D I/I
Base polySi (0.2 μm)
Emitter window and SIC1
SiGE SEG and SIC2
Emitter polySi (in situ P-doped)

Ti-salicide
Contact and W plug
1st metal and via
MIM capacitor
2nd to 4th metal

Figure 7.4   *Process sequence for 0.2 μm SOI/HRS SEG SiGe HBT/CMOS. A 0.3 μm*
CMOS process module was inserted after formation of the trench iso-
lations [After K. Washio et al., IEEE Trans. Electron Devices, Vol. 49,
2002(271–278). ©2002 IEEE]

electrodes exhibited high frequency and high-speed capabilities with $f_{max} = 180$ GHz
and an ECL-gate delay of 6.7 ps, along with good controllability, reliability and
high yield.

Hall *et al.* [67] have reported SiGe heterojunction transistors fabricated on bonded
wafer SOI substrates. The bonded wafer substrates incorporate polySi filled, deep
trenches for isolation. A novel selective and non-selective low-pressure chemical
vapour deposition growth process was used for the epitaxial layers. Experimental transistors exhibited good uniformity across the wafers and collector currents were found
to be ideal, showing the expected enhancement for the SiGe devices compared to Si.

SiGe on thick SOI is an attractive technological option for its low collector-to-
substrate capacitance, high-quality passive elements, and high immunity to substrate
noise [68]. However, these advantages come with the drawbacks of self-heating due to
the low thermal conductance of the buried oxide, and high wafer cost. Investigations
on the impact of self-heating in SiGe-on-SOI devices on transistor performance and
effect of introduction of thermal vias to reduce temperature rise have been performed
by Armstrong and Gamble [69].

A vertical bipolar transistor on SOI has been proposed and demonstrated by Cai
et al. [70, 71]. The transistor operates on the principle that the collector region is
fully depleted so that the charge carriers travel laterally towards the collector reach
through and contact after traversing the intrinsic base layer. The SOI silicon layer
thickness is comparable to that used in SOI CMOS, and no subcollector layer or deep
trench isolation are required. The transistor has been demonstrated in a polySi emitter
SiGe-base npn implementation on SOI with a 140 nm silicon layer. The fabricated
npn bipolar transistors exhibit a BV$_{ceo}$ of 4.2 V and a peak f_T of over 60 GHz.

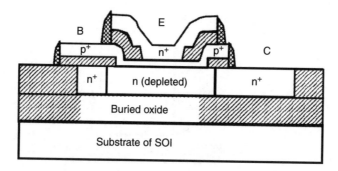

Figure 7.5 Schematic of a vertical bipolar transistor on thin SOI substrate [After J. Cai et al., VLSI Tech. Symp. Dig., 2002(172–173). ©2002 IEEE]

The challenge for SOI BiCMOS integration arises from a fundamental device architectural difference between bipolar and CMOS. Unlike CMOS transistors where current flows in the silicon layer from source to drain, in high-speed vertical bipolar transistors including all SiGe HBTs, electrons flow vertically from the emitter on the top to the collector on the bottom. In order to have low resistance access to the collector from a contact electrode on the top, a subcollector layer on the order of 1–$2~\mu$m thick with heavy doping is needed underneath the collector, which makes it too thick to build on thin silicon film SOI.

The proposed SiGe HBT on SOI structure is shown schematically in Figure 7.5 for an npn transistor. The collector region directly underneath the base is fully depleted during operation. Collector terminal current is carried by the n and n^+ regions surrounding the fully depleted collector. The emitter and base regions can be built in the same way as in a regular vertical bipolar transistor process. Figure 7.6 shows the electron flow path from emitter to collector in forward active mode, and the energy band diagram along the electron path. An electron injected from the emitter flows vertically through the base layer and, instead of continuing vertically down to a subcollector layer as in a bulk transistor, travel laterally towards the collector reach through and contact after traversing the intrinsic base layer.

7.4 Complementary SiGe HBT technology

Currently, SiGe technology development is exclusively based on npn SiGe HBTs due to the larger minority electron mobility in the base and the valence band offset in SiGe strained layers is generally more conducive to npn SiGe HBT designs. Although a complementary (npn-pnp) bipolar technology generally gives much greater design flexibility and higher performance than the npn-only technology, the design of high-performance pnp SiGe HBTs remains a big challenge [72, 73]. The valence band offset in SiGe translate into a conduction that greatly enhances minority electron transport in the device, thereby significantly boosting transistor performance over a similar npn Si BJT. It has been shown that the electric field effects associated with the

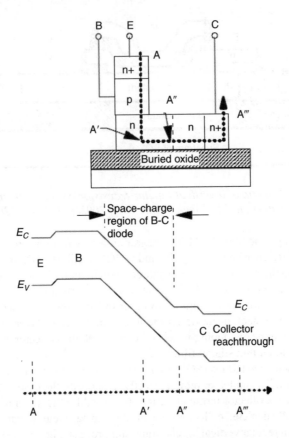

*Figure 7.6   Schematic illustrating the electron path from emitter to collector (top)
and the energy band diagram along the electron path (bottom) [After
J. Cai et al., VLSI Tech. Symp. Dig., 2002(172–173). ©2002 IEEE]*

Ge profile in SiGe HBTs can be tailored to alter the local electric field distribution in
the base-collector space-charge region. For a pnp SiGe HBT, on the other hand, the
valence band offset directly results in a valence band barrier, even at low injection,
which strongly degrades minority hole transport and thus limits frequency response.
Careful optimisation to minimise these hole barriers in pnp SiGe HBTs can yield
impressive device performance compared to Si pnp BJTs [74,75]. The inherent design
differences between npn and pnp SiGe HBTs may be outlined as follows: (1) npn and
pnp profile design optimisation, (2) single Ge profile design for both npn and pnp
transistors, (3) if graded base design preferable to a box-like profile design for pnp
HBTs and (4) if Ge retrograding is required to obtain acceptable SiGe pnp HBT
performance.

The first SiGe HBT pnp transistors were grown with a n-type MBE deposited
epitaxial base and a single crystalline boron-doped emitter deposited by UHVCVD.
pnp SiGe base transistors were also fabricated in the same time frame as the first

npn SiGe base transistors. Epitaxial-base pnp transistors were particularly difficult to fabricate because it required achieving abrupt N type dopant profiles reaching peak concentrations. The MBE n-type base films, as the MBE boron doped layer before, had very high defect densities, which caused considerable leakage and poor Gummel characteristics. However, higher f_T was measured on the SiGe-base pnp transistors than the control homojunction Si-base pnp transistors.

The valence band offset in strained SiGe directly results in a valence band barrier in a pnp SiGe HBT, even at low injection, which strongly degrades minority hole transport, and thus limits frequency response [74, 75]. Careful optimisation to minimise these hole barriers in pnp SiGe HBTs is thus required. It has been shown that the backside Ge profile shape strongly influences high-injection heterojunction barrier effect in SiGe HBTs, which produces premature roll-off of β and f_T at high current density [72, 73]. The backside Ge retrograde has been shown to alter the electric field distribution in the base-collector space-charge region and therefore affects both impact ionization and apparent neutral base recombination (NBR) in SiGe HBT [72]. Impact ionization and apparent NBR are critical to the breakdown voltage and output conductance of SiGe HBTs and are thus key design parameters.

7.5 SiGeC HBT technology

As discussed above, the SiGe alloy-based HBTs have already extended the performance of Si-based devices and the $Si_{1-x}Ge_x$ heterojunction bipolar transistors have reached the level of maturity needed for volume production. However, the $Si_{1-x}Ge_x$ on Si (001) system has some severe limitations. Examples include, the existence of a critical layer thickness for perfect pseudomorphic growth, which depend on the amount of Ge in the SiGe layer. In addition, the major band offset between Si and strained-SiGe is located in the valence band, that is, this system is better suited for hole channel than for electron channel devices.

A major problem in fabricating SiGe devices is the ability to retain the narrow as-grown profile of the boron doping within the base layer during post-epitaxial processing. Heat treatment and transient enhanced diffusion (TED) due to annealing implantation damage can result in the boron profile diffusing into the adjacent silicon regions. This may result in undesirable conduction band barriers and thereby degrade the device performance. TED can be strongly reduced by incorporating carbon into the SiGe base layer. This use of carbon also allows significantly greater flexibility in process design, with a wider latitude in process margins over standard SiGe.

The out-diffusion of boron from the SiGe base as a result of transient enhanced diffusion following extrinsic base implant is the main factor that limits the achievable f_T of SiGe HBTs. When boron diffuses out of the SiGe base, the emitter/base and collector/base junctions are created in silicon instead of SiGe. This leads to the formation of parasitic energy barriers in the conduction band [76], which degrade both gain and f_T.

Research on highly supersaturated, carbon containing alloys on Si substrates started only a few years ago. These new materials alleviate some of the limitations of

SiGe on Si (001). $Si_{1-y}C_y$ and $Si_{1-x-y}Ge_xC_y$ layers can be grown pseudomorphically on Si (001), using MBE, or different chemical vapour deposition techniques. Low carbon concentrations can also be used to suppress diffusion of dopants, such as boron, strain manipulation, thermal stability, and also carbon effects the band structure and charge carrier transport. These unique properties have already found its first device application, the SiGeC heterojunction bipolar transistors [77]. Although the process started with MBE, the process was later transferred to CVD. SiGeC HBTs now offer high-performance wireless solutions, with the potential to integrate digital and analog functions on a single chip.

Research on SiGeC has shown that carbon is very effective in suppressing transient enhanced boron diffusion [78] when present in the base at a concentration of $\sim 1 \times 10^{20} cm^{-3}$ [79, 80]. This is illustrated in Figure 7.7, which shows SIMS boron profiles for SiGe and SiGeC wafers implanted with 2×10^{12} cm^{-2} P at 190 keV, 8×10^{11} cm^{-2} P at 110 keV for a pedestal collector and 6×10^{12} cm^{-2} at 100 keV for a low-doped emitter. The wafers were given an anneal of 15 min at 800 °C and 30 s at 1000 °C. It can be seen that there is considerable broadening of the boron profile in the SiGe wafer, whereas in the SiGeC wafer the profile is very sharp.

Figure 7.8 shows the historical trend in peak f_T in both SiGe and SiGeC technologies, from the first self-aligned device demonstration in 1994 until 2003. IHP has demonstrated the advantages of implementing C-doped SiGe HBTs in a BiCMOS technology with respect to manufacturability, RF performance, modular process integration, and low process complexity. It has been shown that C doping allows the fabrication of the HBT layers before the essential CMOS steps start, paving the way

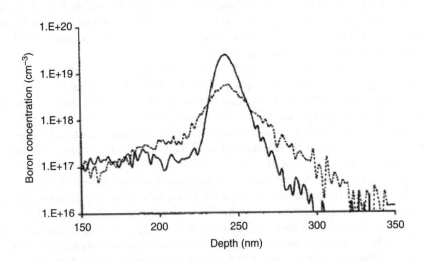

Figure 7.7   A typical Boron SIMS profiles for an SiGeC wafer (solid line) and an SiGe wafer (dotted line) illustrating the suppression of transient enhanced boron diffusion

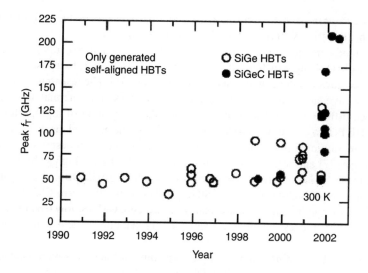

Figure 7.8 Reported cut-off frequency for SiGe and SiGeC HBTs

for a truly modular HBT integration in different CMOS platforms. Integration of high-performance RF LDMOS transistors and other BiCMOS components have also been accomplished.

This IHP process offers f_T (f_{max}) values of up to 300 (350) GHz, and gate delays down to 3.26 ps [58]. The HBTs use a SiGeC base layer of thickness of less than 30 nm, highly doped with B to attain an internal base sheet resistance of less than 2 kΩ. This layer and a Si low-doped emitter layer are deposited in an integrated epitaxy process using a single wafer RPCVD system (ASM Epsilon 2000). C doping is performed in situ during epitaxy using methylsilane. The C-doped SiGe layers forming the base of the HBTs are grown by LPCVD epitaxy, with low defect density and excellent profile control. Incorporation of C in SiGe allows higher B doping levels in very thin SiGe layers without out-diffusion from the SiGe, even after post-epitaxial implants. Such implants were applied to dope the external base regions of the transistors, and for selective doping of the low-doped emitters (LDE) and collectors (LDC). Substantial flexibility is available in IHP technology to trade-off gain, BV_{ceo}, f_T and f_{max} simply by varying the base B dose and the conditions for the LDC implantation. Besides comparable 1/f noise, the Si/SiGeC HBTs show Early voltage-current gain products of more than 20000 V. The excellent quality of the SiGeC base layers is demonstrated by wafer yields of more than 80 per cent for 16 k transistor arrays.

IHP has produced SiGeC HBTs with an f_{max} of 95 GHz and an f_T of 80 GHz for 40 Gbit/s fibre optics applications. Delays per stage of only 16 ps have been demonstrated for ring oscillators containing SiGeC HBTs. IHP has demonstrated the first modular integration of SiGeC HBTs into a 250 nm epi-free, dual-gate CMOS-compatible platform. This module fully uses the advantages of carbon, including suppressed boron diffusion and increased thermal and processing stability. In the following, the focus will be on SiGeC technology from IHP.

The modular integration of high-performance bipolar transistors in a standard CMOS with no changes to the CMOS process is a promising way to realise highly integrated low-power systems in a cost effective fashion [31]. The researchers from IHP, Germany have shown that a strong C-induced suppression of thermal and transient enhanced diffusion of boron enables the modular integration of high-performance SiGeC HBTs into a state-of-the-art CMOS platform. The physical mechanism behind the suppression of B diffusion in C-rich Si and SiGe has been described in several references ([77] and references therein), and explain why the high stability of base B profiles in SiGeC HBTs is important for a modular BiCMOS process. Successful integration of high-performance SiGeC HBTs with a 0.25 μm CMOS platform using only four additional mask layers beyond the core CMOS flow (see Figure 7.9). This BiCMOS process exhibited excellent d.c. parameters and a high yield of SiGeC HBTs.

The availability of both npn- and pnp-type bipolar transistors with matched performance in a complementary BiCMOS process (CBiCMOS) is a promising route to applications simultaneously requiring low voltage, low power, and high speed. IHP has developed a high-speed CBiCMOS process. CBiCMOS process flow and transistor cross sections from IHP are shown in Figures 7.10 and 7.11, respectively. A total of ten mask steps are used for the complementary bipolar module, five for the pnp and four for the npn, and one common mask step for structuring npn emitter poly and pnp base poly. This process offers pnp SiGeC HBTs with peak f_T/f_{max} values of 80 GHz/120 GHz at 2.6 V BV_{ceo}.

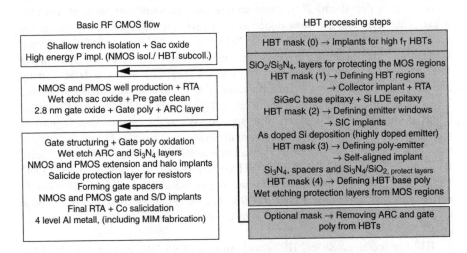

Figure 7.9　HBT-before-CMOS BiCMOS process flow from IHP for fabricating several types of SiGeC HBTs [After D. Knoll et al., IEEE IEDM Tech. Dig., 2001(499–502). ©2001 IEEE]

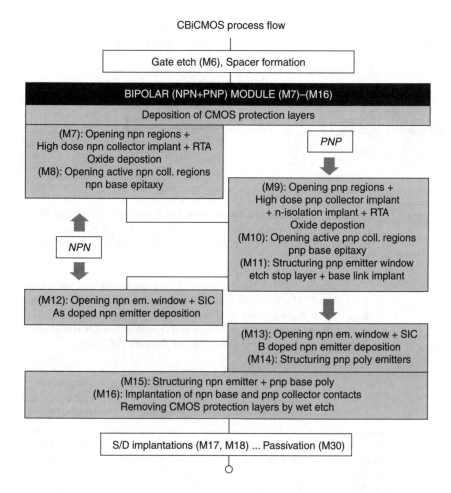

Figure 7.10   Flow of the complementary BiCMOS process [After B. Heinemann et al., IEEE IEDM Tech. Dig., 2003(117–120). ©2003 IEEE]

Table 7.5 compares transistor parameters obtained with the IHP standard SiGe HBT technology and those for C containing HBTs. Device parameters available for IBM's SiGe HBT BiCMOS are also included [19, 81, 82]. The essential difference in comparison with the SiGe-only transistors is the higher base boron doping that can be used without out-diffusion from the SiGe layer. As a result, the SiGeC HBTs reach higher f_{max}, lower noise figures, and higher Early voltages than the C-free devices. Note that the SiGeC devices, like the C-free HBTs, use a much simpler collector construction than applied by the best-in-class, but surpass its performance. Knoll *et al.* [83] have demonstrated that the presence of C has no deleterious effects on d.c. parameters, low frequency noise, or junction leakage.

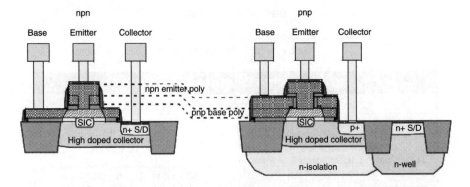

*Figure 7.11   Schematic cross-section of the npn HBT structure (left) and the pnp
HBT structure (right) of the complementary BiCMOS process [After
B. Heinemann et al., IEEE IEDM Tech. Dig., 2003(117–120). ©2003
IEEE]*

*Table 7.5   Comparison of IHP SiGe and SiGeC HBT parameters with published
data from IBM's SiGe HBT BiCMOS. The SiGeC HBTs show higher
f_{max}, lower noise figures, and higher Early voltages than the SiGe-
only devices principally as a result of a higher base boron doping
that can be used without out-diffusion from the SiGe layer*

	SiGe HBTs IHP standard technology	SiGeC HBTs IHP technology	SiGe HBTs IBM BiCMOS
Collector well construction	Implanted, epi-free	Implanted, epi-free	Epitaxially, buried subcoll
Peak f_T (GHz)	45	50	47
Peak f_{max} (GHz)	45	90	65
Current gain	90	200–250	100
Early volt. (V)	15	>100	65
BV_{ceo}(V)	3.1	2.7	3.4
NF_{min}(dB)	<3	<2	2
Gain(dB)	>4	>8	–
Ring osc.	16	19	FI/FO = 1
Delay(ps)	(CML-RO)	(CLM-RO)	(ECL-RO)

7.6   TCAD: SiGe/SiGeC BiCMOS process

In the fabrication of SiGe-based transistors, a low-temperature budget is needed. This
limitation is due to diffusion of dopants to the adjacent layers and also strain relaxation
after thermal annealing. Both these undesired effects degrade the high-frequency
performance of SiGe HBTs. A self-aligned transistor design is the preferable way

to decrease the junction capacitances and an industrial approach is the integration of selective implanted collector (SIC) layers.

Silvaco and SimuCad provide well-integrated simulation software for technology development essential for SiGe/SiGeC. In the following, an example of TCAD application towards SiGe/SiGeC technology development is presented. One can use ATHENA/SUPREM4 to model the full SiGe/SiGeC process flow with a range of sophisticated and experimentally verified models that include the following effects:

- retardation of boron diffusivity in germanium and carbon containing films;
- effects of carbon on interstitial diffusivity and recombination behaviour;
- interstitial injection during epitaxial processes resulting in enhanced dopant diffusion;
- fully coupled diffusion models;
- smart epitaxial simulation automatically depositing polycrystalline films on non-crystalline regions of the device;
- polySi grain boundary diffusion models.

A trench isolated double polySiGe HBT has been simulated using the process simulator ATHENA. The device structure after oxide growth, deposition, etch, epitaxy, implantation and diffusion, as shown in Figure 7.12, is then passed to ATLAS

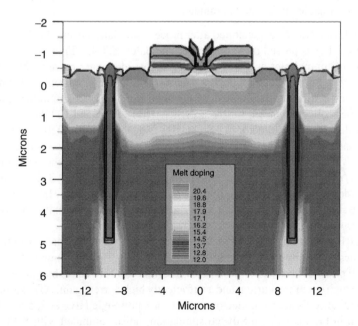

Figure 7.12   A 0.5 μm deep-trench isolated double polySiGe HBT with an emitter area of 40 μm² simulated using the 2D process simulator ATHENA [Source: Silvaco International]

for the device's electrical analysis at various bias points. In a SiGe HBT, the presence of a narrow bandgap SiGe layer in the base region where most of the bandgap offset occurs in the valence band, creates an energy well for holes. The base is generally highly doped to increase high-frequency performance whilst still retaining a reasonable gain. For this reason, the diffusion of boron in SiGe is of great interest and has been discussed in Chapter 3.

The incorporation of high boron levels in the base region along with the Ge-induced effects makes prediction of boron diffusion through the SiGe layer difficult. Most practical devices do not use uniform Ge content in the base, as this results in very thin base region before strain related defects occur. To reduce strain and increase basewidth, most devices have base regions with 'graded' germanium profiles. The effect of carbon incorporation to reduce the Boron out-diffusion has been discussed (see Figure 7.7).

ATLAS/BLAZE seamlessly takes devices created in ATHENA/SUPREM4 and provides the following outputs for the virtual analysis of SiGe HBTs:

- d.c. I–V, Gummel and breakdown curves;
- a.c. analysis for accurate extraction of f_T/f_{max} and capacitance data;
- s-parameter, y-parameter, and z-parameter calculations;
- Transient analysis for large-signal inputs;
- Analysis of critical parameters that impact device reliability;
- 1/f, trap and diffusion-induced noise calculations;
- User-definable models and parameters.

The device structure, generated after process simulation using ATHENA (shown in Figure 7.12), is passed on to the device simulator ATLAS. Figure 7.13 shows the simulated Gummel plot and it is seen that almost ideal base current characteristics are observed. For wireless applications gain verses collector current plot at different frequencies is of interest. Figure 7.14 obtained from ATLAS simulation shows the gain verses collector current for the polySiGe HBT at 2, 5 and 10 GHz.

The design of the collector in a SiGe HBT is dictated by conflicting requirements to simultaneously achieve high breakdown voltage BV_{ceo}, low base-collector capacitance, low base-collector signal delay τ_{bc}, and a high value of the knee current density at which f_T decreases. Impact ionization models available in ATLAS allow simulation of breakdown voltages and recombination in the base of a SiGe HBT. Figures 7.15 and 7.16 show the breakdown characteristics and recombination in the base of a SiGe HBT, respectively.

For RF applications, understanding of the noise in SiGe HBTs is very important as it provides the ability to investigate and optimise the noise behaviour at the operating point. Of prime concern to RF designers is the trade-off between achieving a low noise figure, a high operating current and a sufficiently high current gain. ATLAS simulated minimum noise figure vs. collector current for a polySiGe HBT at 2, 5 and 10 GHz is shown in Figure 7.17. In Silvaco simulation, noise combined with S-PISCES or BLAZE allows analysis of the small-signal noise generated within SiGe HBTs and may be used to extract figures of merit that are essential for optimising the circuit design.

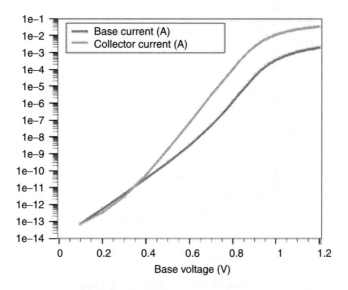

Figure 7.13 d.c. characteristics showing Gummel plot of SiGe HBT [Source: Silvaco International]

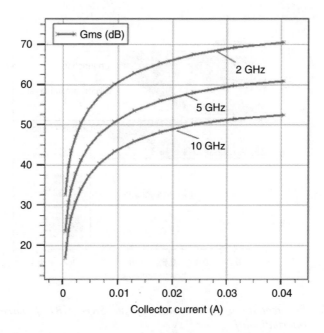

Figure 7.14 Gain vs. collector current for the polySiGe HBT at 2, 5 and 10 GHz [Source: Silvaco International]

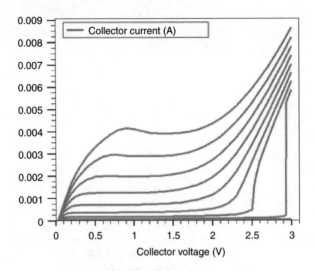

Figure 7.15   Simulation of breakdown voltage using impact ionization model in a SiGe HBT. I–V characteristics optimised for high f_T shows a consequent low breakdown voltage [Source: Silvaco International]

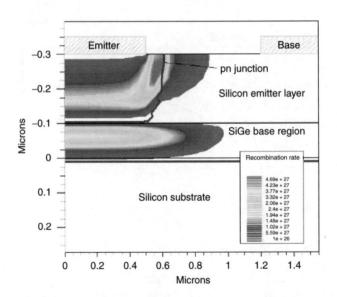

Figure 7.16   Recombination in the base of an SiGe HBT [Source: Silvaco International]

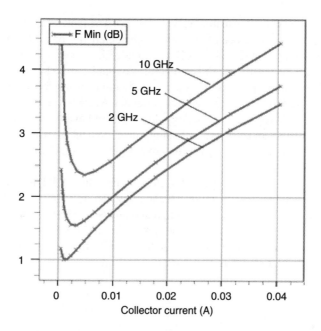

Figure 7.17   Minimum noise figure vs. collector current for the polySiGe HBT at 2, 5 and 10 GHz at a collector voltage of 1 V [Source: Silvaco International]

7.7   Summary

A retrospective look at the development and production of SiGe HBTs reveals that significant challenges have been overcome to make SiGe now a mature technology. In this chapter, key developments including the SiGe epitaxial growth process, overall integration methods, and elimination of yield-limiting defects to enable production of highly integrated products have been presented. Optimising the interaction of the EPI layer with its surroundings has enabled the achievement of high bipolar device yield. Integration of HP SiGe HBT with CMOS logic and a host of passive devices has created technologies that are well suited for a broad range of HP communication products. Modular integration approaches, such as the base-after-gate approach, has simplified the development of SiGe HBT technologies. It is observed that use of a modular integration approach enables a quick migration to next generation and derivative technologies easily. Enhancements have been achieved primarily in cost reduction, increased integration, and application specific voltage/power requirements.

Vertical SiGe HBTs compatible with SOI CMOS have been discussed. The unique feature of collector voltage pinning in thin Si film was discussed in depth, which gives rise to high breakdown voltage, high Early voltage, and low collector capacitance. The SOI device is promising for a better f_T-BV_{ceo} trade-off than that from conventional collector scaling in bulk devices. The fabricated devices show the anticipated strong dependence of d.c. and RF characteristics on SOI substrate bias.

IBM's and IHP's SiGe and SiGeC BiCMOS technologies that have driven the requirements for the most advanced communication applications have been discussed. As a TCAD example, process and device simulation results for a trench isolated double polySiGe HBT using the process simulator ATHENA and device simulator ATLAS towards SiGe/SiGeC technology development have been presented. Use of TCAD tools for the SiGe/SiGeC technology development is expected to offer the process and device designers significant advantages in time-to-market, cost, power and performance prediction.

In contrast to the nearly exponential trend in npn SiGe HBT performance over time, towards complementary bipolar technology, careful analysis of the inherent profile design and differences between npn and pnp SiGe HBTs design is necessary for Ge profile design of high-performance pnp SiGe HBTs. Research areas that need more emphasis for passive elements are modelling at RF frequencies, accurate SPICE models and characterisation as a function of temperature.

References

1 C. K. Maiti, and G. A. Armstrong, *Applications of Silicon-Germanium Heterostructure Devices*. Inst. of Physics Pub., UK, 2001.

2 A. J. Joseph, D. L. Harame, B. Jagannathan, D. Coolbaugh, D. Ahlgren, J. Magerlein, *et al.*, 'Status and direction of communication technologies – SiGe BiCMOS and RFCMOS', *Proc. IEEE*, vol. 93, pp. 1539–1558, 2005.

3 J.-S. Rieh, B. Jagannathan, H. Chen, K. T. Schonenberg, D. Angell, A. Chinthakindi, *et al.*, 'SiGe HBTs with cut-off frequency of 350 GHz', in *IEEE IEDM Tech. Dig.*, pp. 771–774, 2002.

4 R. Lodge, 'Advantages of SiGe for GSM RF front ends', *Electron. Eng.*, vol. 71, pp. 18–19, 1999.

5 R. Gotzfried, F. Beisswanger, S. Gerlach, A. Schuppen, H. Dietrich, U. Seiler, *et al.*, 'RFIC's for mobile communication systems using SiGe bipolar technology', *IEEE Microwave Theory Tech.*, vol. 46, pp. 661–668, 1998.

6 A. Rylyakov, and T. Zwick, '96 GHz static frequency divider in SiGe bipolar technology', in *IEEE GaAs IC Symp. Tech. Dig.*, pp. 288–290, 2003.

7 A. R. Alvarez, *BiCMOS Technology and Applications*. Kluwer Academic Publishers, Boston, 1989.

8 IEEE, 'Special issue on integrated circuits technologies for RF circuit applications', *IEEE Trans. Electron Devices*, vol. 52, pp. 1231–1488, 2005.

9 D. C. Ahlgren, G. Freeman, S. Subbanna, R. Groves, D. Greenberg, J. Malinowski, *et al.*, 'A SiGe HBT BiCMOS technology for mixed signal RF applications', in *IEEE BCTM Proc.*, pp. 195–197, 1997.

10 G. Freeman, D. Ahlgren, D. R. Greenberg, R. Groves, F. Huang, G. Hugo, *et al.*, 'A 0.18 μm 90 GHz f_T SiGe HBT BiCMOS, ASIC-compatible, copper

interconnect technology for RF and microwave applications', in *IEEE IEDM Tech. Dig.*, pp. 569–572, 1999.

11 A. Joseph, D. Coolbaugh, M. Zierak, R. Wuthrich, P. Geiss, Z. He, *et al.*, 'A 0.18 μm BiCMOS technology featuring 120/100 GHz (f_T/f_{MAX}) HBT and ASIC-compatible CMOS using copper interconnect', in *IEEE BCTM Proc.*, pp. 143–146, 2001.

12 D. C. Ahlgren, B. Jagannathan, S.-J. Jeng, P. Smith, D. Angell, H. Chen, *et al.*, 'Process variability analysis of a Si/SiGe HBT technology with greater than 200 GHz performance', in *IEEE BCTM Proc.*, pp. 80–83, 2002.

13 B. A. Orner, Q. Z. Liu, B. Rainey, A. Stricker, P. Geiss, P. Gray, *et al.*, 'A 0.13 μm BiCMOS technology featuring a 200/280 GHz (f_T/f_{max}) SiGe HBT', in *IEEE BCTM Proc.*, pp. 203–206, 2003.

14 S. A. St. Onge, D. L. Harame, J. S. Dunn, S. Subbanna, D. C. Ahlgren, G. Freeman, *et al.*, 'A 0.24 μm SiGe BiCMOS mixed-signal RF production technology featuring a 47 GHz f_t HBT and 0.18 μm L_{eff} CMOS', in *IEEE BCTM Proc.*, pp. 117–120, 1999.

15 N. Feilchenfeld, L. Lanzerotti, D. Sheridan, R. Wuthrich, P. Geiss, D. Coolbaugh, *et al.*, 'High performance, low complexity 0.18 μm SiGe BiCMOS technology for wireless circuit applications', in *IEEE BCTM Proc.*, pp. 197–202, 2002.

16 L. Lanzerotti, N. Feilchenfeld, D. Coolbaugh, J. Slinkman, P. Gray, D. Sheridan, *et al.*, 'A low complexity 0.13 μm SiGe BiCMOS technology for wireless and mixed signal applications', in *IEEE BCTM Proc.*, pp. 237–240, 2004.

17 M. Racanelli, and P. Kempf, 'SiGe BiCMOS technology for RF circuit applications', *IEEE Trans. Electron Devices*, vol. 52, pp. 1259–1270, 2005.

18 D. N. Ngoc, D. L. Harame, J. C. Malinowski, S. J. Jeng, K. T. Schonenberg, M. M. Gilbert, *et al.*, 'A 200 mm SiGe-HBT BiCMOS technology for mixed signal applications', in *IEEE BCTM Proc.*, pp. 89–92, 1995.

19 D. Ahlgren, M. Gilbert, D. Greenberg, S.-J. Jeng, J. Malinowski, D. Nguyen-Ngoc, *et al.*, 'Manufacturability demonstration of an integrated SiGe HBT technology for the analog and wireless marketplace', in *IEEE IEDM Tech. Dig.*, pp. 859–862, 1996.

20 D. L. Harame, K. M. Newton, R. Singh, S. L. Sweeney, S. E. Strang, J. B. Johnson, *et al.*, 'Design automation methodology and rf/analog modeling for rf CMOS and SiGe BiCMOS technologies', *IBM J. Res. and Dev.*, vol. 47, pp. 139–175, 2003.

21 D. J. Harame, J. H. Comfort, J. D. Cressler, E. F. Crabbe, J. Y. C. Sun, B. S. Meyerson, *et al.*, 'Si/SiGe epitaxial-base transistors-part I: Materials, physics, and circuits', *IEEE Trans. Electron Devices*, vol. 42, pp. 455–467, 1995.

22 D. J. Harame, J. H. Comfort, J. D. Cressler, E. F. Crabbe, J. Y. C. Sun, B. S. Meyerson, *et al.*, 'Si/SiGe epitaxial-base transistors-part II: Process integration and analog applications', *IEEE Trans. Electron Devices*, vol. 42, pp. 469–482, 1995.

23 D. L. Harame, and B. S. Meyerson, 'The early history of IBM's SiGe mixed signal technology', *IEEE Trans. Electron Devices*, vol. 48, pp. 2555–2567, 2001.

24 IEEE, 'Special issue on silicon germanium – advanced technology, modeling and design', *Proc. IEEE*, vol. 93, p. 1519, 2005.

25 S. S. Iyer, G. L. Patton, S. S. Delage, S. Tiwari, and J. M. C. Stork, 'Silicon-germanium base heterojunction bipolar transistors by molecular beam epitaxy', in *IEEE IEDM Tech. Dig.*, pp. 874–876, 1987.

26 B. S. Meyerson, 'UHV/CVD growth of Si and Si-Ge alloys: Chemistry, Physics and Device Applications', *Proc. IEEE*, vol. 80, pp. 1592–1608, 1992.

27 G. Freeman, J.-S. Rieh, B. Jagannathan, Y. Zhijian, F. Guarin, and A. Joseph, 'Device scaling and application trends for over 200 GHz sige HBTs', in *Proc. IEEE RFIC Symp.*, pp. 6–9, 2003.

28 C. P. Moreira, E. Kerherve, P. Jarry, A. A. Shirakawa, and D. Belot, 'Dual-mode RF receiver front-end using a 0.25-μm 60-GHz f_T SiGe:C BiCMOS7RF technology', in *SBCCI. Symp. Dig.*, pp. 88–93, 2004.

29 P. Deixler, A. Rodriguez, W. De Boer, H. Sun, R. Colclaser, D. Bower, *et al.*, 'QUBiC4X: An $f_T/f_{max} = 130/140$ GHz SiGe:C-BiCMOS manufacturing technology with elite passives for emerging microwave applications', in *IEEE BCTM Proc.*, pp. 233–236, 2004.

30 S. Van Huylenbroeck, A. Sibaja-Hernandez, A. Piontek, L. J. Choi, M. W. Xu, N. Ouassif, *et al.*, 'Lateral and vertical scaling of a QSA HBT for a 0.13 μm 200 GHz SiGe:C BiCMOS technology', in *IEEE BCTM Proc.*, pp. 229–232, 2004.

31 D. Knoll, B. Heinemann, R. Barth, K. Blum, J. Borngraber, J. Drews, *et al.*, 'A modular, low-cost SiGe:C BiCMOS process featuring high-F_T and high-BV$_{CEO}$ transistors', in *IEEE BCTM Proc.*, pp. 241–244, 2004.

32 B. Heinemann, R. Barth, D. Bolze, J. Drews, P. Formanek, O. Fursenko, *et al.*, 'A complementary BiCMOS technology with high speed npn and pnp SiGe:C HBTs', in *IEEE IEDM Tech. Dig.*, pp. 117–120, 2003.

33 D. L. Harame, D. Ahlgren, D. Coolbaugh, J. Dunn, G. Freeman, J. Gillis, *et al.*, 'Current status and future trends of SiGe BiCMOS technology', *IEEE Trans. Electron Devices*, vol. 48, pp. 2575–2594, 2001.

34 L. D. Lanzerotti, A. St. Amour, C. W. Liu, and J. C. Sturm, 'Si$_{1-x-y}$Ge$_x$C$_y$/Si heterojunction bipolar transistors', in *IEEE IEDM Tech. Dig.*, pp. 930–932, 1994.

35 F. Sato, T. Hashimoto, T. Tatsumi, and T. Tashiro, 'Sub-20 ps ECL circuits with high-performance super self-aligned selectively grown SiGe base (SSSB) bipolar transistors', *IEEE Trans. Electron Devices*, vol. 42, pp. 483–488, 1995.

36 T. F. Meister, H. Schafer, M. Franosch, W. Molzer, K. Aufinger, U. Scheler, *et al.*, 'SiGe base bipolar technology with 74 GHz f_{max} and 11 ps gate delay', in *IEEE IEDM Tech. Dig.*, pp. 739–742, 1995.

37 B. Jagannathan, M. Khater, F. Pagette, J.-S. Rieh, D. Angell, H. Chen, *et al.*, 'Self-aligned SiGe NPN transistor with 285 GHz f_{max} and 207 GHz f_t in a manufacturable technology', *IEEE Electron Device Lett.*, vol. 23, pp. 258–260, 2002.

38 P. Deixler, A. Rodriguez, W. De Boer, H. Sun, R. Colclaser, D. Bower, *et al.*, 'QUBiC4X: A $f_T/f_{max} = 130/140$ GHz SiGe:C-BiCMOS manufacturing

technology with elite passives for emerging microwave applications', in *IEEE BCTM Proc.*, pp. 236–239, 2004.

39 K. Washio, E. Ohue, K. Oda, M. Tanabe, H. Shimamoto, T. Onai, *et al.*, 'A selective-epitaxial-growth SiGe-base HBT with SMI electrodes featuring 9.3-ps ECL-gate delay', *IEEE Trans. Electron Devices*, vol. 46, pp. 1411–1416, 1999.

40 B. S. Meyerson, 'Low-temperature silicon epitaxy by ultrahigh vacuum chemical vapor deposition', *Appl. Phys. Lett.*, vol. 48, pp. 797–799, 1986.

41 R. A. Johnson, M. J. Zierak, K. B. Outama, T. C. Bahn, A. J. Joseph, C. N. Cordero, *et al.*, 'A 1.8 million transistor CMOS ASIC fabricated in a SiGe BiCMOS technology', in *IEEE IEDM Tech. Dig.*, pp. 217–220, 1998.

42 D. L. Harame, E. F. Crabbe, J. D. Cressler, J. H. Comfort, J. Y.-C. Sun, S. R. Stiffler, *et al.*, 'A high-performance epitaxial SiGe-base ECL BiCMOS technology', in *IEEE IEDM Tech. Dig.*, pp. 19–22, 1992.

43 D. L. Harame, J. M. C. Stork, B. S. Meyerson, T. N. Nguyen, and G. J. Scilla, 'Epitaxial-base transistors with ultra-high vacuum chemical vapor deposition (UHV/CVD) epitaxy/enhanced profile control for greater flexibility in device design', *IEEE Electron Device Lett.*, vol. 10, pp. 156–158, 1989.

44 A. Neugroschel, C.-T. Sah, and M. S. Carroll, 'Degradation of bipolar transistor current gain by hot holes during reverse emitter-base bias stress', *IEEE Trans. Electron Devices*, vol. 43, pp. 1286–1290, 1996.

45 J. H. Comfort, G. L. Patton, J. D. Cressler, W. Lee, E. F. Crabbe, B. S. Meyerson, *et al.*, 'Profile leverage in self-aligned epitaxial Si or SiGe base bipolar technology', in *IEEE IEDM Tech. Dig.*, pp. 21–24, 1990.

46 M. Smayling, J. Reynolds, D. Redwine, S. Keller, and G. Falessi, 'A modular merged technology process including submicron CMOS logic, nonvolatile memories, linear functions, and power components', in *IEEE CICC Proc.*, pp. 24.5.1–24.5.4, 1993.

47 P. G. Y. Tsui, P. V. Gilbert, and S.-W. Sun, 'A versatile half-micron complementary BiCMOS technology for microprocessor-based smart power applications', *IEEE Trans. Electron Devices*, vol. 42, pp. 564–570, 1995.

48 E. Bayer, W. Bucksch, K. Scoones, K. Wagensohner, J. Erdeljac, and L. Hutter, 'A 1.0 μm linear BiCMOS technology with power DMOS capability', in *IEEE BCTM Proc.*, pp. 137–1414, 1995.

49 S. Subbanna, J. Johnson, G. Freeman, R. Volant, R. Groves, D. Herman, *et al.*, 'Prospects of silicon-germanium-based technology for very high-speed circuits', in *IEEE MTT-S Dig.*, pp. 361–364, 2000.

50 J. Dunn, D. Harame, S. St. Onge, A. Joseph, N. Feilchenfeld, K. Watson, *et al.*, 'Trends in silicon germanium BiCMOS integration and reliability', in *Proc. Int. Reliability Phys. Symp.*, pp. 237–242, 2000.

51 G. Freeman, J. Kierstead, and W. Schweiger, 'Electrical parameter data analysis and object-oriented techniques in semiconductor process development', in *IEEE BCTM Proc.*, pp. 81–88, 1996.

52 M. Racanelli, P. Ma, and P. Kempf, 'SiGe BiCMOS technology for highly integrated wireless transceivers', in *IEEE 25th GaAs IC Symp. Dig.*, pp. 183–186, 2003.

53 K. Washio, 'SiGe HBT and BiCMOS technologies for optical transmission and wireless communication systems', *IEEE Trans. Electron Devices*, vol. 50, pp. 656–668, 2003.

54 M. Racanelli, and P. Kempf, 'SiGe BiCMOS technology for communication products', in *IEEE CICC Proc.*, pp. 331–334, 2003.

55 J. S. Dunn, D. C. Ahlgren, D. D. Coolbaugh, N. B. Feilchenfeld, G. Freeman, D. R. Greenberg, *et al.*, 'Foundation of rf CMOS and SiGe BiCMOS technologies', *IBM J. Res. and Dev.*, vol. 47, pp. 101–138, 2003.

56 D. Harame, A. Joseph, D. Coolbaugh, G. Freeman, K. Newton, S. M. Parker, *et al.*, 'The emerging role of SiGe BiCMOS technology in wired and wireless communications', in *Proc. IEEE Int. Caracas Conf. Devices, Circuits and Systems*, pp. D052–1–D052–17, 2002.

57 J.-S. Rieh, B. Jagannathan, H. Chen, K. Schonenberg, S.-J. Jeng, M. Khater, *et al.*, 'Performance and design considerations for high speed SiGe HBTs of $f_T/f_{max} = 375$ GHz/210 GHz', in *Int. Conf. Indium Phosphide and Related Materials*, pp. 374–377, 2003.

58 M. Khater, J.-S. Rieh, T. Adam, A. Chinthakindi, J. Johnson, R. Krishnasamy, *et al.*, 'SiGe HBT technology with $f_{max}/f_T = 350/300$ GHz and gate delay below 3.3 ps', in *IEEE IEDM Tech. Dig.*, pp. 247–250, 2004.

59 S. Subbanna, R. Groves, B. Jagannathan, D. Greenberg, G. Freeman, R. Volant, *et al.*, 'Using silicon-germanium mainstream BiCMOS technology for X-band and LMDS (25–30 GHz) microwave applications', in *IEEE MTT-S Symp. Dig.*, pp. 401–404, 2002.

60 B.-U. Klepser, and W. Klein, 'Ramp-up of first SiGe circuits for mobile communications: positioning of SiGe vs. GaAs and silicon', in *GaAs IC Symp.*, pp. 37–40, 1999.

61 U. Konig, 'SiGe and GaAs as competitive technologies for RF-applications', in *IEEE BCTM Proc.*, pp. 87–92, 1998.

62 M. Mouis, and A. Chantre, 'VLSI integration of SiGe epitaxial base bipolar transistors', in *Proc. URSI ISSSE Symp.*, pp. 50–55, 1998.

63 C. Kermarrec, T. Tewksbury, G. Dawe, R. Baines, B. Meyerson, D. Harame, *et al.*, 'SiGe HBT's reach the microwave and millimeter-wave frontier', in *IEEE BCTM Proc.*, pp. 155–162, 1994.

64 M. Mastrapasqua, P. Palestri, A. Pacelli, G. K. Celler, M. R. Frei, P. R. Smith, *et al.*, 'Minimizing thermal resistance and collector-to-substrate capacitance in SiGe BiCMOS on SOI', *IEEE Electron Device Lett.*, vol. 23, pp. 145–147, 2002.

65 F. Sato, T. Hashimoto, H. Tezuka, M. Soda, T. Suzaki, T. Tatsumi, *et al.*, 'A 60-GHz f_T super self-aligned selectively grown SiGe-base (SSSB) bipolar transistor with trench isolation fabricated on SOI substrate and its application to 20-Gb/s optical transmitter IC's', *IEEE Trans. Electron Devices*, vol. 46, pp. 1332–1338, 1999.

66 K. Washio, E. Ohue, H. Shimamoto, K. Oda, R. Hayami, Y. Kiyota, *et al.*, 'A 0.2-μm 180-GHz-f_{max} 6.7-ps-ECL SOI/HRS self-aligned SEG SiGe HBT/CMOS technology for microwave and high-speed digital applications', *IEEE Trans. Electron Devices*, vol. 49, pp. 271–278, 2002.

67 S. Hall, A. C. Lamb, M. Bain, B. M. Armstrong, H. Gamble, H. A. W. El Mubarek, *et al.*, 'SiGe HBTs on bonded wafer substrates', *Microelectronic Eng.*, vol. 59, pp. 449–454, 2001.

68 S. Ueno, M. Watanabe, T. Kato, T. Shinohara, K. Mikami, T. Hashimoto, *et al.*, 'A single-chip 10 Gb/s transceiver LSI using SiGe SOI/BiCMOS', in *IEEE ISSCC Tech. Dig.*, pp. 82–83, 2001.

69 G. A. Armstrong, and H. S. Gamble, 'Simulation of self heating effects in heterojunction bipolar transistors fabricated in wafer bonded SOI substrates', in *Silicon-on-Insulator Technology and Devices IX Editor P. L. Hemment, The Electrochemical Society Proc. Series, Pennington, NJ*, vol. 99-3, pp. 249–254, 1999.

70 J. Cai, A. Ajmera, C. Ouyang, P. Oldiges, M. Steigerwalt, K. Stein, *et al.*, 'Fully-depleted-collector polysilicon-emitter SiGe-base vertical bipolar transistor on SOI', in *VLSI Tech. Symp. Dig.*, pp. 172–173, 2002.

71 J. Cai, M. Kumar, M. Steigerwalt, H. Ho, K. Schonenberg, K. Stein, *et al.*, 'Vertical SiGe-base bipolar transistors on CMOS-compatible SOI substrate', in *IEEE BCTM Proc.*, pp. 215–218, 2003.

72 G. Zhang, J. D. Cressler, L. Lanzerotti, R. Johnson, Z. Jin, S. Zhang, *et al.*, 'Electric field effects associated with the backside Ge profile in SiGe HBTs', *Solid State Electron.*, vol. 46, pp. 655–659, 2002.

73 G. Zhang, J. D. Cressler, G. Niu, and A. Pinto, 'A comparison of npn and pnp profile design tradeoffs for complementary SiGe HBT technology', *Solid State Electron.*, vol. 44, pp. 1949–1954, 2000.

74 D. L. Harame, J. M. C. Stork, B. S. Meyerson, E. F. Crabbe, G. L. Patton, G. J. Scilla, *et al.*, 'SiGe-base PNP transistors fabricated with n-type UHV/CVD LTE in a "No D_t" process', in *Dig. Symp. on VLSI Technol.*, pp. 47–48, 1990.

75 D. L. Harame, B. S. Meyerson, E. F. Crabbe, C. L. Stains, J. M. Cotte, J. M. C. Strok *et al.*, '55 GHz polysilicon-emitter graded SiGe-base PNP transistor', in *Proc. Symp. VLSI Technology*, pp. 71–75, 1991.

76 Md. R. Hashim, R. F. Lever, and P. Ashburn, '2D simulation of the effects of transient enhanced boron out-diffusion from base of SiGe HBT due to an extrinsic base implant', *Solid State Electron.*, vol. 43, pp. 131–140, 1999.

77 H. J. Osten, *Carbon-Containing Layers on Silicon – Growth, Properties and Applications*. Trans-Tech Publications, Switzerland, 1999.

78 L. D. Lanzerotti, J. C. Sturm, E. Stach, R. Hullullull, T. Buyuklimanli, and C. Magee, 'Suppression of boron transient enhanced diffusion in SiGe heterojunction bipolar transistors by carbon incorporation', *Appl. Phys. Lett.*, vol. 70, pp. 3125–3127, 1997.

79 H. J. Osten, G. Lippert, D. Knoll, R. Barth, B. Heinemann, H. Rucker, *et al.*, 'The effect of carbon incorporation on SiGe heterobipolar transistor performance and process margin', *IEEE IEDM Tech. Dig.*, pp. 803–806, 1997.

80 I. M. Anteney, G. Lippert, P. Ashburn, H. J. Osten, B. Heinemann, G. J. Parker, *et al.*, 'Characterization of the effectiveness of carbon incorporation in SiGe for the elimination of parasitic energy barriers in SiGe HBTs', *IEEE Electron Device Lett.*, vol. 20, pp. 116–118, 1998.

81 S. Subbanna, G. Freeman, D. Ahlgren, D. Greenberg, D. Harame, J. Dunn, *et al.*, 'Integration and design issues in combining very-high-speed silicon-germanium bipolar transistors and ULSI CMOS for system-on-a-chip applications', in *IEEE IEDM Tech. Dig.*, pp. 845–848, 1999.

82 G. Niu, J. D. Cressler, U. Gogineni, and D. L. Harame, 'Collector-base junction avalanche multiplication effects in advanced UHV/CVD SiGe HBTs', *IEEE Electron Device Lett.*, vol. 19, pp. 288–290, 1998.

83 D. Knoll, B. Heinemann, H. J. Osten, K. E. Ehwald, B. Tillack, P. Schley, *et al.*, 'Si/SiGe:C heterojunction bipolar transistors in an Epi-free well, single-polysilicon technology', in *IEEE IEDM Tech. Dig.*, pp. 703–706, 1998.

Chapter 8
MOSFET: compact models

The primary goal of compact model development for circuit simulation is to provide physics-based, scalable models that are fully integrated into a design kit environment. Progress in MOS device physics, model development and model implementation process has qualitatively changed the capabilities of compact models precisely at a time when the rapid expansion of RF MOSFET applications is imposing the most stringent demands on the new generation of MOSFET models. A MOSFET compact model predicts the output current (I_{ds}) and its derivatives (g_m and g_{ds}) as a function of temperature, bias voltage, channel length and device width.

Si MOSFETs have so far been primarily used in digital circuits and at frequencies below 1 GHz. By optimisation of the technology and device geometry n-channel Si MOSFETs with cut-off frequencies beyond 150 GHz have been realised. The feasibility of using CMOS for implementing RF circuits for wireless communications has been demonstrated [1–3]. One of the major obstacles for use of CMOS in RF circuits is the lack of good MOS models that are valid up to RF. The rapid growth of RF CMOS and mixed-signal integrated circuit (IC) design has increased the need for accurate, robust models for simulating analog circuit designs.

There are two major problems faced in simulating MOS circuits, particularly MOS-based analog circuits. The first is the development of models which are suitable for the new fabrication technologies for short-channel devices. The second problem is the specification of model parameters. Although most MOS models are physics-based, some parameters do not have well-defined values and there are other parameters for which the physical values do not produce the best fit to actual device characteristics. So, it is generally necessary to extract model parameters from measured transistor data obtained for a given fabrication process. MOSFET analytical models are based on either a regional approach, surface potential formulation, or semi-empirical equations. Models based on the regional approach use a different set of equations to describe the device behaviour in different regions. Where weak and strong inversion regions are generally bridged by using a non-physical curve fitting. Two different types of compact MOSFET models can be distinguished on the basis of different modelling approach: threshold-voltage based models [4–6] and surface-potential-based models [7, 8]. For a comprehensive review on high-frequency modelling technique of MOSFETs for radio frequency integrated circuit (RFIC) design, covering intrinsic and extrinsic components, equivalent circuits, parameter extraction, noises and distortion behaviours for RF circuit applications, the reader may refer to a paper by Cheng *et al.* [9].

The demands for advanced models, which can describe nanoscale silicon MOSFET devices for analog and mixed-signal applications and account for the physical effects on small geometry devices, have led to enormous efforts in the development of advanced physics-based compact models. For recent research activities in developing new models for MOSFETs, active and passive devices for circuit applications and ESD protection, which are critical to realising the RF systems-on-a-chip (SoCs), the reader may refer to the special issue on advanced compact models and 45-nm modelling challenges by the *IEEE Transactions on Electron Devices* editors [10].

In this chapter, currently available MOSFET models and the impact of the new modelling paradigm on MOSFET circuit simulation with particular attention to RF issues, non-quasistatic effects and symmetric surface-potential based models are discussed. New models from Philips, MOS Model 11, EKV, HiSIM and PSP will be introduced, and a comparison with the well-established BSIM model will be carried out.

8.1   Charge-based MOSFET models

The MOSFET models that are currently available can be divided into three generations. The first generation of MOSFET models treats the device as almost ideal describing the MOSFET from very simple physically based parameters. The solution adopted was the threshold voltage (V_t) based formulation that assumes the surface potential is a very simple function of the input voltage; constant for V_g above V_t and a linear function of gate voltage below V_t. This results in separate solutions for different regions of MOSFET operation requiring smoothing functions to connect the regions.

Originally, most of the model development was done to achieve a good fit for d.c. characteristics while the main requirement for the capacitance model was good continuity rather than good fit. For MOSFETs, the level 1, 2 and 3 models are generally referred to as the first generation models. The Shichman–Hodges model (also known as Level 1) is the original MOSFET model developed and is applicable mainly to long channel devices. Attention was given to physically accurate representation without equal consideration to mathematical representation, which often creates numerical problems during circuit simulation. The C–V portion of the model is the Meyer model, which is not based on charge conservation model. Level 2 model addresses several short channel effects such as the velocity saturation. However, the mathematical implementation of the model is complicated, leading to convergence problems. For C–V calculation, either the Meyer model of Level 1 or the Ward–Dutton model can be used. The Ward–Dutton model [11] is a charge-based model, which forms the backbone of all present models. Semi-empirical Level 3 is regarded as a simplified version of Level 2. This model has proven to be robust and is popular for digital circuit design. However, the model is not scalable.

In the second generation of models, the parameters in the model are empirical in their character. The BSIM model places less emphasis on the exact physical formulation of the device, but instead relies on empirical parameters and polynomial equations to handle various physical effects. This generally leads to improved circuit simulation behaviour compared to previous models, although its accuracy degrades in submicron FETs. Several empirical parameters without clear physical meanings are incorporated in the model. This approach gains the advantage of improved convergence properties, but at the cost of complicating the parameter extraction process, as well as weakening the link between model parameters and fabrication process.

BSIM2 is an extension to BSIM, with comprehensive modifications which make it suitable for analog circuit design. Although BSIM2 is an improvement upon BSIM, in terms of model accuracy as well as convergence behaviour in circuit simulation, it still breaks the transistor operation into several regions. This leads to discontinuity in the first derivative in I–V and C–V characteristics, a result that can cause numerical problems in simulation. With the help of smoothing functions, BSIM3 adopts a single equation to describe device characteristics in various operating regions. This eliminates the discontinuity in the I–V and C–V characteristics. BSIM3 has evolved through several versions. BSIM3v1 and BSIM3v2 contain many mathematical problems that have largely been solved by the third version, BSIM3v3. Although BSIM3v3 is a de facto public-domain industry standard MOS model, it has limitations when approaching for high frequency analog circuits exceeding several GHz. An enhanced version specifically for analog circuits, BSIM4, offers several improvements over BSIM3, through improved noise modelling, and in the incorporation of extrinsic parasitics. Interested readers may refer to an excellent review on compact noise models for MOSFETs by Jindal [12]. MOSFET models included in circuit simulators can be classified into three categories: analytical models, table lookup models and empirical models. Most of the models in current use are analytical. Further details of a range of different models can be found in Reference 13.

Until recently threshold-voltage-based MOSFET compact models were considered standard and attempted to accommodate the increasing demands of the design community by a rapid increase in the number of model parameters without changing the regional nature of the model. Interests in analog and mixed-signal IC designs require different operation regions of compact models for the circuit designs and simulations [14].

For example BSIM3v3 includes the following physical phenomena [15]:

- short and narrow channel effects on threshold voltage;
- non-uniform doping effect (in both lateral and vertical directions);
- mobility reduction due to vertical field;
- bulk charge effect;

- velocity saturation;
- drain-induced barrier lowering;
- channel length modulation (CLM);
- substrate current-induced body effect (SCBE);
- subthreshold conduction;
- source/drain parasitic resistances;
- poly-gate depletion and quantum mechanical effects.

A new variant on the BSIM family, BSIM5 uses a single set of equations to calculate charges throughout all the bias regions. It incorporates short-channel, non-uniform doping and other effects to accurately model subtle details of the device behaviour. Special attention is paid to higher-order physics such as accurate current saturation and quantum mechanical effect. It is fully symmetric and smooth. Velocity saturation, velocity overshoot and source-velocity limit are modelled in a unified way. BSIM5 shows that the classical velocity saturation underestimates the drain current for 45 nm technology, while the hydrodynamic model overestimates the current because source-velocity limit starts to take effect. The BSIM5 core model is being extended to model non-classical devices such as ultra-thin-body SOI and multi-gate devices including FinFET.

A summary of various MOSFET models developed during the last four decades is shown in Figure 8.1. It can be seen from the figure that the number of parameters required to fit a model is becoming very large. For example, BSIM3v3 requires several hundred parameters for a complete model [16]. Automated extraction program of model parameter through a dedicated program such as ICCAP requires significant experience and specialist knowledge. Perhaps the ultimate, but yet unrealised goal

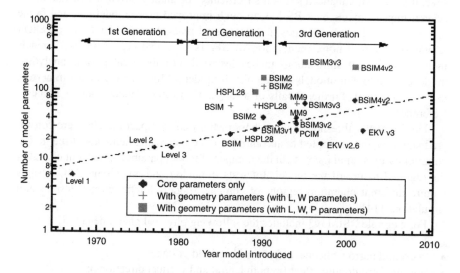

Figure 8.1   Different MOS models and the number of d.c. model parameters vs. the year of the model introduction

in TCAD compact modelling is a versatile model with as few parameters as possible (e.g., <30), and good accuracy in all regions of operation. Saijets *et al.* [17] have evaluated several mainstream MOS models.

A fundamental requirement of accurate MOSFET models is symmetry with respect to source–drain interchange or 'Gummel symmetry' [7]. Some of the fundamental problems of V_t-based models such as BSIM3, include the use of source-referenced threshold voltage, producing a singularity in the I–V characteristic. Unphysical description of the moderate inversion region leads to erroneous results for the g_m/I_d ratio (critical for analog design) and inconsistent modelling of charges and currents, producing negative transcapacitances [16]. The use of source-referenced threshold voltage and asymmetric bulk charge linearisation in standard V_t-based models not only violates the Gummel symmetry but also prevents the modelling of common current division circuits.

In most V_t-based MOSFET models reported in the literature a trade-off between accuracy and complexity has resulted in a regional approach, where different sets of equations are used for different regions of device operation. Continuity of the current and its derivatives with respect to bias at the transition points is ensured by using so-called smoothing functions. The above continuity is beneficial for the convergence properties of the circuit simulation program, which uses the Newton–Raphson method to solve circuit equations. The main drawback of regional models is the imprecise description of drain current I_d and transconductance g_m in the transition region between subthreshold and superthreshold, the so-called moderate inversion region. The above-mentioned region becomes particularly important with the continuous down-scaling of MOS transistors toward submicron dimensions and the subsequent decrease in supply voltage.

There is now a consensus that threshold-voltage-based models of bulk MOSFETs need to be replaced by more physical surface-potential-based models [18, 19] or inversion-charge based models [20]. One approach is to find the density of the inversion charge at the two ends of the channel and formulate the model outputs in terms of these charge densities. These are 'charge based models' meaning that both conductance and capacitance calculations are based on charge. This approach has been used in EKV, ACM and BSIM5 models.

In another approach, one solves for the input equation surface potential at the two ends of the channel. The terminal charges, currents and derivatives are then calculated from the surface potential and hence are called surface potential models. SP, MOS Model 11 and HiSIM are based on this approach. Progress in development of these models has brought about qualitative improvement in ability to simulate CMOS circuits and is reviewed below.

8.2 Surface-potential based MOSFET models

Continued scaling of MOS transistors has lead to much competition in the quest for the best TCAD compact model for MOSFETs [4–8, 21]. The continuing shrinking of CMOS transistors to the 100 nm and below has led to the need to incorporate small-geometry electrical effects, which were previously considered to be second-order, into increasingly sophisticated SPICE models for these devices. Rapid growth in low-power, low-voltage CMOS imposes tighter requirements on model accuracy. The aggressive reduction of the power supply voltage implies that threshold-voltage-based approach to compact modelling is close to its limits and the next generation advanced compact MOSFET models should be surface-potential based. Since the source–drain overlap regions are critical in modern scaled MOSFETs the surface-potential-based approach is essential as it allows the physical modelling of these regions. For most low-voltage analog applications, a MOSFET is typically biased within approximately 300 mV of its threshold, and hence the use of regional models is no longer appropriate.

Modelling of short-channel MOSFETs requires a precise description of physical effects such as mobility reduction, series resistance, velocity saturation, channel length modulation, static feedback, self-heating and drain-induced barrier lowering. With the aggressive scaling of oxide thickness, modelling of tunnelling currents is essential. A surface-potential-based formulation enables a particularly simple implementation of this effect.

As a result, considerable attention has been drawn to charge sheet MOSFET models based on surface potential formulations. These so-called surface-potential based models are inherently single-piece, and give an accurate and continuous description of current and its derivatives in all operation regions. The charge sheet MOSFET model is based on surface potential formulations and describes the drain current in the whole range of applied voltages using one single equation. Models based on surface potential formulation are inherently continuous and require the solution of an implicit equation. Surface potentials at drain and source are given by an implicit relation based on the terminal voltages, and as a consequence this can only be solved iteratively, resulting in long computation times, which is undesirable for VLSI circuit simulation. In order to reduce computation time, an explicit, yet accurate relation between surface potential and terminal voltages is preferable.

A big advantage of a complete surface-potential-based model is that the overall model consistency is automatically preserved through the surface potential. Therefore, the number of model parameters can be drastically reduced in comparison with conventional models. This parameter reduction comes without any loss in reproduction accuracy of measurement data.

The above effects have been incorporated in the surface-potential-based MOSFET model [21], and it has been found to give an accurate description of the drain current in all practical operation regions, even in the moderate inversion region. An explicit approximate relation for surface potential as a function of terminal voltages has been developed [21], which results in an accurate description of all operation regions. The use of the explicit relation for ϕ_s does not result in a noticeable decrease in accuracy when compared to implicit ϕ_s-based models or in an increase in simulation time when compared to V_t-based models. In other words the new model combines the advantages of both regional and single-piece models.

Traditionally, the development of surface-potential-based models has been inhibited by the inherent complexity of this approach. However, reformulation of the charge sheet model using a symmetric linearisation method [22] resulted in a model formulation which is no more complex than less accurate V_t-based or inversion charge based models while producing results that are numerically equivalent to the original form of the surface-potential-based model. Furthermore, the difficult problem of computing the surface potentials as functions of the terminal voltages has been solved by improving the convergence of the numerical algorithms and more recently using analytical approximations accurate to within 10 nV. The reported results were limited to the range adequate in bulk MOSFET models but a simple extension of the work in [23, 24] produces analytical approximations valid for extreme forward biases of the source/drain junction occasionally encountered in SOI devices. Typical results shown in Figures 8.2 and 8.3 are accurate not only for surface potential but also for a much more demanding task of evaluating the transcapacitances. While not using the threshold voltage as a key variable, surface-potential-based models are capable of a very detailed modelling of V_t as a function of geometry. Also, surface-potential-based modelling of $V_t(L)$ dependence has been shown to be superior to the traditional description.

8.2.1 SP model

The surface-potential-based compact MOSFET model (SP 'core' model) has been developed at the Pennsylvania State University. This model solves several long-standing problems of surface-potential based modelling and provides a basis for a comprehensive compact MOSFET model [18]. In the SP model, accurate analytical

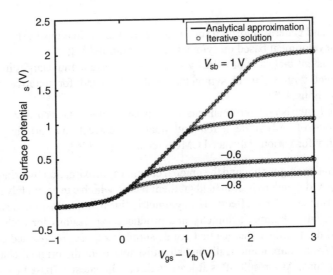

Figure 8.2   Surface potential computed using new analytical approximation and numerical iteration for different back bias; $t_{ox} = 3$ nm and $N_{sub} = 10^{17}$ cm^{-3} [After G. Gildenblat et al., Proc. 11th Intl. Conf. Electronics, Circuits, and Systems, 2004 (638–641). ©2004 IEEE]

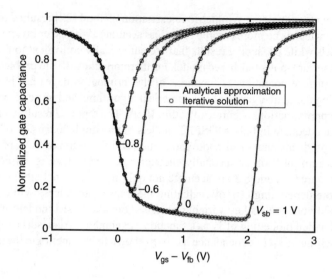

Figure 8.3   *Normalized C–V curves computed using new analytical approximation and numerical iteration for different back bias; $t_{ox} = 3$ nm and $N_{sub} = 10^{17}$ cm^{-3} [After G. Gildenblat et al., Proc. 11th Intl. Conf. Electronics, Circuits, and Systems, 2004(638–641). ©2004 IEEE]*

approximation has been reported for the surface potential of MOS transistors. These include

accurate and generic algorithm for surface potential evaluations [23];
model development based on symmetric linearisation [23];
self-consistent inclusion of velocity saturation in current and charges in a manner consistent with Gummel symmetry, without a need for iterative numerical computations, [25];
simplified quasistatic terminal charge consistent with d.c. current, analytical inclusions of polySi depletion and quantum corrections, offering a model flexibility for various advanced CMOS technologies [26].

SP formulation goes beyond the gradual channel approximation by using the bias- and geometry-dependent lateral-field gradient factor, which can accurately model the charge-sharing and DIBL effects. The symmetric linearisation method developed in SP preserves the Gummel symmetry and produces expressions for both the drain current and the terminal charges that are as simple as those in V_t-based or charge-based models, and are numerically indistinguishable from the original charge-sheet model equations. As a result SP is at least as fast as the latest V_t-based models.

Analytical inclusions of polySi depletion and quantum corrections are accomplished through developing the non-iterative approximations for surface potentials by perturbation method and the simplified model formulations by the symmetrically linearised surface-potential-based approach. The physical content and computational efficiency of the model is significantly increased while the number of model parameters is reduced without sacrificing the quality of the fit of experimental data. Surface potential models use a single expression for drain current and give a physics-based and accurate description in all operation regions including the moderate inversion and the accumulation region. As discussed above the charge-based model and the V_t-based model are special simplified cases of the surface-potential-based model. For a comprehensive review on PSP model, the reader may refer to a recent paper by Gildenblat *et al.* [27].

8.2.2 HiSIM

In the surface-potential-based modelling, the surface potential is the measure of all device characteristics, which are very much dependent on technology. HiSIM (Hiroshima-university STARC IGFET Model) is another MOSFET model for circuit simulation, based on the complete drift-diffusion approximation with channel-surface potentials [28]. As the simulation is based on the drift-diffusion approximation and a channel-surface-potential description, the modelling allows the MOSFET characteristics to be closely based on their physical origins. HiSIM obtains the potentials by solving the Poisson equation iteratively both at the source side and drain side. An accuracy of 10 pV has been achieved with faster simulation time than some threshold-voltage-based models [29].

In HiSIM major phenomena that appear in advanced MOSFETs are demonstrated to be modelled closely based on their physical origins, and thus resulting in a small number of model parameters. HiSIM implements charge and capacitance equations that maintain the gradual-channel approximation and provide a solution in the saturation condition by extending the surface potential beyond the pinch-off point. The most important practical advantage of HiSIM is the reduced number of model parameters (about 70 model parameters), and one parameter set is enough for all gate lengths. All model parameters have a clear, direct relation to the MOSFET physics resulting in easily extractable parameters with a clearly defined physical meaning [29]. For a comprehensive recent review on HiSIM, the reader may refer to a paper by Miura-Mattausch *et al.* [30].

Simulated surface potential values by HiSIM are shown in Figure 8.4. Once the surface-potential values at source and drain side are known, all device characteristics are calculated without any additional parameter. Since this is the core approach of the surface-potential-based modelling and at the same time the core of the device physics, a one-to-one correspondence between circuit simulation model and device physics is readily obtainable. Also higher-order phenomena observed such as noise have been shown to be determined by the potential gradient along the channel, which again highlights the strength of the concept of surface-potential-based modelling.

8.2.3 MOS model 11

MOS Model 11 (MM11) is a compact MOSFET model intended for digital, analog and RF circuit simulation in modern CMOS technologies from Philips and

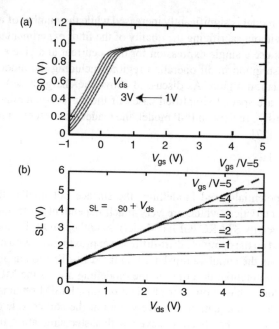

Figure 8.4   *Simulated surface potential values with HiSIM at (a) source side as a function of V_{gs} and (b) drain side as a function of V_{ds}. The short-channel effect is incorporated in the values [After H. J. Mattausch et al., Proc. Solid-State and Integrated-Circuit Technology, 2001(861–866). ©2004 IEEE]*

is a successor to the original widely used Philips MOS Model 9 (MM9). MM9 employs smoothing functions to achieve continuity in device characteristics. The model is accurate for circuit simulation of sub-quarter micron technologies. MM11 is a symmetrical, surface-potential based model, which accurately describes distortion behaviour, making it highly suitable for digital and analog as well as RF circuit design. Being surface-potential based, it includes accurate descriptions of the currents and charges, especially the first-order derivatives (transconductance, conductance, capacitances) and the equations are valid in all regions (accumulation, depletion and weak, moderate and strong inversion). In MM11, a linearisation is performed around the average of source and drain surface potentials, which results in simpler expressions without a loss of accuracy.

In MM11, special emphasis has been given to distortion modelling. For an accurate description of distortion, the model should accurately describe the drain current and its higher-order derivatives. Parameter extraction in MM11 is relatively straightforward. And the public domain model is available in commercial circuit simulators such as SPECTRE, HSPICE and ADS. MOS Model 11 incorporates bias-dependent source/drain partitioning of gate leakage current, quantum mechanical effects and poly-depletion.

MM11 contains improved expressions for mobility reduction. velocity saturation and various conductance effects. In addition, the MM11 approach also ensures that the model symmetry with respect to source–drain interchange is maintained. Note that this approach is very similar to the symmetric linearisation method used in SP, which made it easy to merge the best features of MM11 and SP into one model called PSP.

The PSP model [31], from Pennsylvania State University has been jointly developed by Philips and Pennsylvania State University. This surface potential model has evolved from the MOS Model 11 and SP [32]. It includes 'universal' dependence on vertical effective field and the deviations from the universality associated with Coulomb scattering. The PSP model was recently selected to be the industry standard for future nanometer chip design by the Compact Model Council, USA.

The main features of PSP are as follows:

- physical surface-potential-based formulation in both intrinsic and extrinsic model modules;
- physical and accurate description of the accumulation region;
- inclusion of all relevant small-geometry effects;
- modelling of the halo implant effects, including the output conductance degradation in long devices;
- Coulomb scattering and non-universality in the mobility model;
- non-singular velocity-field relation enabling the modelling of RF distortions including intermodulation effects;
- complete Gummel symmetry;
- quantum-mechanical corrections;
- correction for the polySi depletion effects;
- GIDL/GISL model;
- surface-potential-based noise model including channel thermal noise, flicker noise and channel induced gate noise;
- advanced junction model including trap-assisted tunnelling, band-to-band tunnelling and avalanche breakdown;
- stress model.

8.2.4 EKV model

On the basis of a new approach to analytical MOSFET modelling, the Enz-Krummenacher-Vittoz EKV model was first published in 1995 [33]. EKV model is unique in its use of substrate-referencing, as opposed to source-referencing of the preceding models. This fundamental philosophical change allows the EKV model a greater scope of fundamentally eliminating the asymmetry problems unavoidable in the source-referencing models. Referencing the voltages to the substrate not only preserves and exploits the symmetry of the device in the model, but also clearly distinguishes between effects that do not affect this symmetry property, such as all the effects related to the transverse field (mobility reduction due to the transverse field, poly depletion, quantum effect and non-uniform doping in the transverse direction) and the effects related to the longitudinal direction, such as velocity saturation and non-uniform doping in the longitudinal direction.

In the following, the main features of the public domain EKV 3.0 MOSFET model are discussed. The EKV model was initially developed for the design of very low-power analog ICs, with the objective of having a simple analytical model valid in all modes of inversion. It was important to correctly handle the weak inversion part since many micropower circuits were designed based on the MOS transistor operating in this region. Since its introduction, the model has become popular owing to its simplicity and the relatively few parameters.

The EKV model is a physics-based and design-oriented MOSFET model. On the basis of the surface-potential approach, a linearisation technique is at the heart of the analytical framework. Even though early EKV model used the linearisation of the inversion charge with respect to the channel voltage to derive the drain current, it was actually not a charge-based model.

EKV also redefined the forward and reverse currents extending it to all modes of inversion. The basic concept of normalized forward current (also called inversion factor) and reverse current were also defined and used for circuit design optimisation. The moderate inversion region is covered by an empirical interpolation function after proper normalization. Designing circuits operating in the ever more important moderate and weak inversion regions of operation of the MOS transistor is a challenge, even for RF circuits. A proper normalization of drain current leads to a level-of-inversion-centred analysis of MOS physics, providing many guidelines to the designer for the trade-offs in circuit design. The EKV model approach is therefore not only a circuit simulation model, but it also provides a whole framework for advanced analog IC design. A rigorous derivation of the charge-based EKV model and a detailed modelling of the inversion charge linearisation can be found in Reference 34. A comparison of various MOS models is shown in Table 8.1.

Table 8.1   Comparison of various MOS models

	BSIM4	MOS9	MOS11	EKV	HiSIM
Modelling method	Analytical	Analytical	Analytical	Analytical	Iterative
Inversion	V_t-based	V_t-based	ψ-based	hybrid	ψ-based
d.c. current	drift	drift	drift-diff.	drift	drift-diff.
Referencing	Dynamic	Source	Source	Bulk	Source
Symmetry	Yes	No	Yes	Yes	Yes
Induced gate noise	Yes	Yes	Yes	Ignored	Yes
Tunnelling	Yes	Ignore	Yes	Ignored	Yes

8.3 Model evaluation

A judgment on the quality of each MOSFET model described above is sometimes necessary. It should be noted that the quality of fit is not the only criterion for choosing a compact model. Ease of parameter extraction, correlation of parameters, number of parameters, redundancy of parameters and the model behaviour should also be considered. A comparison of models is necessary to show weaknesses and strengths of each particular model so that the user community can decide which model is appropriate for their use. However, it may be noted that the quality requirements might be different for digital and analog applications and/or specific operating conditions. Test conditions are given for some parameters. For some parameters, the test conditions/algorithms will vary among the foundries. In those cases, the foundry-specific test conditions and algorithm should be used. The following list provides a guideline in the evaluation of the quality of a compact model:

- short channel effects;
- narrow width effects;
- number of regional models and scalability;
- lateral and vertical non-uniform doping effects;
- mobility reduction with gate and substrate bias;
- output conductance;
- velocity saturation;
- subthreshold conduction;
- linear region fit accuracy;
- saturation region fit accuracy;
- transition from linear to saturation region;
- continuous derivatives;
- substrate current modelling;
- junction area, field and gate sidewall, overlap capacitances;
- charge conservation and high frequency characteristics;
- physical values for physical model parameters;
- source/drain parasitic effects;
- temperature parameters;
- leakage currents;
- noise parameters.

Saijets *et al.* [17] have evaluated the existing mainstream MOS models at radio frequencies. The strengths and weaknesses of the models were also compared. The four mainstream RF CMOS models chosen for comparison are EKV [34], BSIM3v3 [16], Philips MOS Model 9 [35] and the old classic Berkeley Spice Level 3 MOSFET model [36]. In this evaluation study, identical equivalent circuits, including the parasitic series resistances and zero-bias capacitances were assumed for the extrinsic a.c. models and the models were compared by comparing their s-parameter fits. The authors first extracted a complete scalable d.c. model and also the a.c. parameters were extracted using both capacitance measurements and s-parameter fitting. Both d.c. and a.c. extractions were made by using the APLAC circuit simulator and by programs written in APLAC description language. In a.c. measurements the n-MOSFET devices were considered as two-port having the gate as the input and the drain as the output.

It was found that the real differences of the a.c. model behaviour results from differences in their d.c. and active charge models. It also became evident that a more complex MOSFET model is necessary and more development of parameter extraction techniques is required. Both d.c. and a.c. extraction may very easily fall into local numerical minima that can represent non-physical values, complicating the model optimisation.

8.4   Modelling of SOI MOSFETs

In general an SOI model can be classified into three categories: partially depleted (PD), ideal fully depleted (FD) and dynamic depleted (DD). Partially depleted has been the most popular form of SOI technology because of its ease in manufacturability and compatibility with bulk CMOS processing. Ultra-thin film PD-SOI technology offers better short-channel control over bulk technology [37]. A fully depleted model is similar to a partially depleted model, but without a node for the floating body. It has to handle transition from FD to PD because all SOI devices can operate in accumulation by applying appropriate gate bias [38].

An SOI CMOS SPICE model is similar to a traditional CMOS SPICE model with a fifth node to model the floating body potential, an optional sixth node to model the resistance of a body tied transistor and a seventh node (for self-heating effects). Existing SPICE models currently offered in commercial electrical simulators are BSIM3SOI/BSIMPD (University of Berkeley), SOISPICE (University of Florida) and STAG (University of Southampton).

BSIMProPlus [39] is considered to be the most efficient and accurate SPICE model parameter extraction tool on the market for digital, analog, mix-signal and RF modelling application especially in advanced process technologies. BSIMProPlus provides accurate and fast extraction routines for I–V, C–V, temperature-model and RF small equivalent circuit model parameter extractions.

A recent enhancement variant of the BSIMSOI model is BSIMPD [40], which includes all major physics in SOI MOSFETs, and has been tested extensively within IBM on its state-of-the-art high-speed SOI technology. The following features are significant:

- Floating body simulation in both I–V and C–V. The body potential is determined by the balance of all the body current components.
- An improved parasitic bipolar current model. This includes enhancements in the various diode leakage components, second order effects (high-level injection and Early effect), diffusion charge equation and temperature dependence of the diode junction capacitance.

- An improved impact-ionization current model. The contribution from BJT current is also modelled.
- A gate-to-body tunnelling current model, which is important to thin-oxide SOI technologies.
- Enhancements in the threshold voltage and bulk charge formulation of the high positive body bias regime.
- An external body node (the 6th node) facilitates the modeling of distributed body-resistance.
- Self-heating. An external temperature node (the 7th node) facilitates the simulation of thermal coupling.
- A unique SOI low-frequency noise model, including a new excess noise resulting from the floating body effect.

Since the issues discussed in Section 1 also apply to SOI MOSFETs, surface-potential-based approach has been extended to modelling of SOI devices. A variant of SP model, SP–SOI has been formulated within the framework of the bulk SP model [18]. This assures that all small geometry effects are physically modelled and standard benchmark tests are automatically satisfied. SP–SOI also has the following additional features:

- internal floating body;
- improved analytical approximation of surface potential which allows high forward body bias;
- self-heating model;
- backgate charge.

A simplified circuit representation of SP–SOI is shown in Figure 8.5. The internal body potential is determined by the junction current, impact ionization, GIDL/GISL currents, and capacitive coupling under transient operation. Unlike the standard bulk and SOI models, SP models are automatically symmetric, provided one uses symmetric rather than the source based linearisation and the drift velocity model is non-singular.

Typical device characteristics are shown in Figure 8.6 for three device sizes, using the same set of parameters. A careful reproduction of the drain conductance in bulk SP model has been retained in SP–SOI, including impact ionization effects. For negative gate biases the drain current is caused primarily by the tunnelling currents in the source-drain overlap regions and is accurately reproduced by the model.

8.5 Modelling of heterostructure MOSFETs

Strain engineering in MOSFETs is receiving serious attention due to many advantages of strain on carrier mobility and therefore drive current of MOSFETs and has been discussed in Chapter 4. Due to the introduction of strain in the MOSFET channel, the effective mass of carriers decreases and results in an increase in carrier mobility. Strained-$Si_{1-x}Ge_x$- and strain-compensated $Si_{1-x-y}Ge_xC_y$-channel MOSFETs are the most investigated heterostructure MOSFETs [41].

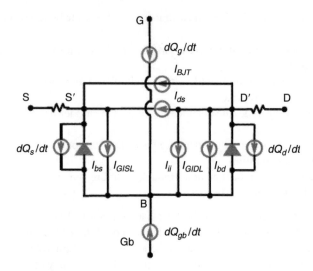

*Figure 8.5   Simplified circuit representation of SP–SOI model [After W. Wu et al.,
IEEE CICC Proc., 2005(819–822). ©2005 IEEE]*

For compact modelling of strained-Si heterostructure MOSFETs, a single-piece
charge model in the entire accumulation, depletion, and strong inversion regions
for strained-Si MOSFETs that is scalable in terminal biases and strained-Si parameters with physically derived flat-band voltages has been developed and validated
with numerical simulation [42]. The capacitance–voltage characteristics of strained-
Si n-MOSFETs shows a 'plateau' in the accumulation region [43], which is unique
to heterojunction devices. The physical understanding can be easily investigated
by numerical simulations. However, there are so far no complete analytical equations to physically model these effects in the entire region of MOSFET operations.
Chandrasekaran *et al.* [42] have presented a physics-based strained-Si n-MOSFET
charge model that is scalable for terminal biases and strained-Si parameters, such
as Ge mole fraction x, strained-Si layer thickness, and doping. The authors have
derived physical equations of the flat-band voltages for the strained-Si/SiGe device
and applied them to the regional surface-potential and charge solutions that could be
added to form a unified solution. The authors have used the unified regional charge-
based approach, where regional solutions are physically derived and the transitions
across regions are controlled by the smoothing parameters. Being physics-based,
the model can be used to extract physical parameters from measured strained-Si
devices.

For the case of a strained-Si channel (under biaxial tensile strain realised using the
lattice mismatch of an underlying relaxed-SiGe layer [41]), the energy band structure
is modified and the inversion carrier transport properties are enhanced, depending on
the Ge content. The enhancement follows from improved velocity saturation and overshoot effects, as well as increased mobility. The improved overshoot has been shown
to result from a strain-induced increase in the carrier energy relaxation time. Further,

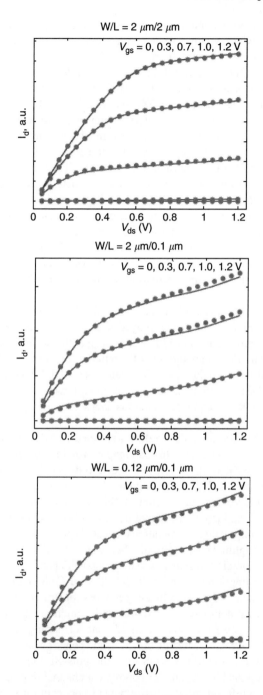

Figure 8.6 *Comparison of SP–SOI (solid lines) with measured data (symbols) for three different geometries using the same model parameter set [After W. Wu et al., IEEE CICC Proc., 2005(819–822). ©2005 IEEE]*

the band structure modification results in a narrowed energy bandgap, dependent on Ge mole fraction, which yields a reduced threshold voltage for the heterostructure MOSFET relative to the Si counterpart.

Strained-Si on SOI (sSOI) MOSFETs are susceptible to the local thermal heating generated in the channel because of the low thermal conductivity of buried oxide (BOX). Even with a 100-nm-thin BOX, SOI MOSFETs d.c. I–V characteristics, from which the SPICE simulation model parameters are extracted, suffer current loss due to self-heating effect (SHE). However, for most logic circuits, SHE is insignificant since the average power consumption per device is low and the switching time is much shorter than the thermal time constant. Thus, using the conventional methods of extracting the parameters from d.c. I–V data, SPICE simulation underestimates the device current due to SHE, which is present in the d.c. data but absent in logic circuits. Therefore, for accurate logic circuit simulation, one needs to obtain SHE-free I–V data and then extract SHE-free model parameters to close the gap between the d.c. I–V data and circuit speed. The pulse I–V measurement technique is a good way to directly obtain I–V data free from self-heating, but it is difficult, requiring the fast time resolution setup and careful consideration of the coupling.

Process/physics-based compact model, UFPDB [44], which is unified for bulk-Si and PD-SOI MOSFETs, has been employed to model strained-Si MOSFETs [45]. UFPDB has been implemented in SPICE3, to project the performance advantage of PD-SOI CMOS with floating bodies over the bulk-Si counterpart as the technologies are scaled to 100 nm or below [46]. The small set of UFPDB parameters relate directly to the MOSFET structure and the pertinent device physics, and hence can be well estimated without requiring numerous data sets of measured electrical device characteristics.

UFPDB (Ver. 2.0) was upgraded to allow a strained-Si/SiGe option. A new parameter, based on the relation between the bandgap narrowing and the Ge content in the underlying relaxed-SiGe buffer layer, was added to define the narrowed bandgap in the strained-Si channel and depletion region. The reduced bandgap necessitates an increase in threshold voltage for I_{off} control. Simulations were based on this control being implemented via increased channel doping in both n- and p-channel devices. Unloaded CMOS ring oscillator simulations revealed speed enhancement in the range 14–16 per cent for optimal Ge content of x between 0.2 and 0.3.

The simulation prediction suggests that strained-Si channel technology is worth pursuing as the scaling limits of bulk-Si and PD-SOI CMOS are approached. The simulations also quantified the significant speed loss for strained-Si/SiGe CMOS due to the increased areal components of source/drain junction capacitance, especially when the control was effected via increased channel doping. This loss would be avoided in strained-Si/SiGe CMOS on thin SOI, and hence an additional speed advantage (~ 8 per cent) of PD-SOI CMOS over the bulk-Si counterpart is intimated. Further, SOI would enable the simpler control via channel doping in both devices, while retaining a significant speed enhancement (13 per cent for $x = 0.20$) due to the strained-Si/SiGe channels.

8.6 RF MOS modelling

The RF properties of deep-sub-micron silicon MOSFETs are very promising in terms of performance, scalability, and system-on-chip perspective. Till the mid 1990s, most RF circuits and systems have been implemented with either compound semiconductor transistors, such as GaAs MESFETs, HEMTs, HBTs, or silicon BJTs and SiGe HBTs. Until recently, the achievable microwave properties of silicon MOSFETs have been inferior to those of other high-frequency transistors. However, during the last decade, the continuous downscaling of CMOS has made it a candidate for RF applications [47, 48]. An 0.18 μm CMOS technology exhibits excellent potential for RF applications with f_T and f_{max} of almost 50 GHz, and 0.35 dB of NF_{min} at 2 GHz [3]. As the gate length shrinks below 100 nm, f_T will exceed 100 GHz. The improved RF performance of silicon-based technologies over the years and their potential use in telecommunication applications increased the research in RF modelling of MOS transistors. Chang *et al.* [49] have reported the state-of-the-art 90-nm foundry RF CMOS technology with discussion on various integrated passive elements and user-friendly design kit for RF IC design.

RF circuitry consists primarily of four RF building blocks, amplifiers, filters, mixers, and oscillators. The design of these building blocks requires accurate small-signal as well as large-signal RF MOSFET models. Much attention has been paid to model the small-signal RF behaviour of MOS transistors, using compact models, for example BSIM3v3 [50], MM9 [51], as well as equivalent circuit-based models [52, 53]. However, large-signal RF MOSFET models are equally important in the design of mixers and power amplifiers.

All of the requirements for a MOSFET model in low-frequency application, such as continuity, accuracy, and scalability of the d.c. and capacitance models should be maintained in an RF model that has additional requirements:

(a) predict bias dependence of small-signal parameters at high-frequency operation;
(b) simulate intermodulation distortion;
(c) predict HF noise in low noise amplifiers (LNA);
(d) include the non-quasistatic (NQS) effect.

A modified model based on s-parameter measurements to enhance the RF performance of EKV model has been reported [53] by adding new capacitance equations using subcircuits, as shown in Figure 8.7. The combination of the d.c. and RF parts of the model predicts the non-linear high frequency performance up to 10 GHz over various bias points. Once the equivalent circuit element values of the linear model are known, the next step is to find equations that fit the extracted capacitances over all bias points. This modelling procedure is shown in Figure 8.8.

The MOSFET equivalent circuit elements are extracted from s-parameters at all gate voltages and drain voltages that occur in large-signal operation. The accuracy of the equivalent circuit-based model is validated by large-signal RF measurements using a non-linear network measurement system. Both the equivalent circuit-based

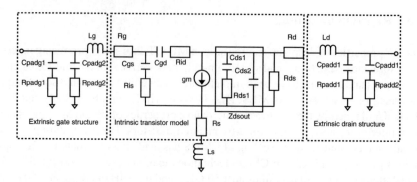

Figure 8.7   Small-signal equivalent circuit of a MOSFET for the extraction of the linear intrinsic elements [After C. E. Biber et al., IEEE MTT-S Symp. Dig., 1997(865–868). ©1997 IEEE]

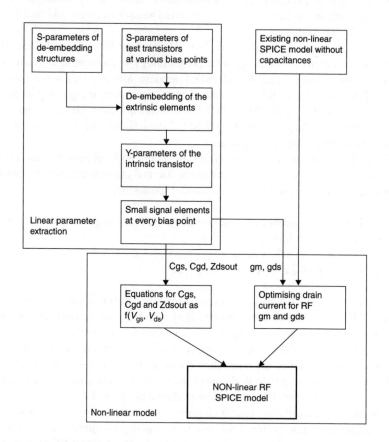

Figure 8.8   Flow chart of the modelling process used to obtain a non-linear RF SPICE model [After C. E. Biber et al., IEEE MTT-S Symp. Dig., 1997(865–868). ©1997 IEEE]

model and the BSIM3v3 model meet the requirements to describe accurately the RF large-signal behaviour of MOS transistors [4, 54].

In RF MOSFETs, the influence of the distributed substrate resistance becomes significant as the operation frequency increases [50, 55–58]. At low frequencies, the impedance of the junction capacitance is so large that the substrate resistance may not be seen from the drain terminal. However, with increasing frequency, the impedance of the junction capacitance reduces and the effects of the substrate resistance start to be seen. At high frequencies, combined with the fact that the MOSFET is a four-terminal device, the signals in RF MOSFETs are coupled through the substrate R–C network in a complex way. In strong inversion the substrate signal coupling mainly affects the small-signal output characteristics, which are important for RF design.

RF CMOS models with the substrate components, including the substrate resistance and drain/source junction capacitance, have been published [50, 52, 56, 58–60]. HF characteristics of substrate resistance in MOSFETs has been studied by Cheng and Matloubian [61] at different bias conditions for a 0.35 μm BiCMOS technology in the frequency range up to 10 GHz. A strong bias dependence of the real part of admittance, y_{22}, mainly contributed by the channel conductance, was observed and a very weak bias dependence of substrate resistance was found after de-embedding the measured admittance to remove the influence of channel resistance and gate-to-drain capacitance. The measured y_{22} vs. frequency (after two-step de-embedding) for a single-finger device is given in Figure 8.9, in which both real and imaginary parts of the measured y_{22} are functions of applied biases, especially at higher gate bias. Figure 8.10 gives the equivalent circuit for the device at the given measurement conditions. It is known that the measured y_{22} includes at least the contributions from the gate resistance, drain series resistance, source series resistance, channel resistance, gate-to-source capacitance, gate-to-drain capacitance, and gate-to-bulk capacitance besides the substrate components (SCs). To understand the HF behaviour of the SCs, the contributions of other components such as gate resistance and gate capacitance should be de-embedded from the measured y_{22} shown in Figure 8.9.

8.7 Large-signal MOSFET models

A large-signal FET model generally consists of charge sources and current sources. These sources are referred to as the device's state functions. The device state functions can be represented either by functional descriptions or by look-up tables. Different methods exist to determine their electrical properties. At present, RF large-signal models are mostly generated from d.c., C–V, and/or s-parameter measurements. Schreurs *et al.* [62, 63] extracted RF large-signal model parameters directly from large-signal RF measurements, using the nonlinear network measurement system (NNMS). However, the NNMS is not a widespread measurement system [64]. Functional descriptions in large-signal models can be subdivided into three categories: (a) compact models (e.g. BSIM3, MM9, and EKV), (b) empirical models, and (c) neural network models [63].

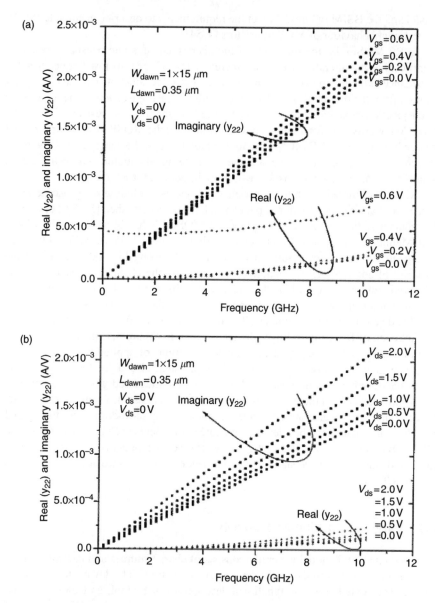

Figure 8.9   *(a) Measured y$_{22}$ data at different gate biases show strong gate-bias dependence and (b) measured y$_{22}$ data at different drain biases show drain-bias dependence [After Y. Cheng and M. Matloubian, IEEE Electron Device Lett., Vol. 21, 2000(604–606). ©2000 IEEE]*

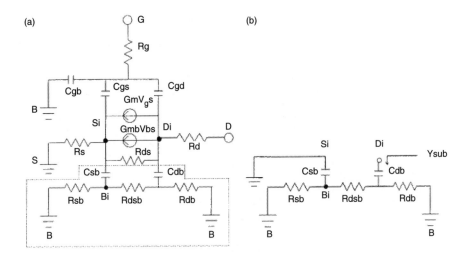

Figure 8.10   *(a) HF equivalent circuit of a MOSFET and (b) the substrate network [After Y. Cheng and M. Matloubian, IEEE Electron Device Lett., Vol. 21, 2000(604–606). ©2000 IEEE]*

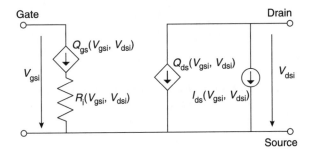

Figure 8.11   *Large-signal MOSFET model. The magnitude of the current source and the charge sources depends on the bias conditions and on the element values of the modified equivalent circuit [After E. P. Vandamme et al., Solid State Electron., Vol. 46, 2002(353–360)]*

The large-signal equivalent circuit-based, RF MOSFET model developed by Vandamme *et al.* [54] is shown in Figure 8.11, which contains a current source at the drain terminal $I_{ds}(V_{gsi}, V_{dsi})$, two charge sources, one at the gate terminal $Q_{ds}(V_{gsi}, V_{dsi})$ and one at the drain terminal $Q_{ds}(V_{gsi}, V_{dsi})$ and a resistance $R_i(V_{gsi}, V_{dsi})$, stemming from non-quasistatic effects. Contrary to MESFETs, the current source at the gate can be omitted for a MOSFET, because no d.c. current flows through the gate (gate tunnelling currents only start playing a role for a gate-oxide thickness below 2 nm, i.e. for 0.1 μm technologies and below). The magnitude of

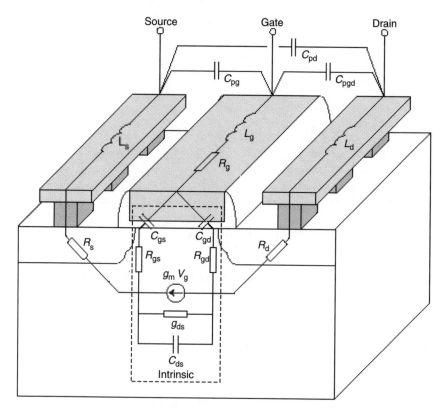

Figure 8.12   Cross-section of a MOSFET with main network elements determining its electrical behaviour (V_g is the voltage across C_{gs}) [After E. P. Vandamme et al., Solid State Electron., Vol. 46, 2002(353–360)]

the charge source and the current source depends on the intrinsic gate and drain voltages, V_{gsi} and V_{dsi}, respectively, as well as on the electrical properties and geometry of the MOSFET.

The MOS transistor, shown schematically in Figure 8.12 can be divided into an intrinsic part and an extrinsic part. The extrinsic elements are bias-independent and consist of the series inductors and series resistors in the gate, source- and drain leads, denoted by L_g, L_s, L_d and R_g, R_s, R_d, respectively, and the parallel coupling capacitors between the pads of gate and source C_{pg}, gate and drain C_{pgd} and drain and source C_{pd}.

The extrinsic elements are extracted from the s-parameters, biased at various V_{gs} while keeping $V_{ds} = 0$ V. However, before the extrinsic element values can be extracted, the measured s-parameters need to be de-embedded; that is, the influence of pad parasitics and interconnections in the RF test structure must be taken into account. Therefore an improved three-step de-embedding procedure [65], based on Reference 66 is required. Next, the de-embedded s-parameters are converted to

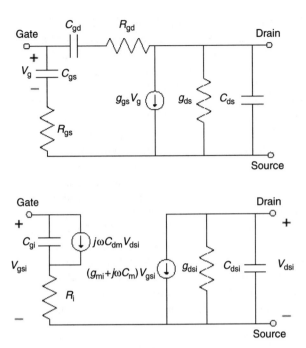

Figure 8.13   Equivalent circuit (top) and modified equivalent circuit (bottom) of the intrinsic part of a MOSFET. The modified equivalent circuit is used for large-signal modelling [After E. P. Vandamme et al., Solid State Electron., Vol. 46, 2002(353–360)]

y- or z-parameters, using a conversion table [67], from which the extrinsic element values are extracted. At $V_{ds} = 0$ V, it can be assumed that $R_{gs} = R_{gd}$, $C_{gs} = C_{gd}$ and $g_m = 0$ S [68], which simplifies the equivalent circuit.

The intrinsic part of the MOS transistor, shown in Figure 8.13, consists of the gate–source and gate–drain capacitors and resistors, C_{gs}, C_{gd} and R_{gs}, R_{gd}, respectively; the transconductance, g_m, the output conductance, g_{ds} and the drain–source junction capacitance, C_{ds} (source and bulk are grounded). For large-signal modelling it is desirable to eliminate the coupling between gate and drain through C_{gd} and R_{gd} [64] This eliminates consistency problems with gate–source and gate–drain charges and leads to a simpler large-signal model with only two internal terminal voltages, V_{gsi} and V_{dsi}. A modified equivalent circuit for the intrinsic part of the MOSFET, is shown in Figure 8.13, with an additional current source and two additional transcapacitances C_m and C_{dm} to model the coupling between gate and drain. It can be shown that, at least up to 50 GHz, this circuit behaves electrically in the same manner as the conventional intrinsic equivalent circuit.

Excellent RF characteristics of CMOS in an integrated BiCMOS process and highly scaled MOS devices with a thin gate oxide have been reported in an advanced industrial 0.25 μm BiCMOS process, offering four levels of Al and a full menu of active and passive devices [69]. The process uses an essential industrial-standard, qualified CMOS platform, and offers SiGeC HBTs with different speed/breakdown voltage and 2.5 V CMOS transistors for digital applications, and isolated NMOS devices for RF applications. Accurate modelling and parameter extraction for RF CMOS have become essential for the development of RF Si CMOS-based circuits. A sub-circuit extension of BSIM3v3 model of a sub-micron MOSFETs for small-signal RF circuit simulation have been reported [70].

For RF applications, models for CMOS transistors have been developed on the basis of BSIM3v3. The improvements of high-frequency behaviour have been modelled using a sub-circuit, which contains an intrinsic MOSFET, and lumped elements for substrate, gate, source, and drain. The substrate and gate resistances significantly affect the model accuracy in the high-frequency range. In the following, the techniques of extraction and optimisation of model parameters for the n-MOSFETs in the frequency range from 0.1 GHz to 40 GHz are discussed.

The equivalent sub-circuit for a RF MOSFET is shown in Figure 8.14. The sub-circuit contains an intrinsic MOSFET (BSIM3v3) and lumped elements for sub-strate, gate, source, and drain. The BSIM3v3 model consists of a non-quasistatic model and a capacitance model, which makes it the ideal for RF simulations.

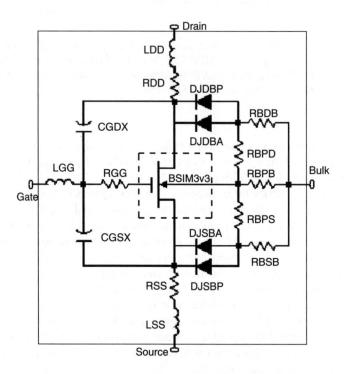

Figure 8.14   RF MOS equivalent circuit using BSIM3v3 [After B. Senapati et al., Proc. IWPSD 2003(818–820)]

In the RF BSIM3v3 macro model, there is a resistance network for the substrate, which is described by the following resistances, namely, RBPD, RBPS, RBDB, RBSB, and RBPB. Together with the resistance network, the internal drain-bulk junction diode and source-bulk junction diodes of the BSIM3v3.2 model are replaced by the external elements, namely, DJDBA, DJDBP, DJSBA, and DJSBP for area and perimeter, respectively. RSS and RDD are resistances of the source and drain, respectively. In order to model high-frequency response, extra lumped elements must be considered for the accurate prediction of the high-frequency characteristics. Thus the inductors LGG, LSS, and LDD and capacitors CGDX and CGSX that are parasitic elements existing between gate, source, or drain are included.

The RF MOSFET sub-circuit model was implemented in the SPICE3 simulator. The model parameter extraction and optimisation were performed from the measured experimental data of the MOSFET using ICCAP [71]. Two-port s-parameters for the frequency range from 0.1 to 40 GHz are measured for different bias conditions using a HP 8510 network analyser and GSG probes. Different control biases are applied from an HP-4142B source measure unit. Model parameters are extracted after open and short de-embedding the probe-pad parasitic.

Figure 8.15 shows the drain currents (I_d) vs. drain voltage (V_d) for different gate biases. The drain currents are in good agreement with the experimental results for low, medium, and high gate biases. Figure 8.16 shows the current gain ($|h_{21}|$) and power gain (G_{max}) of the MOSFET for frequency ranges 0.1 to 40 GHz at a $V_d = 2.5$ V and $V_g = 1.25$ V. Figure 8.17 shows the s-parameters of the MOSFET for frequency

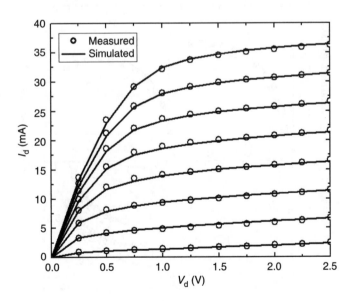

Figure 8.15 Measured and simulated drain current vs. drain voltage for a gate voltage sweep from 0.75 to 2.5 V [After B. Senapati et al., Proc. IWPSD 2003(818–820)]

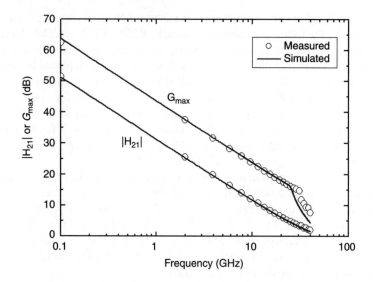

Figure 8.16   Measured and simulated | h_{21} | vs. frequency at a V_g = 1.25 V and V_d = 2.5 V [After B. Senapati et al., Proc. IWPSD 2003(818–820)]

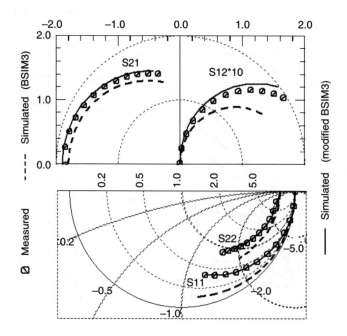

Figure 8.17   Measured and simulated s-parameters at a V_g = 1.25 V and V_d = 2.5 V for a frequency sweep from 0.1 GHz to 21 GHz [After B. Senapati et al., Proc. IWPSD 2003(818–820)]

ranges 0.1 to 21 GHz at a $V_d = 2.5$ V and $V_g = 1.25$ V. Measured data are presented by open circle, the modified BSIM3 model by a solid line and BSIM3 model by a dotted line indicating the relative improvement achieved in the modified BSIM3 model.

8.8 Summary

As the mainstream MOS technology is scaled into the deep sub-micron regime, development of a truly physical and predictive compact model for circuit simulation that covers geometry, bias, temperature, d.c., a.c., RF, and noise characteristics becomes a major challenge. Well-constructed models, however, without accurate model parameters for the used process, may lead to wrong prediction in circuit simulation. Model parameter extraction, in particular from the device characterisation is crucial and remains as an important practical issue that needs serious attention.

The surface-potential is based approach is being developed using solutions of several long-standing problems of compact modelling. These include symmetric linearisation method enabling extremely simple yet accurate expressions for the drain current and terminal charges, non-iterative computation of the surface potential and extension of the model formulation beyond the gradual channel approximation. Approaches to MOSFET compact models have been described. A big advantage of a complete surface-potential-based model is that the overall model consistency is automatically preserved through the surface potential. Therefore, the number of model parameters can be drastically reduced in comparison with conventional models.

The modelling of partially-depleted SOI MOSFETs using the third-generation surface-potential-based models has been discussed. An SP-based model has been shown to be an alternative to the more traditional threshold-voltage-based SOI models. Modelling issues for heterostructure MOSFETs are outlined, which need further attention. A subcircuit model for RF CMOS valid under different bias conditions and RF frequency range up to 20 GHz has been described. The usefulness and accuracy of the n-MOSFET model are discussed. A method to extract the important parameters and fine tuning of the parameters from d.c. and RF measurements is also discussed.

References

1 Q. Huang, F. Piazza, P. Orsatti, and T. Ohguro, 'The impact of scaling down to deep submicron on CMOS RF circuits', *IEEE J. Solid-State Circuits*, vol. 33, pp. 1023–1036, 1998.

2 B. Razavi, 'CMOS technology characterization for analog and RF design', *IEEE J. Solid-State Circuit*, vol. 34, pp. 268–276, 1999.

3 E. Morifuji, H. S. Momose, T. Ohguro, T. Yoshitomi, and H. Kimijima, 'Future perspective and scaling down roadmap for RF CMOS', in *IEEE VLSI Tech. Symp. Dig.*, pp. 163–164, 1999.

4 Y. Cheng, M.-C. Jeng, Z. Liu, J. Huang, M. Chan, K. Chen, *et al.*, 'A physical and scalable I-V model in BSIM 3v3 Analog/Figital Circuit Simulation', *IEEE Trans. Electron Devices*, vol. 44, pp. 277–287, 1997.

5 M. Shur, T. A. Fjeldly, T. Ytterdal, and K. Lee, 'Unified MOSFET Model', *Solid State Electron.*, vol. 35, pp. 1795–1802, 1992.

6 T. Skotnicki, C. Denat, P. Senn, G. Merckel, and B. Hennion, 'A New analog/digital CAD model for sub-halfmicron MOSFETs', in *IEEE IEDM Tech. Dig.*, pp. 165–168, 1994.

7 K. Joardar, S. K. Gullapalli, C. McAndrew, M. E. Bumham, and A. Wild, 'Improved MOSFET model for circuit simulation', *IEEE Trans. Electron Devices*, vol. 45, pp. 134–148, 1998.

8 M. Miura-Mattausch, U. Feldmann, A. Rahm, M. Bollu, and D. Savignac, 'Unified Complete MOSFET Model for analysis of digital and analog circuits', *IEEE Trans. Computer-Aided Design*, vol. CAD-15, pp. 1–7, 1996.

9 Y. Cheng, M. J. Deen, and C.-H. Chen, 'MOSFET modeling for RF IC Design', *IEEE Trans. Electron Devices*, vol. 52, pp. 1286–1303, 2005.

10 IEEE, 'Special issue on advanced compact models and 45-nm modeling challenges,' *IEEE Trans. Electron Devices*, vol. 53, pp. 1957–2202, 2006.

11 D. E. Ward and R. W. Dutton, 'A charge-oriented model for MOS transistor capacitances', *IEEE J. Solid State Circuits*, vol. SC-13, pp. 703–707, 1978.

12 R. P. Jindal, 'Compact noise models for MOSFETs', *IEEE Trans. Electron Devices*, vol. 53, pp. 2051–2061, 2006.

13 D. P. Foty, *MOSFET Modeling with SPICE—Principles and Practice*. Prentice-Hall, 1997.

14 Y. P. Tsividis and G. Masetti, 'Problems in precision modeling of the MOS transistor for analog applications', *IEEE Trans. Computer-Aided Design*, vol. CAD-3, pp. 72–79, 1984.

15 Y. Cheng, M. Chan, K. Hui, M. Jeng, Z. Liu, J. Huang, *et al.*, 'BSIM3v3 manual.' University of California/Berkeley, 1996.

16 W. Liu, *MOSFET Models for Spice Simulation, Including BSIM3v3 and BSIM4*. John Wiley & Sons, NY, 2001.

17 J. Saijets, M. Andersson, and M. A. Berg, 'A comparative study of various MOSFET models at radio frequencies', *Analog Integrated Circuits and Signal Processing*, vol. 33, pp. 5–17, 2002.

18 G. Gildenblat, H. Wang, T.-L. Chen, X. Gu, and X. Cai, 'SP: An advanced surface-potential-based compact MOSFET model', *IEEE J. Solid State Circuits*, vol. 39, pp. 1394–1406, 2004.

19 G. Gildenblat, X. Cai, T. Chen, X. Gu, and H. Wang, 'Reemergence of the surface-potential-based compact MOSFET models', in *IEEE IEDM Tech. Dig.*, pp. 863–866, 2003.

20 M. Bucher, C. Lallement, C. Enz, and F. Krummenacher, 'Accurate MOS modelling for analog circuit simulation using the EKV model', in *IEEE Intl. Symp. Circuits and Systems*, pp. 703–706, 1996.

21 R. van Langevelde and F. M. Klassen, 'An explicit surface-potential-based MOSFET model for circuit simulation', *Solid State Electron.*, vol. 44, pp. 409–418, 2000.

22 H. Wang, T.-L. Chen, and G. Gildenblat, 'Quasi-static and non-quasi-static compact MOSFET models based on symmetrically linearization of the bulk and inversion charges', *IEEE Trans. Electron Devices*, vol. 50, pp. 2262–2272, 2003.

23 T.-L. Chen and G. Gildenblat, 'Analytical approximation for the MOSFET surface potential', *Solid State. Electron.*, vol. 45, pp. 335–339, 2001.

24 W. Wu, T. Chen, G. Gildenblat, and C. C. McAndrew, 'Physics-Based Mathematical Conditioning of the MOSFET Surface Potential Equation', *IEEE Trans. Electron Device*, vol. 51, pp. 1196–1200, 2004.

25 G. Gildenblat, T.-L. Chen, and P. Bendix, 'Analytical approximation for perturbation of MOSFET surface potential by polysilicon depletion layer', *Electron. Lett.*, vol. 35, pp. 1974–1976, 1999.

26 G. Gildenblat, T.-L. Chen, and P. Bendix, 'Closed form approximation for the perturbation of MOSFET surface potential by quantum-mechanical effects', *Electron. Lett.*, vol. 36, pp. 1072–1074, 2000.

27 G. Gildenblat, X. Li, W. Wu, H. Wang, A. Jha, R. van Langevelde, *et al.*, 'PSP: An advanced surface-potential-based MOSFET model for circuit simulation', *IEEE Trans. Electron Devices*, vol. 53, pp. 1979–1993, 2006.

28 H. J. Mattausch, M. Miura-Mattausch, H. Ueno, S. Kumashiro, T. Yamaguchi, K. Yamashita, *et al.*, 'HiSIM: The first complete drift-diffusion MOSFET model for circuit simulation', in *Proc. Solid-State and Integrated-Circuit Technology*, pp. 861–866, 2001.

29 M. Miura-Mattausch, H. J. Mattausch, T. Ohguro, T. Lizuka, M. Taguchi, S. Kumashiro, *et al.*, 'MOSFET model HiSIM based on surface-potential description for enabling accurate RF-CMOS design', *J. Semicond. Technol. Sci.*, vol. 4, pp. 133–139, 2004.

30 M. Miura-Mattausch, N. Sadachika, D. Navarro, G. Suzuki, Y. Takeda, M. Miyake, *et al.*, 'HiSIM2: Advanced MOSFET model valid for RF circuit simulation', *IEEE Electron Device Lett.*, vol. 53, pp. 1994–2007, 2006.

31 G. Gildenblat, C. McAndrew, H. Wang, W. Wu, D. Foty, L. Lemaitre, *et al.*, 'Advanced compact models: gateway to modern CMOS design', in *Proc. 11th Intl. Conf. Electronics, Circuits and Systems*, pp. 638–641, 2004.

32 G. Gildenblat and T. L. Chen, 'Overview of an advanced surface-potential based mosfet model (SP)', in *Proc. Nanotech*, pp. 657–661, 2002.

33 C. C. Enz, F. Krummenacher, and E. A. Vittoz, 'An analytical MOS transistor model valid in all regions of operation and dedicated to low-voltage and low-current applications', *Analog Integrated Circuits and Signal Processing*, vol. 8, pp. 83–114, 1995.

34 J.-M. Sallese, M. Bucher, F. Krummenacher, and P. Fazan, 'Inversion charge linearization in MOSFET modeling and rigorous derivation of the EKV compact model', *Solid State Electron.*, vol. 47, pp. 677–683, 2003.

35 R. M. D. A. Velghe, D. B. M. Klaassen, and F. M. Klaassen, 'MOS model 9.' Unclassified report NL-UR 003/94, Philips Electronics N.V., 1994.

36 P. Antognetti and G. Massobrio, *Semiconductor Device Modeling with SPICE*. McGraw Hill Book Co., NY, 1987.

37 S. K. H. Fung, M. Khare, D. Schepis, W. Lee, S. H. Ku, H. Park, *et al.*, 'Gate length scaling accelerated to 30 nm regime using ultra-thin film PD-SOI technology', in *IEEE IEDM Tech. Dig.*, pp. 629–632, 2001.

38 D. Sinitsky, S. Tang, A. Jangity, F. Assaderaghi, G. Shahidi, and C. Hu, 'Simulation of SOI devices and circuits using BSIM3SOI', *IEEE Electron Devices Lett.*, vol. 19, pp. 323–325, 1998.

39 Cadence Design Systems, 'BSIMProPlus Datasheet', 2003.

40 P. Su, S. K. H. Fung, S. Tang, F. Assaderaghi, and C. Hu, 'BSIMPD: A partial-depletion SOI MOSFET model for deep-submicron CMOS designs', in *IEEE CICC Proc.*, pp. 197–200, 2000.

41 C. K. Maiti, and G. A. Armstrong, *Applications of Silicon-Germanium Heterostructure Devices*. Inst. of Physics Pub., UK, 2001.

42 K. Chandrasekaran, X. Zhou, S. B. Chiah, W. Shangguan, and G. H. See, 'Physics-based single-piece charge model for strained-Si MOSFETs', *IEEE Electron Devices Lett.*, vol. 52, pp. 1555–1562, 2005.

43 C. K. Maiti, L. K. Bera, and S. Chattopadhyay, 'Strained-Si Heterostructure field effect transistors', *Semicond. Sci. Technol.*, vol. 13, pp. 1225–1246, 1998.

44 J. G. Fossum, 'UFSOI MOSFET models (ver. 7.0) user's guide, Univ. of Florida, Gainesville', 2002.

45 J. G. Fossum and W. Zhang, 'Performance Projections of Scaled CMOS Devices and Circuits with Strained Si-on-SiGe Channels', *IEEE Trans. Electron Devices*, vol. 50, pp. 1042–1049, 2003.

46 M. M. Pelella and J. G. Fossum, 'On the performance advantage of PD/SOI CMOS with floating bodies', *IEEE Electron Devices Lett.*, vol. 49, pp. 96–104, 2002.

47 J. N. Burghartz, M. Hargrove, C. S. Webster, R. A. Groves, M. Keene, K. A. Jenkins, *et al.*, 'RF potential of a 0.18-μm CMOS logic device technology', *IEEE Trans. Electron Devices*, vol. 47, pp. 864–870, 2000.

48 T. Manku, 'Microwave CMOS—device physics and design', *IEEE J. Solid State Circuits*, vol. 34, pp. 277–285, 1999.

49 C.-S. Chang, C.-P. Chao, J. G. J. Chern, and J. Y.-C. Sun, 'Advanced CMOS technology portfolio for RF IC applications', *IEEE Trans. Electron Devices*, vol. 52, pp. 1271–1285, 2005.

50 W. Liu, R. Gharpurey, M. C. Chang, U. Erdogan, R. Aggarwal, and J. P. Mattia, 'RF MOSFET modeling accounting for distributed substrate and channel resistances with emphasis on the BSIM3v3 SPICE model', in *IEEE IEDM Tech. Dig.*, pp. 309–312, 1997.

51 R. R. J. Vanoppen, J.A.M.Geelen, and D. B. M. Klassen, 'The high-frequency analogue performance of MOSFETs', in *IEEE IEDM Tech. Dig.*, pp. 173–176, 1994.

52 S. H.-M. Jen, C. C. Enz, D. R. Pehlke, M. Schroter, and B. J. Sheu, 'Accurate modeling and parameter extraction for MOS transistors valid up to 10 GHz', *IEEE Trans. Electron Devices*, vol. 46, pp. 2217–2227, 1999.

53 C. E. Biber, M. L. Schmatz, and T. Morf, 'Improvements on a MOSFET model for non-linear RF simulations', in *IEEE MTT-S Int. Microwave Symp. Dig.*, pp. 865–868, 1997.

54 E. P. Vandamme, D. Schreurs, C. van Dinther, G. Badenes, and L. Deferm, 'Development of a RF large signal MOSFET model, based on an equivalent circuit, and comparison with the BSIM3v3 compact model', *Solid State Electron.*, vol. 46, pp. 353–360, 2002.

55 M. B. Das, 'High-Frequency Network Properties of MOS Transistors Including the Substrate Resistivity Effects', *IEEE Trans. Electron Devices*, vol. ED-16, pp. 1049–1069, 1969.

56 J.-J. Ou, X. Jin, I. Ma, C. Hu, and P. R. Gray, 'CMOS RF modeling for GHz communication IC', in *IEEE IEDM Tech. Dig.*, pp. 94–95, 1998.

57 S. F. Tin, A. A. Osman, K. Mayaram, and C. Hu, 'A simple subcircuit extension of the BSIM3v3 model for CMOS RF design', *IEEE J. Solid State Circuits*, vol. 35, pp. 612–624, 2000.

58 D. R. Pehlke, M. Schroter, A. Burstein, M. Matloubian, and M. F. Chang, 'High-frequency application of MOS compact models and their development for scalable RF model libraries', in *IEEE CICC Conf. Dig.*, pp. 219–222, 1998.

59 Y. Cheng, M. Schroter, C. Enz, M. Matloubian, and D. Pehlke, 'RF modeling issues of deep-submicron MOSFETs for circuit design', in *Proc. Solid-State and Integrated Circuit Technology*, pp. 416–419, 1998.

60 S. F. Tin and K. Mayaram, 'Substrate network modeling for CMOS RF circuit simulation', in *IEEE CICC Proc.*, pp. 583–586, 1999.

61 Y. Cheng and M. Matloubian, 'On the high-frequency characteristics of substrate resistance in RF MOSFETs', *IEEE Electron Device Lett.*, vol. 21, pp. 604–606, 2000.

62 D. M. M.-P. Schreurs, J. Verspecht, S. Vandenberghe, and E. Vandamme, 'Straightforward and accurate nonlinear device model parameter-estimation method based on vectorial large-signal measurements', *IEEE Trans. Microwave Th. and Tech.*, vol. 50, pp. 2315–2319, 2002.

63 D. Schreurs, S. Vandenberghe, G. Carchon, B. Nauwelaers, E. Vandamme, G. Badenes, *et al.*, 'Evaluation of non-linear modelling techniques for MOSFETs based on vectorial large-signal measurements', in *IEEE ISCAS Symp. Proc.*, pp. 429–432, 2000.

64 P. Jansen, D. Schreurs, W. De Raedt, B. Nauwelaers, and M. Van Rossum, 'Consistent small-signal and large-signal extraction techniques for heterojunction FET's', *IEEE Trans. Microwave Th. and Tech.*, vol. 43, pp. 87–93, 1995.

65 E. P. Vandamme, D. M. M.-P. Schreurs, and G. Van Dinther, 'Improved Three-Step De-embedding Method to Accurately Account for the Influence of Pad Parasitics in Silicon On-Wafer RF Test-Structures', *IEEE Trans. Electron Devices*, vol. 48, pp. 737–742, 2001.

66 H. Cho and D. Burk, 'A Three Step Method for the De-embedding of High Frequency S-parameter Measurements', *IEEE Trans. Electron Devices*, vol. 38, pp. 1371–1375, 1991.
67 G. Gonzalez, *Microwave transistor amplifiers: Analysis and design*. Prentice-Hall, 1997.
68 Y. P. Tsividis, *Operation and Modeling of the MOS Transistor*. McGraw-Hill, 1999.
69 K.-E. Ehwald , D. Knoll, B. Heinemann, K. Chang, J. Kirchgessner, R. Mauntel, *et al.*, 'Modular integration of high-performance SiGe:C HBTs in a deep sub-micron, epi-free CMOS process', in *IEEE IEDM Tech. Dig.*, pp. 561–564, 1999.
70 B. Senapati, K.-E. Ehwald, D. Knoll, and F. Furnhammer, 'An extended BSIM3v3 model for RF MOSFETs', in *Proc. IWPSD*, pp. 818–820, 2003.
71 Hewlett-Packard Company, San Francisco, CA, *IC-CAP User's Manual — High Frequency Model Tutorials*, 1997.

Chapter 9
HBT: compact models

The demand for high-speed wireless and communication applications has pushed the development of SiGe and SiGeC epitaxial base HBT processes [1]. The corresponding circuit design activities essentially depend on accurate compact models of SiGe HBTs in all relevant modes of the transistor operation. Unlike the high-performance digital CMOS that can afford a few years between products, the analog and mixed-signal applications involving SiGe demands significantly reduced product cycle time. Thus, there is an intimate relationship between the TCAD, process development and circuit designers.

For many advanced high-speed and RF circuits, the impact of substrate parasitic elements on bipolar transistor electrical performances cannot be neglected. These parasitic elements influence strongly the output characteristics. A compact model consists of lumped elements such as resistors, diodes, capacitors and dependent current sources connected together with minimum number of nodes to form an equivalent circuit to represent a semiconductor device. All lumped elements can be calculated directly from layout and technological data. Development of reliable device models is an essential part in the circuit design for analog and mixed-signal applications. Accurate device characterisation also plays an important and critical role in model development and circuit benchmarking.

The combination of narrow basewidth, due to epitaxial growth techniques, has resulted in a tremendous performance boost of SiGe-based bipolar transistors. In order to capture the resulting new electrical and physical effects occurring in such technologies, improved compact models are needed for predicting accurately the circuit performance. A compact model should incorporate all physics relevant to future technology generations with a small number of model parameters compared to models used in industry today [2–7].

In general, two strategies for developing compact models are possible: (1) constructing numerical tables with interpolation formulae that have little physical basis to a physical model and (2) deriving physics-based compact mathematical expressions. Table models are easier to construct but are very much limited in use. The term

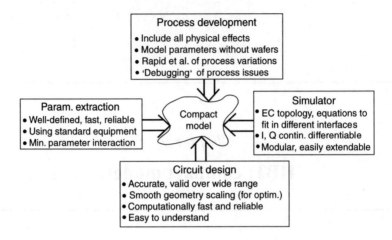

Figure 9.1   Requirements for compact model development [After M. Schroter, CMC Meeting presentation]

physics-based refers to the relation of the model equations to the underlying physics describing the operation of a device. It should be mentioned that a successful compact model and its development have to satisfy many requirements from different directions (see Figure 9.1). Such a model development requires the background in respective transistor theory, process information and device simulation. Preceding all these requirements is the necessity to create carefully controlled device measurement test structures that allow for accurate extraction of the desired device and circuit parameters.

Robust measurement and modelling of high-speed semiconductor devices, particularly at the extreme levels of performance found in modern SiGe devices, require extreme care in getting reliable data, and importantly, a model to fit the data. The characterisation of high-performance SiGe HBTs, and associated passives, requires care owing to the demanding requirements associated with these devices. Circuit measurement is particularly demanding for applications utilizing SiGe devices owing to the high-frequency and low-noise characteristics of these designs [8,9]. Frequently, off-the-shelf measurement solutions do not exist to satisfy the needs.

Compact models should be fully symmetric, model all intrinsic capacitances and, as such, be capable of using a small number of extraction parameters to model analog and RF performance of the devices representing the most advanced technology nodes. Further, it is highly desirable that this model can be used for the extrapolation of future technologies and device parameter variation based on technology fluctuations. The development of sophisticated compact models takes 10–20 years before they actually become suitable for production circuit design.

Compact models are judged on their success in predicting large-signal behaviour and this is where the real difference is found between the various models. The small-signal behaviour may be found by the linearisation of the large-signal circuit for a given bias point. A sound compact transistor model should have the following attributes:

- physics-based formulation for the prediction of electrical behaviour;
- accurate and valid over a wide operational range;
- scalability for various transistor configurations;
- fast and numerically stable;
- provide model parameters even if wafers are not available (predictive);
- suitable to assist process development;
- capable for the rapid evaluation of impact of process variations.

As the field of compact modelling is enormous, ranging from bipolar (vertical and lateral, homo- and heterojunctions) to MOSFET (lateral and vertical, MESFET and HEMT) models, the scope of this chapter will be limited to treating only the vertical bipolar models (as lateral bipolar device are less frequently used), with emphasis on their applicability for RF circuit design. Although the available models were developed mainly for the homojunction devices, extensions to heterojunction devices (viz., SiGe-base) are possible and are considered in this chapter. Compact circuit models usually consist of a set of model equations that are either empirical or derived from device physics, or a combination of both. Historically, one can distinguish the models as follows:

- Physical models (generally computed from doping profile etc.) that express all the compact-model elements in physical parameters. The advantages are that they are accurate and suitable for yield calculations but take long calculation time.
- Compact models (extracted) are defined by topology and element values/equations (analytical equation for every circuit component). The advantages are that they are fast, accurate if parameters are correct and to some extent suitable for yield calculations. The disadvantages are that they are inaccurate, if model does not fit the device or parameters are inaccurate.
- Database models are obtained from bulk measurements. Although they are accurate and fast, they require a large number of measurements, are not accurate outside the measured regions and are also not scalable. Model complexity is defined by number of nodes and elements in the equivalent circuit along with the element equations and number of parameters. Currently the following transistor models are available and are listed below according to their complexity:
- Ebers-Moll (EM);
- Gummel-Poon (GP);
- Extended Gummel-Poon;
- VBIC;
- MEXTRAM;
- HICUM.

The goal of this chapter is to provide an overview on vertical bipolar inter-company (VBIC), high current model (HICUM) and most exquisite transistor model (MEXTRAM) and their associated internal circuits necessary to deploy the model for design purposes. These include (1) the basic operating principle, (2) how the physical effects in advanced transistors are covered, (3) the geometry scaling methods used, (4) the parameter determination methodology, (5) how a hierarchy of models can be generated in an efficient way and (6) future research directions.

9.1   Advanced models

The first compact bipolar model was published in 1954 by Ebers and Moll [10]. It was originally only a static (d.c.) model. Extension of the basic EM model has resulted in a simple but fast compact model and is suitable for the simulation of large digital circuits. Since a number of physical effects occur in integrated circuits that are not included in the EM model, it was replaced by the GP model in 1970 [11, 12] and is based on the integral charge control relation (ICCR), an extension of the Moll-Ross relation [13], which includes the integral charge control concept. The charge control formulation of the GP model accounts for the key physical parameters and mechanisms that control bipolar junction transistor (BJT) behaviour in an intuitive and consistent manner, so that a number of second-order effects such as the forward and reverse Early effect, and high injection are naturally included [14].

Several variants of the GP model have been proposed, and one of the well-known variants found in most circuit simulators today is the SPICE Gummel-Poon (SGP) model [15–17]. The equivalent circuit for an integrated vertical npn transistor in the SGP model found in most commercial simulators is shown in Figure 9.2. Modified versions of this model can also be found in many commercial simulators such as H-SPICE (Meta-software), SmartSPICE (Silvaco), PSPICE (Microsim), MSPICE (Mentor Graphics) and SPECTRE. Additional elements such as capacitors and resistors are added to the existing equivalent circuit to account for parasitics. Because the depletion and diffusion charge components are not used in the definition of

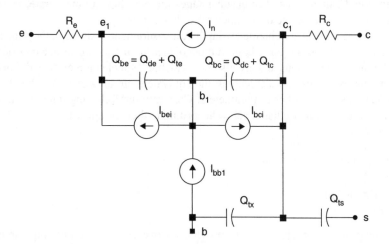

Figure 9.2   Topology of the equivalent circuit diagram for SGP model [After H. C. de Graaff, Proc. ESSDERC, 1997]

the main current. In, the d.c. and a.c. characteristics in the GP model are more or less decoupled. This gives a greater freedom in the parameter extraction, but results in a non-physical parameter set that is not suitable for proper scaling of the transistor [16].

Continued scaling of bipolar devices has resulted in a high performance but, at the same time, has caused increased concern for problems such as quasi-saturation, avalanche multiplication and self-heating effects, so SGP is not sufficiently accurate for state-of-the-art scaled bipolar devices [18, 19]. With the complicated transistor models used in the circuit simulators today, device characterisation with respect to circuit models becomes a non-trivial task. To meet the needs for accurate modelling of advanced devices, several compact bipolar models such as Most EXquisite TRansistor Model (MEXTRAM) largely depending on the modelling of the total stored charge from Philips [20] and HIgh CUrrent Model (HICUM) from the Ruhr University at Bochum [21, 22] are being developed independently and consistently by dedicated developers in conjunction with bipolar process development.

In the United States, efforts have been made to formulate new standard models for MOS and bipolar. The Compact Model Council (CMC), mostly consisting of US semiconductor companies and CAD suppliers, has been founded to 'promote the international, non-exclusive standardization of compact model formulations, the model interfaces and their availability in circuit simulators'. The VBIC model for silicon bipolar junction transistors has been developed as an industry standard and public domain replacement for the SGP model [23]. Presently there are two compact models recommended by CMC: MEXTRAM and HICUM.

In the following, the focus will be on the state-of-the-art compact models for vertical bipolar transistors. The four most widely known models (SGP, MEXTRAM, HICUM and VBIC) will be reviewed and compared with each other with respect to the modelled physical effects and to their performance. Comparison will be made in terms of physical effects such as main current and charges, quasi-saturation, collector-base depletion charge, emitter-base current crowding, influence of extrinsic regions, non-quasistatic effects, avalanche and thermal modelling. A performance comparison will also be made with an emphasis on the applicability to RF circuit design. SiGe HBTs non-linear distortion, geometrical scaling, statistical modelling and thermal modelling for self-heating will receive explicit attention.

9.2 VBIC model

In 1994, efforts were taken to develop a new industry-standard model for vertical bipolar transistors. The outcome has been the development of a new compact VBIC model [23, 24], which is an extension of the model developed by Kull *et al.* [25].

The VBIC model has been designed to be a more complex, improved version of SGP model with the following additional features:

- current-dependent base resistance modelling;
- modelling of the collector epilayer resistance;

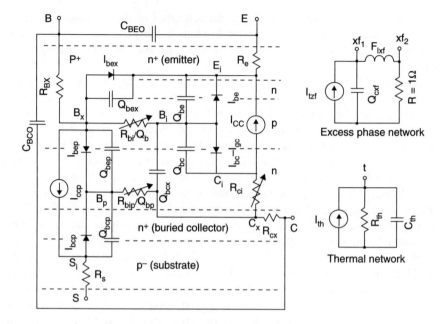

Figure 9.3   *Equivalent network of a npn transistor in the VBIC model. Thermal network is also shown [After B. Senapati and C. K. Maiti, IEE Proc. Circuits, Devices and Systems, Vol. 149, 2002(129–135). ©2002 IEEE]*

- self-heating of a bipolar transistor;
- forward and reverse Early effect based on the junction charges;
- high-injection effects in the base;
- excess phase modelled as a separate subcircuit;
- parasitic substrate pnp transistor;
- weak avalanche injection.

Figure 9.3 shows the VBIC equivalent network for a npn transistor. The dotted lines separate the different active and passive regions of the transistor. The parasitic substrate pnp is modelled by a simplified GP transistor. Quasi-saturation is modelled after Reference 25 with the elements RCI, Q_{bcx} and a modified Q_{bc} and by including the elements I_{bex} and Q_{bex}, along with RBI, I_{be} and Q_{be}. In contrast to the GP model, the base current is not connected to the collector current. In the VBIC model, both ideal and non-ideal base currents are calculated from separate saturation currents and ideality factors. Consequently, the temperature dependence of each current mechanism is modelled separately. Excess phase is modelled with a second-order network that implements the Weil–McNamee approximation [26]. A weak avalanche current Igc is also included for the b–c junction [27].

The VBIC model includes two depletion charge models, the standard SGP model and a single-piece model, in which the capacitance is adjusted to a constant value for a forward-bias junction. The base-emitter depletion charge is partitioned between

Q_{be} and Q_{bex}. The base-collector charge element (Q_{bc}) includes a depletion charge and a diffusion charge term and is modelled together with Q_{bcx}, the charge associated with the base push-out into the collector. Finally, to model capacitances associated with extrinsic parasitics, the constant capacitances CBEO and CBCO are included. Note that in Figure 9.3 the intrinsic base resistances, RBI and RBIP, are modulated by the normalized base charges, Q_b and Q_{bp}, respectively, and the intrinsic collector resistance, RCI, is modulated by V_{bci}.

The Early effect approximation in the SGP model is known to cause inaccuracies in output conductance modelling for modern BJTs [28], so an Early effect model based on the junction depletion charges significantly improves output conductance modelling. However, this means that existing SGP model parameter sets need to have the Early voltages adjusted to work reasonably with VBIC [29]. VBIC includes noise models similar to those of the SGP model, with shot, thermal and 1/f components. BJT electrical behaviour varies with temperature, and so VBIC has temperature mappings defined for its model parameters. The mappings are similar to, but improved on, the temperature mappings for the SGP model. The resistances have a temperature variation [29]

$$R(T_2) = R(T_1)(T_2/T_1)^{XR} \tag{9.1}$$

where the temperatures are in degrees K, and there is a separate XR parameter for each doping type region. This follows from an empirical model of the temperature dependence of mobility [30]. The transport saturation current varies as

$$I_s(T_2) = I_s(T_1)(r_T^{XIS} \exp(-E_A(1 - r_T)/V_{t\beta}))^{1/NF} \tag{9.2}$$

where $r_T = T_2/T_1$ and XIS and E_A are parameters. This is similar to the SGP model and is based on the variation of n_{ie}^2 with temperature. The other saturation currents have similar temperature mappings. This allows fitting of β_F and β_R over bias and temperature. The forward transport current (I_{cc}) of the intrinsic transistor is modelled with the forward (I_{tfi}) and reverse (I_{tri}) current components as

$$I_{cc} = \frac{1}{Q_b}(I_{tfi} - I_{tri})$$

$$I_{tfi} = IS\left(\exp\left[\frac{qV_{bei}}{NF \cdot kT}\right] - 1\right); \quad I_{tri} = IS\left(\exp\left[\frac{qV_{bci}}{NR \cdot kT}\right] - 1\right) \tag{9.3}$$

where $V_{bei} = V_{bi} - V_{ei}$ and $V_{bci} = V_{bi} - V_{ci}$ are the branch voltages. The normalized base charge, Qb, is modelled using depletion and diffusion charge components, forward and reverse Early voltages (VEF,VER) and knee currents (IKF, IKR) as

$$q_1 = 1 + \frac{q_{je}}{VER} + \frac{q_{jc}}{VEF}$$

$$q_2 = \frac{I_{tfi}}{IKF} + \frac{I_{tri}}{IKR} \tag{9.4}$$

$$q_b = 0.5\left(q_1 + \sqrt{q_1^2 + 4q_2}\right)$$

where $q_{je}(V_{bei},PE,ME)$ and $q_{jc}(V_{bci},PC,MC)$ are the base-emitter and base-collector charges, respectively.

The base current (I_b) is divided into base-emitter current (I_{be}) and base-collector current (I_{bc}). The total base-emitter current (I_{bt}) is further partitioned between (I_{be}) and (I_{bex}) to model the distributed nature of the base and are given by

$$I_{bt} = \text{IBEI}\left(\exp\left[\frac{qV_{bei}}{NEI \cdot kT}\right] - 1\right) + \text{IBEN}\left(\exp\left[\frac{V_{bei}}{NEN \cdot kT}\right] - 1\right) \tag{9.5}$$

$$I_{be} = \text{WBE} \cdot I_{bt}; \quad I_{bex} = (1 - \text{WBE}) \cdot I_{bt}$$

where WBE is a model parameter that varies from 1 to 0. The base-collector current (I_{bc}) includes both the ideal and non-ideal components and is given by

$$I_{bc} = \text{IBCI}\left(\exp\left[\frac{qV_{bci}}{NCI \cdot kT}\right] - 1\right) + \text{IBCN}\left(\exp\left[\frac{qV_{bci}}{NCN \cdot kT}\right] - 1\right) \tag{9.6}$$

and also a weak avalanche current component (I_{gc}) is given by Reference 31

$$I_{gc} = (I_{cc} - I_{bc})\text{AVC1}(PC - V_{bci})\exp[-\text{AVC2}(PC - V_{bci})^{ME-1}] \tag{9.7}$$

where AVC1 and AVC2 are model parameters. The intrinsic collector current (I_{epi}) is modelled with the enhanced Kull's quasi-saturation model [25] with the elements: intrinsic collector resistance (RCI), epitaxial charge (QCO) and epi-doping factor (GAMM). The transport current (I_{ccp}) of the parasitic transistor can be divided into two components: forward and reverse. Owing to the distributed nature of the parasitic components, the forward component of I_{ccp} can further be split into $V_{bep}(= V_{bx} - V_{bp})$ and $V_{bci}(= V_{bi} - V_{ci})$ with model parameters WSP and $(1-\text{WSP})$ as

$$I_{tfp} = \text{ISP}\left(\text{WSP}\exp\left[\frac{qV_{bep}}{NFP \cdot kT}\right] + (1 - \text{WSP})\exp\left[\frac{qV_{bci}}{NFP \cdot kT}\right] - 1\right)$$

$$I_{trp} = \text{ISP}\left(\exp\left[\frac{qV_{bcp}}{NFP \cdot kT}\right] - 1\right) \tag{9.8}$$

$$I_{ccp} = \frac{1}{Q_{bp}}(I_{tfp} - I_{trp})$$

The normalized base charge, Q_{bp}, of the parasitic transistor includes only high-level forward injection and is given by

$$Q_{bp} = 0.5\left(1 + \sqrt{1 + 4\frac{I_{tfp}}{\text{IKP}}}\right) \tag{9.9}$$

The base-emitter current (I_{bep}) and the base-collector current (I_{bcp}) of the parasitic pnp transistor are modelled similarly as in case of the intrinsic npn transistor described above. The parasitic base current (I_{rbp}) is further controlled by the parasitic base resistance (RBIP). The intrinsic base resistances (RBI) and parasitic resistance (RBIP) are modulated by the normalized base charges Q_b and Q_{bp}, respectively.

Excess phase is modelled with a second-order network in the VBIC model as a separate option. A self-heating model (released as a separate option for VBIC)

Table 9.1 *VBIC model parameter list for a transistor at room temperature [After B. Senapati, 'Modelling and Characterization of SiGe-HBTs for RF Applications', PhD Thesis, Indian Institute of Technology, Kharagpur, 2001.]*

Basic-effects	Parameter	Measurements
Basic d.c.	IS, NF, NR, IBEI, NEI, IBCI, NCI	GP ($I_{C,B}-V_{be}$, $I_{E,B}-V_{bc}$)
β-degradation	IBEN, NEN, IBCN, NCN, IKF, IKR	GP ($I_{C,B}-V_{be}$, $I_{E,B}-V_{bc}$)
Junction capacitor	CJE, PE, ME, AJE	Capacitance–Voltage (C–V)
	CJC, PC, MC, FC, AJC	
Early effects	VEF, VER	$I_{C,E}-V_{CE,EC}$, C–V
Parasitic transistor	ISP, NFP, IBEIP, IBENP, IBCIP	GP($I_{S,B}-V_{bc}$)
	NCIP, IBCNP, NCNP, IKP	(Parasitic pnp)
Parasitic capacitance	CJEP, CJCP, MS, PS, AJS	C–V (Parasitic pnp)
Extrinsic capacitance	CBEO, CBCO	C–V
Resistors	RCX, RBX, RBI, RBP, RE, RS	GP ($I_{C,B}-V_{be}$), ($V_{ce}-I_b$)
Partitioning	WSP, WBE	Optimisation
Quasi-saturation	RCI, HRCF, QCO, GAMM, VO	High current (I_c-V_{ce})
Avalanche	AVC1, AVC2	High bias (I_c-V_{ce})
Self-heating	RTH, CTH	High current (I_c-V_{ce})
Transit time	TF, QTF, XTF, VTF	High frequency (f_T-I_c)
	ITF, TR, TD	

includes the thermal resistance RTH and capacitance CTH, along with the thermal power source Ith, which couples, the power generated in the BJT to the thermal network. The local temperature rise at node t is linked to the electrical model through the temperature mappings of the model parameters. A complete VBIC model parameter list, grouped according to basic effects, is given in Table 9.1. The terminal currents of the transistor are given by

$$I_c = I_{cc} - I_{bc} + I_{gc} + I_{rbp}$$
$$I_b = I_{be} + I_{bex} + I_{bc} - I_{gc} + I_{bep} + I_{ccp}$$
$$I_e = -I_{cc} - I_{be} - I_{bex}$$
$$I_s = -I_c - I_b - I_e$$

(9.10)

In comparison to the GP model, the base currents are no longer directly related to the collector current, and an alternate formulation of the normalized base charge Qb has been used in order to obtain dynamic Early voltages. The shortcomings of the VBIC model are the following:

- circuit structure that involves many (7) internal nodes and circuit elements;
- a clear and unified parameter extraction being difficult for the full model;
- incorrect epilayer model;
- limitations in the modelling of lightly doped epilayers;

- modelling of $f_T(Ic)$ at the onset of quasi-saturation;
- incorrect modelling of the Early effect in the presence of high injection;
- no a.c. current crowding;
- incorrect modelling of distortion at high frequency and current levels.

By not using the advanced features the VBIC can be made to be similar to the SGP model, and the existing extraction techniques can be applied for some of the VBIC parameters. Cao *et al.* [32] have developed such a methodology to extract the parameters of Si-BJTs for the VBIC model.

9.3   MEXTRAM model

MEXTRAM was introduced by De Graaff and Kloosterman at Philips in 1985 as Level 501 [20]. This was followed by Level 502 in 1987 [33], Level 503 in 1994 and Level 504 in 2000 [34]. MEXTRAM bipolar transistor model has been placed in the public domain since 1994 [35,36]. MEXTRAM is employed for intensive physical modelling of the quasi-saturation phenomena including the base widening, Kirk effect [14], hot-carrier behaviour in the epilayer, and advance modelling of distortion effects [37,38]. In the Level 504 model, special attention is paid to SiGe transistors, and self-heating effects are included in the description using a one-pole thermal equivalent circuit similarly to the approach implemented in VBIC. Significantly for advanced technology devices MEXTRAM can model effects of a graded Ge profile in SiGe HBTs [39].

Several physical phenomena such as current gain, output conductance, base push-out, cut-off frequency, noise behaviour and temperature scaling have been included in MEXTRAM. It is suitable for digital and analog circuit design and has demonstrated accuracy in a wide variety of applications [40–44]. MEXTRAM is also based on a reliable, robust and unambiguous transistor parameter extraction method [45–47]. Most parameters of the MEXTRAM model can be extracted directly from depletion capacitance or terminal currents measurements at different temperatures. The evolution of the equivalent circuit for MEXTRAM is shown in Figures 9.4, 9.5, 9.6, and 9.7.

A new version (Level 504) of the MEXTRAM compact modelling procedure for bipolar transistors has been released by the Philips Research Laboratories. Besides other improvements, the new MEXTRAM version is now capable of modelling of SiGe transistors. The formulations for a graded base SiGe HBT and neutral base recombination have been incorporated in the MEXTRAM 504 [44], which have been implemented in a number of commercial simulators. MEXTRAM provides a sophisticated model for modelling avalanche effects in the b-c junction. MEXTRAM uses base–charge partitioning to describe the non-quasistatic effects. A full parameter extraction using the compact MEXTRAM model has also been reported [42,48].

Thermal phenomena of bipolar transistors are addressed in MEXTRAM by a simple thermal impedance and a power dissipation source. A set of

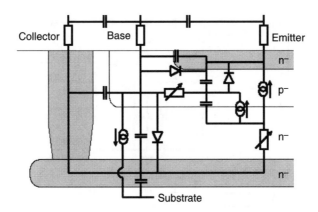

Figure 9.4   *Equivalent circuit without any optional element [Source: Mextram Manual]*

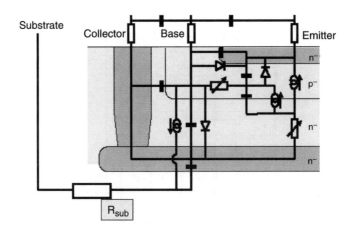

Figure 9.5   *Equivalent circuit with the substrate resistance [Source: Mextram Manual]*

temperature-scaling parameters is extracted along with the corresponding electrical model parameters in the straight forward procedure. The physics-based modelling of the MEXTRAM parameters also provides the opportunity for the geometrical scaling [41,49]. MEXTRAM has been used in various SiGe HBTs applications including high-performance SiGe HBT integration [42], low-noise amplifiers [50–52] and SiGe HBTs phototransistors [43]. MEXTRAM uses an additional collector node to model the epilayer in terms of a separate network element, but without making use of the possibilities of partitioning the epilayer charge similarly to Kull's epilayer model. Sheridan *et al.* [53] have employed MEXTRAM (Level 504) parameter extraction for the IBM self-aligned high-breakdown 0.32 μm BiCMOS SiGe graded-base test technology.

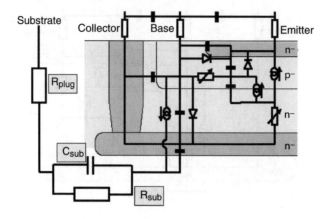

Figure 9.6   Equivalent circuit with the substrate resistance including capacitances [Source: Mextram Manual]

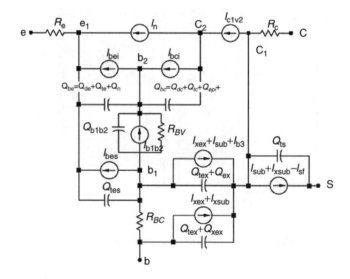

Figure 9.7   Topology of the equivalent circuit diagram for MEXTRAM model [After H. C. de Graaff, Proc. ESSDERC, 1997]

9.4   The HICUM model

HIgh CUrrent bipolar compact transistor Model (HICUM) is the result of bipolar modelling research conducted at the Ruhr University-Bochum, Germany by Schroter and Rein [21, 54–56]. HICUM development resulted from the experience that the SGP model is not accurate enough for high-speed large-signal transient applications and the required high collector current densities. HICUM was originally designed to improve on the standard GP model in forward operation or weak saturation. The

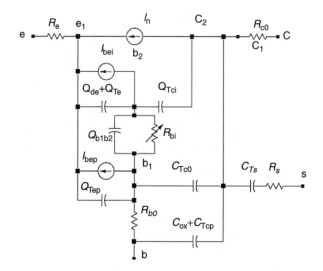

Figure 9.8   Topology of the equivalent circuit diagram for HICUM model [After H. C. de Graaff, Proc. ESSDERC, 1997]

model was subsequently extended to take account of the parasitic substrate transistor, saturation and quasi-saturation, avalanche and tunnelling currents, and self-heating effects resulting in the equivalent circuit shown in Figure 9.8 HICUM now employs an advanced transistor model for the bipolar transistors with main emphasis on circuit design for high frequency ECL-type circuits for fibre-optic applications.

Since its first appearance, the model has always been updated to accommodate the changes brought to bipolar fabrication techniques. The model has been extended to high-speed small-signal applications since 1989 as well as to SiGe HBTs since 1993. Extensions for graded-base SiGe HBTs have been derived using the generalized integral charge control relation (GICCR), which permits modelling of HBTs with (graded) bandgap differences within the junctions. HICUM uses an implementation of the Winkel's delayed charge control model to describe non-quasistatic effects. HICUM allows one to model reverse-bias tunnelling currents in the e-b diode, a feature that could be of interest if the e-b diode is used as a varactor diode. Base-emitter tunnelling model is also available (e.g., for simulation of varactor leakage). Simple parasitic substrate transistor is included in the equivalent circuit and parallel RC network takes into account the frequency-dependent coupling between the buried layer and the substrate terminal. An equivalent circuit for HICUM is presented in Figure 9.8. HICUM is based on an extended and generalized integral charge control relation (GICCR) [7, 57]. However, in contrast to the (original) GP model as well as the SGP model and its variants, in HICUM the GICCR concept is applied consistently without inadequate simplification and additional fitting parameters (such as the Early voltage). However, the equations governing the HICUM model are much more complicated and numerous than available in the SGP model. Understanding them requires a more in-depth look at the physical properties of the BJT. The

HICUM model has been extended to simulate high-frequency small-signal operation of BJTs [22] and advanced SiGe heterojunction bipolar transistors [7, 58–60], and further developments are still ongoing following the advancements in bipolar transistor technology and design [61,62]. Details of HICUM model equations can be found in Reference 63. The current dependence of the internal b-c depletion capacitance is neglected in HICUM for simplicity, which is justified to a large extent because of small current levels. In its present form, HICUM employs as many as 87 transistor-specific model parameters and is designed for geometric scalability, and a separate program called TRADICA is available to generate HICUM model parameters from layout data. For a comprehensive review on HICUM for RF bipolar modelling and circuit applications, the reader may refer to a paper by Schroter [64].

The presently available version of HICUM (named HICUM/Level 2) includes many physical effects that are relevant for today's silicon-based processes (including SiGe technologies). As a consequence, its equivalent circuit is fairly complicated and not well suited for rough analytical calculations often performed by circuit designers in the preliminary design phase. Therefore, a simplified version of the model (called HICUM/Level 0) [65] is intended to be offered. The combination of these different levels of complexity during circuit design is also expected to save computational effort and time. The four most widely known models (SGP, MEXTRAM, HICUM and VBIC) are compared below with respect to the modelled physical effects and to their performance in Table 9.2.

Both MEXTRAM and HICUM models have matured for more than a decade now and provide an accuracy that meets most application needs specially for SiGe HBTs. Despite their different appearances, as MEXTRAM introduces an extra collector node, the two models are equivalent to a large extent (see Table 9.2) as is apparent from the following:

- Both models describe the e-b depletion capacitance in essentially the same way.
- Modelling of the Early effect is mostly done in the same way.
- Both models take account of ohmic resistances in series with the base, collector and emitter terminals.
- Both models describe the parasitic pnp transistor in terms of a simple transistor model.
- a.c. emitter current crowding is described in both models in essentially the same way.
- Both models describe self-heating effects.

9.5   Parameter extraction

The accuracy of compact models in circuit simulation depends not only on the correct physical description of various physical phenomena in the device but also on a reliable, robust and unambiguous extraction methodology for model parameters. In order to minimise the correlation between electrical and temperature-dependent parameters, the electrical parameters may be split into parameters extracted at low-injection (not affected by the self-heating) and high-injection condition parameters.

Table 9.2   Physical effects included in various transistor models [After H. C. de Graaff, Proc. ESSDERC, 1997.]

	SGP	MEXTRAM	HICUM	VBIC
High base injection	Yes	Yes	Yes	Yes
Early effect	Early voltage	Depletion charges	Depletion charges	Depletion charges
Quasi-saturation	No	Kirk effect	Kirk effect	Yes
Non-quasistatic effects	Yes	Yes	Yes	Yes
Stored base charges	Yes	Yes	Yes	Yes
Internal R_b	Conductivity modulation	Conductivity modulation	Conductivity modulation	Conductivity modulation
Noise	Yes	Yes	No	Yes
Weak avalanche	Without Kirk effect	With Kirk effect	Yes	Without Kirk effect
Parasitic pnp	No	Yes	No	Yes
Current-dependent C_{Tc}	No	Yes	Yes	No
Self-heating	No	Yes	Yes	Yes
Internal nodes	3	5	4	7
Number of parameters	35	39	36	70/63

Moreover, in order to further reduce the parameter correlation, the electrical and temperature-dependent parameters could be further split into small groups that are extracted sequentially.

An automated measurement system capable of accurately measuring the various characteristics of a device and extracting parameters is very important. During development of a new device design, data from device simulators or measurements on prototypes can be used as input to the extraction algorithm and a prototype compact model can be extracted. Requirements for parameter extraction methodologies are

- well-defined, simple, fast and reliable procedures;
- standard measurement equipment;
- modular – implementation enabling quick adaption to process and model evolution;
- model implementation in CAD systems;
- modular formulation with minimised interactions of parameters with simple equations and equivalent circuit topology.

In the following, VBIC parameter extraction techniques will be illustrated. Specifically, extraction of the Early voltages [66, 67] and the transistor series resistances and other compact model parameters are also covered. A brief overview on the procedures to determine model parameters is discussed, followed by a discussion on a measurement setup that can be used to perform d.c. and a.c. (scattering or s-parameter) measurements, and also the issue of de-embedding in s-parameter measurements. Finally, some comparisons are shown between simulated results obtained from both SGP model and VBIC to measured data for various device characteristics.

The Early voltages are determined by the method described in reference [28] and knee currents from analysis of forward, reverse and substrate d.c. current gains (e.g. the knee current is where this gain drops to half its low-current value). These parameters can further be refined by optimisation. The saturation currents, emission coefficients and temperature parameters for all transport and diode-like currents are estimated from low bias data and then refined using optimisation. Fitting the low bias parameters is straight forward. Resistances can be determined using existing methods. However, there is a degree of indeterminacy in extracting resistances from d.c. data only. Initial extracted resistances are thus refined by simultaneous optimisation of a.c. and d.c. data (including high bias data) [68]. Transit time parameters and excess phase are likewise extracted using existing techniques and then refined as part of the simultaneous d.c. and a.c. optimisation. In the following, some examples to demonstrate VBIC's capabilities for modelling bias, frequency, geometry and temperature-dependent transistor behaviour, as may be applied to SiGe HBTs, are presented.

9.6   Extraction algorithm

A typical algorithm for parameter extraction methods for VBIC is given below [69]. In this algorithm, when a parameter has been extracted, its value is used in all the subsequent steps:

Step 1.   From the low current region of the forward Gummel plot, Ibei, Nei, Iben, Nen, Is, and Nf are extracted.

Step 2.   From the low current region of the reverse Gummel plot, Ibci, Nci, Ibcn, and Ncn are extracted.

Step 3.   From the self-heating part of the d.c. measurement, Rth and Cth are extracted.

Step 4.   From the high current region of the forward Gummel plot, Re is extracted.

Step 5.   From the high current region of the reverse Gummel plot, the Rb + Rc (if needed, Ikr) are extracted.

Step 6.   From the collector resistance part of the d.c. measurement, Rc and, if necessary, Xrc(= Xrb = Xre) are extracted.

Step 7.   From the temperature-dependent parameters part of the d.c. measurement, Xis, Xii and Tnf are extracted.

Step 8.   Steps 4–7 are repeated until the best fits are obtained.

Step 9.   From the bias independent parameters of the cold-capacitor measurements, Cje, Cjc, Cbeo and Cbco (and, if included, Lb, Lc, Le) are extracted.

Step 10. From the bias-dependent parameters of the cold-capacitor measurements, Me, Mc, Pe, Pc and Fc (and if needed Wbe and the distribution of Rb between Rbi and Rbx) are extracted.

Step 11. From the active s-parameter measurement, Tf and the distribution of Rc between Rci and Rcx are extracted. If necessary, the s-parameter measurement can be used to fine-tune the resistances Re, Rb and Rc.

Step 12. If parameters for distributed base are used in Step 10, or some of the resistors have been fine-tuned in Step 11, then Steps 4–11 should be repeated for the best possible fit. Steps 4–6 do not extract new values for the resistors that have been fine-tuned in Step 11.

An automated measurement setup that may be used for the extraction of SiGe HBT device parameters is shown in Figure 9.9. An HP-4145B-based or equivalent system is chosen as it has the capability to measure d.c. characteristics of semiconductor devices down to the 100 μV and 1 pA range, and its high sensitivity. For capacitance–voltage measurements, HP-4061A semiconductor measurement system consists of an HP-4275A multi-frequency LCR meter and a HP-4140B pA meter. The measurement system has the capability to measure capacitance down to 1 fF for frequencies ranging from d.c. to 10 MHz. As device characteristics evaluation is the prime objective, the controller is commonly used to collect the measured data and transmit the data to the host computer for post-processing. High-frequency measurements are performed using HP-8510B network analyser for extraction of transit time.

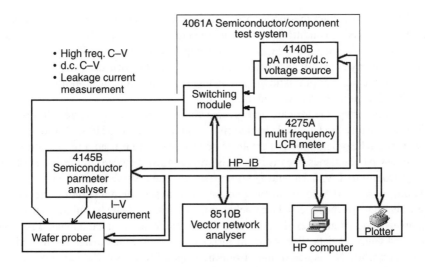

Figure 9.9   Electrical measurement system for device characterisation [After B. Senapati, 'Modelling and Characterization of SiGe-HBTs for RF Applications', PhD Thesis, Indian Institute of Technology, Kharagpur, 2001]

Table 9.3   The different measurement setups. The substrate voltage is normally set to −1 V with respect to the common in the different measurement setups [After B. Senapati, 'Modelling and Characterization of SiGe-HBTs for RF Applications', PhD Thesis, Indian Institute of Technology, Kharagpur, 2001.]

Measurement name	Bias setting	Measured data
Forward-Early	V_{be}, V_{cb}, $V_{cb,max}$	I_c, I_b
Reverse-Early	V_{bc}, V_{be}, $V_{be,max}$	I_e, I_b
Forward-Gummel	V_{bc}, V_{be}	I_c, I_b, I_{sub}
Reverse-Gummel	V_{be}, V_{bc}	I_e, I_b, I_{sub}
RE-flyback	I_c, V_{be}	I_e, V_{ce}
RCC-active	V_{bc}, V_{be}	I_c, I_b, I_{sub}
Output characteristic	I_b 1/4 $I_{b,set}$, 1/2 $I_{b,set}$	I_c, I_b, I_{sub}
	V_{ce}, $V_{cb,max}$	I_c, I_{sub}, V_{be}
Base-emitter		
depletion capacitance	V_{be}–$V_{be,max}$	C_{be}
Base-collector		
depletion capacitance	V_{bc}–$V_{cb,max}$	C_{bc}
Substrate-collector		
depletion capacitance	V_{sc}–$V_{cb,max}$	C_{sc}
s-parameters	V_{cb1}, V_{cb2}, V_{cb3},	I_c, I_b, s-par

In order to extract the model parameters, several measurements are performed at different biasing conditions at room temperature (see Table 9.3). The d.c. characterisation is done using the experimental setup described above. The purpose is to extract as many transistor parameters as possible from d.c. measurements [15,16] and minimise the number of unknown parameters requiring numerical optimisation [68] for fitting them in the VBIC model.

As a first step, the forward and reverse Gummel plots are measured. In order to measure the forward Gummel characteristics of the transistor, a HP-4145B or equivalent that can be programmed as a voltage or current source is used. For measuring the Gummel characteristics, the instrument is programmed to sweep emitter-base voltage (V_{be}) from 0 to 1 V. I_e, I_b and I_c are measured. The reverse junction voltage used should be set at 0 V to avoid the generation of avalanche/tunnelling currents and self-heating. For the reverse Gummel characteristics, the same technique is used but in reverse operation, when the emitter and collector leads are interchanged. From the forward Gummel data (linear part of $\ln(I_c)$ vs. V_{be} curve) estimates for IS and NF are made while NR is determined from the slope of $\ln(I_e)$ vs. V_{bc} curve of reverse Gummel data. The base-emitter ideal (IBEI/NEI) and non-ideal (IBEN/NEN) parameters are estimated from the ideal and non-ideal portions in $\ln(I_b)$ vs. V_{be} curve of the forward Gummel plot, respectively. Similarly, base-collector ideal (IBCI/NCI) and non-ideal (IBCN/NCN) parameters are extracted from the $\ln(I_b)$ vs. V_{bc} curve of reverse Gummel plot.

The forward current gain (β_F) is plotted as the ratio of the measured collector current (I_c) to the base current (I_b) and reverse current gain (β_R) as the ratio of the measured emitter current (I_e) to the base current (I_b). The knee currents are the high injection currents at which the current gain, β, starts to decrease from its peak value. From the forward β vs. I_c and reverse β vs. I_e curves estimates are made for IKF and IKR, respectively. IK refers to a collector current where β drops to half of its peak value. The parasitic pnp transistor parameters are determined from substrate current in the reverse Gummel plot, because the base-collector junction is forward biased and the parasitic transistor is conducting. If there is no substrate terminal available for a device, a p-n diode is generally used to model the parasitic substrate effects associated with the extrinsic base instead of a pnp transistor [25].

The output characteristics of the SiGe HBTs are measured using either forced-voltage (V_{be} = constant) or a forced-current (I_b = constant). In the measurement (V_{ce}) is swept and I_c is measured. For analog design, the Early voltage (VE) is very important. Physically, VE accounts for the amount of basewidth modulation due to change in the collector-base reverse voltage. Experimentally, forward VEF and reverse VER Early voltages are obtained from the low-bias forward and reverse output characteristics and the junction capacitances; base-emitter CJE (V_{be},PE,ME) and collector-base CJC (V_{bc},PE,MC), respectively [28].

As device dimensions shrink, the self-heating effect becomes more important. Although negligible at a very low injection ($I_c < 1\,\mu A$) level, it becomes extremely important at high injection ($I_c > 1$ mA) level. Therefore, VE estimation using forced-V_{be}, at high injection levels, is expected to rapidly degrade due to the self-heating effect, as I_c increases sharply even with a small increase in V_{cb}. On the other hand, forced-I_b output characteristics at high injection will show a negative slope in the I_c vs. V_{ce} characteristics due to an increase in I_s with V_{cb}. Extraction of VE becomes difficult under such conditions. Fortunately, owing to higher thermal conductivity of Si, although self-heating being present at all temperatures, the variation in I_c at low temperatures is small. It has been reported that self-heating has a much stronger degrading effect on VE as opposed to the degradation in VE due to base pushout or barrier effect at high injection [70].

The output characteristics of a bipolar transistor normally exhibit a quasi-saturation and/or high injection effects that are used to extract the epilayer parameters [31, 45–47]. The output characteristics are measured at three different values of the base currents and epilayer parameters (VO, RCI, HRCF and GAMM) are extracted [25]. During the weak avalanche breakdown parameters extraction, the V_{be} voltage is kept constant and V_{cb} is increased until the avalanche breakdown occurred [31]. To avoid the high current effects and internal heating, the base-emitter voltage is set to a low value. Forward weak avalanche breakdown parameters AVC1 and AVC2 have been extracted and optimised [31].

The next step is the inclusion of the finite base, collector and emitter resistances in the VBIC model [71]. The emitter resistance (RE) is determined by stimulating the base with a current in strong saturation and measuring the collector-emitter voltage (V_{ce}). The collector current is kept small (typically $< 1\,\mu A$) and the applied base current is swept up to 10 mA. The inverse of the gradient of characteristics can then be used to obtain RE. The measurement of the collector resistance (RC) is similar to the emitter resistance. Here, current is applied to the base and collector, and the collector-emitter voltage (V_{ce}) is measured. The $I_b - V_{ce}$ characteristics for two collector current values are measured and the collector resistance is determined from the ratio of V_{ce} voltage difference to collector current difference.

The distributed nature of the base resistance makes it difficult to model. Non-linear effects of the base resistance with the base current have been incorporated in the SGP model in a rather simple manner. In VBIC model, the base resistances of the intrinsic and parasitic transistors are bias-dependent and the resistances are modulated by the normalized base charges. The base resistances RBI and RBX are obtained from the Gummel plot [71] and further refined by optimising a fit to the experimental data as described later.

The thermal network shown in Figure 9.3 is used to model self-heating. Self-heating is a process causing an increase in temperature due to thermal dissipation of current flowing through a resistive element. A transistor's electrical response is deeply coupled to its internal (and surrounding) temperatures. A change in temperature will affect currents, which in turn affects the temperature. It is important to model this relation between temperature and electrical behaviour of a transistor, especially when dealing with high currents such as in power amplifier circuits. At a high power level, the heat generated in the device raises the internal junction temperature, which in turn alters current gain (β) or V_{be}. The decrease of β with increasing dissipated power results in a negative differential output resistance (NDR) due to the self-heating in the device. If the ambient temperature is increased, β decreases and $I_c - V_{ce}$ characteristics drop to a low collector current level.

To determine the thermal resistance, Dawson *et al.* [72] and Bovolon *et al.* [73] have suggested a fitting of a linear relationship between β or V_{be} and the junction temperature. However, fitting to such a linear relation has been questioned [74]. As forward current gain of SiGe HBTs (see Figure 9.10) at different temperatures is found to be non-linear, we adopted the extraction technique developed by Marsh *et al.* [74] in which any nonlinearity in β or V_{be} and the junction temperature becomes irrelevant.

In order to extract the thermal resistance parameters measurements are performed at various biasing conditions and different temperatures. The d.c. characterisation of the devices, in common-emitter configuration using a probe station with a thermal chuck having a temperature control of $\pm 0.5\,^\circ$C, is carried out using a HP-4145B semiconductor parameter analyser. To avoid breakdown of the transistor, the base-collector voltage is kept below the breakdown voltage of the transistor.

The mean device junction temperature (T_j) can be expressed as

$$T_j = T_{\text{subs}} + R_{\text{th}}P_{\text{diss}} = T_{\text{subs}} + R_{\text{th}}(V_{\text{CE}}I_c + V_{\text{be}}I_b) \tag{9.11}$$

where R_{th} is the thermal resistance, P_{diss} is the power dissipation and T_{subs} is the substrate temperature. If three substrate temperatures ($T_1 < T_2 < T_3$) are chosen to be equally spaced (ΔT), the mean device junction temperature can be solved using Equation (9.11) for three dissipated power levels, at fixed I_b and I_c and

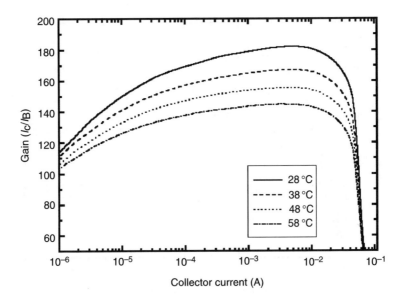

Figure 9.10   Forward current gain of a SiGe HBT at different temperature [After B. Senapati, 'Modelling and Characterisation of SiGe-HBTs for RF Applications', PhD Thesis, Indian Institute of Technology, Kharagpur, 2001]

is given by Reference 74

$$T_j = T_2 + \Delta T \left(\frac{P_2}{P_3} - \frac{P_2}{P_1} \right) / \left(\frac{P_2}{P_3} + \frac{P_2}{P_1} - 2 \right) \tag{9.12}$$

where P_1, P_2 and P_3 are the dissipated powers corresponding to substrate temperatures T_1, T_2 and T_3, respectively.

The mean device junction temperature is calculated using Equation (9.12) at a fixed base current ($I_b = 300\,\mu A$) for $T_1 = 28\,°C$, $T_2 = 43\,°C$, and $T_3 = 58\,°C$ from the output characteristics as shown in Figure 9.11. The thermal resistance (R_{th}) is computed at three dissipated powers P_1, P_2 and P_3 using Equations (9.11) and (9.12). To find the effect of R_{th} on P_{diss}, R_{th} is computed at different dissipated power levels in a similar manner. Thermal resistance is found to increase linearly with power dissipation (see Figure 9.12) and is due to the decreasing thermal conductivity of Si/SiGe with the temperature.

It may be noted that the accuracy of the extracted R_{th} values is influenced by the substrate temperature and the temperature difference (ΔT) used during measurements. It should also be emphasised that the thermal resistance should be extracted under a high self-heating biasing condition such that the three points are obtained in the saturation region of the transistor output characteristics (see Figure 9.11). The thermal resistance of double-mesa SiGe HBTs has been calculated and a dependence on temperature and on dissipated power similar to GaAs has been found [75].

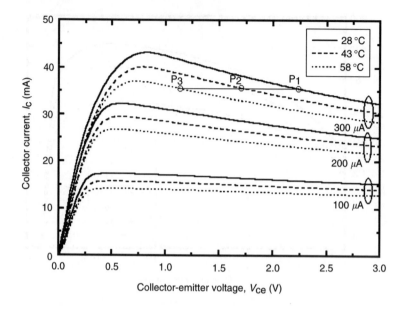

Figure 9.11   Temperature dependence of forward output characteristics of a SiGe
HBT. Self-heating occurs at a high injection level [After B. Senapati,
'Modelling and Characterization of SiGe-HBTs for RF Applications',
PhD Thesis, Indian Institute of Technology, Kharagpur, 2001]

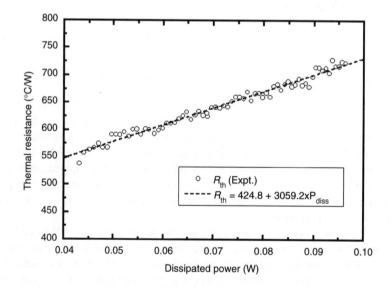

Figure 9.12   Extracted thermal resistance as a function of dissipated power at $I_b =$
300 μA [After B. Senapati, 'Modelling and Characterization of SiGe-
HBTs for RF Applications', PhD Thesis, Indian Institute of Technology,
Kharagpur, 2001]

The junction capacitances are extracted from capacitance-voltage (C–V) measurements at 1 MHz using the setup shown in Figure 9.9. The depletion capacitance data should be taken in a forward bias condition up to a point until the diffusion capacitance kicks in. A rule of thumb is to use a voltage up to a value when the capacitance becomes two to three times the zero-bias capacitance. From $C_{be}(V_{be})$ data, CJE, PE and ME are determined while CJC, PC and MC are determined from $C_{bc}(V_{bc})$ data using a non-linear least-square optimisation [16].

Transit time (τ_F) is commonly determined from cut-off frequency. The cut-off frequency is obtained from s-parameter measurements (up to 10 GHz) in the common emitter configuration by extrapolating $|h_{21}|$. The s-parameter measurements are carried out using an HP-8510B vector network analyser at room temperature. All parasitic components associated with resistance, inductance and capacitance of probes, pads and interconnects have been successfully de-embedded [76, 77]. The input and output RF connections of transistor fixture are made through SMA connectors (50 Ω). Regulated d.c. power supplies are connected through bias-T network for biasing the transistor. As a small increase in V_{be} can cause a large base-emitter current swing that could destroy the transistor, precaution must be taken. After all the connections are made, V_{be} is set to +0.75 V and V_{ce} is then slowly increased. This would help protecting the transistor base-emitter junction by limiting base-emitter current flow. f_T is measured at two V_{ce} values as a function of collector current. The forward transit time (τ_F) is obtained from the intercept of the $1/(2\pi f_T)$ vs. $1/I_c$ curve. The voltage and current-dependent parameters (VTF, ITF) of the transit time are estimated by optimisation.

Several precautions necessary during the measurements are listed below:

- All measurements should be taken from the same location (die) of a chip.
- To generate scalable model parameters, the dimensions of the measured devices have to be known. In most cases, design rules and mask geometries are sufficient. For the emitter-base and base-collector region, however, SEM photos are preferred to obtain as accurate as possible structural data.
- s-parameters are preferred to be provided as raw data, that is, not de-embedded. As a consequence, data of the corresponding OPEN (and SHORT) structures are also required not only for de-embedding but also for checking the quality of the measurements. This way, effort for extracting model parameters on inconsistent or incorrect data can be avoided.

9.7   VBIC model implementation in SPICE

The use of scalable models allows designers to pick the right device for their application without being constrained to a finite library of pre-approved devices. The statistical model also allows the designers to explore the actual design space with a tighter design tolerance. For their own technology development, researchers from IBM have developed a fully scalable, statistical GP model for the SiGe HBTs without sacrificing model accuracy, flexibility or usability [78, 79]. Proper scaling of the emitter area and perimeter with drawn dimensions is crucial to developing a scalable model for an npn transistor. Tight process control significantly simplifies the development of an accurate statistical model.

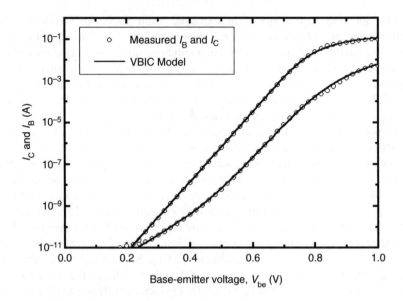

Figure 9.13   Forward Gummel plot of a SiGe HBT: measured and simulated using VBIC model [After B. Senapati, 'Modelling and Characterization of SiGe-HBTs for RF Applications', PhD Thesis, Indian Institute of Technology, Kharagpur, 2001]

Extracted d.c. and a.c. parameters of the SiGe HBTs are implemented in the VBIC model using Fortran pseudo-code program [23, 80, 81]. Figures 9.13 and 9.14 show the forward and reverse Gummel plots of a SiGe HBT, respectively. Measured data is presented by open circle and the VBIC model by a solid line. The extracted ideality factor from the Gummel plot is found to be approximately 1.0, indicating that the device has not suffered from boron out-diffusion into the emitter. Forward and reverse current gain is shown in Figure 9.15. The maximum forward and reverse current gain at room temperature are found to be 170 and 12, respectively.

The forward output characteristics with self-heating is shown in Figure 9.16. Measured data is represented using open circle, the SGP model by a dashed line, and the VBIC model by a solid line. The negative slope in the forward output characteristics at a high level injection shows the importance of self-heating in the transistor, which is dependent on the quiescent operating point. As the device begins to operate at a high power level, the heat generated raises the internal junction temperature, which in turn alters the V_{be}. The GP model fails to predict this important effect, while VBIC captures the shift with an accurate electrothermal model. It is also noted that owing to a very high base doping concentration ($\sim 3 \times 10^{19}/\text{cm}^{-3}$) the basewidth modulation by the base-collector voltage is less pronounced, leading to a high Early voltage of 40 V. The reverse output characteristics of SiGe HBTs is shown in Figure 9.17.

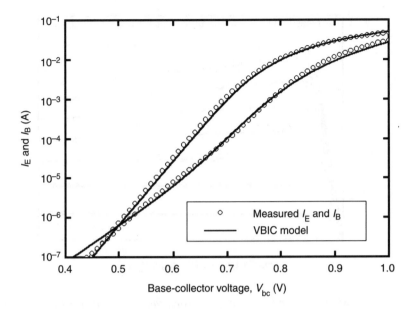

Figure 9.14   Reverse Gummel plot of a SiGe HBT: measured and simulated using VBIC model [After B. Senapati, 'Modelling and Characterization of SiGe-HBTs for RF Applications', PhD Thesis, Indian Institute of Technology, Kharagpur, 2001]

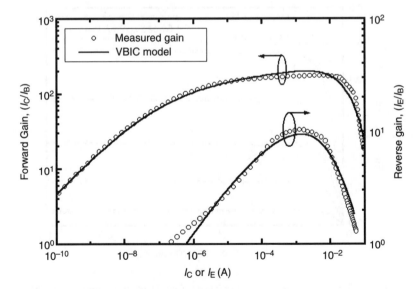

Figure 9.15   Forward and reverse d.c. current gain of a SiGe HBT: measured and simulated using VBIC model [After B. Senapati, 'Modelling and Characterization of SiGe-HBTs for RF Applications', PhD Thesis, Indian Institute of Technology, Kharagpur, 2001]

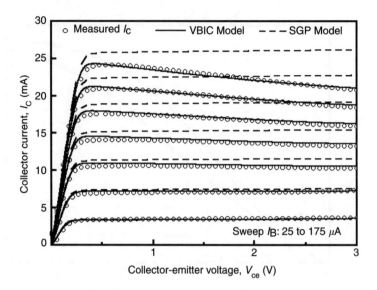

Figure 9.16　Measured and simulated forward output characteristics with self-heating effect for a SiGe HBT. Note that the VBIC model prediction is superior to that of by SGP [After B. Senapati, 'Modelling and Characterization of SiGe-HBTs for RF Applications', PhD Thesis, Indian Institute of Technology, Kharagpur, 2001]

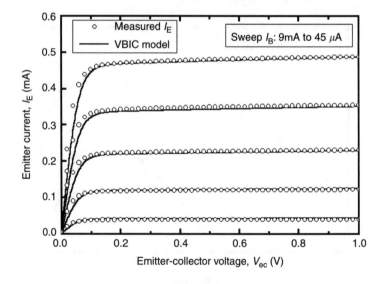

Figure 9.17　Reverse output characteristics: measured and simulated using VBIC model [After B. Senapati, 'Modelling and Characterization of SiGe-HBTs for RF Applications', PhD Thesis, Indian Institute of Technology, Kharagpur, 2001]

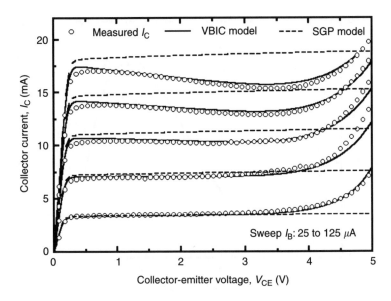

Figure 9.18   *Measured and simulated forward output characteristics including avalanche multiplication and self-heating effects for a SiGe HBT. The VBIC model predicts more accurately the avalanche multiplication and the self-heating effect [After B. Senapati, 'Modelling and Characterization of SiGe-HBTs for RF Applications', PhD Thesis, Indian Institute of Technology, Kharagpur, 2001]*

The forward output characteristics with an avalanche breakdown and self-heating effect is shown in Figure 9.18. It can be seen that the VBIC model predicts more accurately the transition between the saturation and linear region, which makes possible a better prediction of current gain in the linear region. The difference becomes larger as the bias is increased and the device goes into breakdown via avalanche multiplication effect. It is also noted that the quasi-saturation effect is negligible, which reduces the d.c. current gain due to high injection, as does the unity gain cut-off frequency.

Figure 9.19 shows a typical cut-off frequency characteristics for two collector bias condition. Measured data is represented by solid dots and the VBIC by a solid line. It is observed that the VBIC model predicts f_T reasonably well when the device is operated in the linear region and also when the device is pushed into near breakdown. Accurate modelling of these effects is critical for high frequency design where an accurate description of the device bandwidth is required over a broad bias range. Several important VBIC model parameters (extracted and optimised) are listed in Table 9.4.

Parameter extraction is closely related to the process and parameter extraction procedures and list of test structures needed for various parameters extraction are generally made available by the model developer. For example, for HICUM, the parameter determination methodology has been described in more detail in References 82 and 83. Mostly, the parameters can be determined using standard

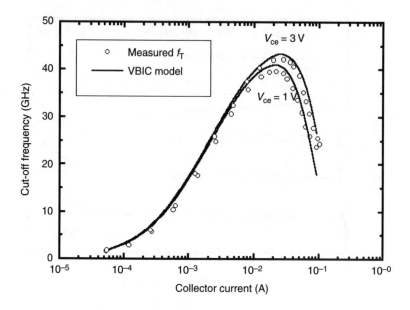

Figure 9.19   Measured and simulated cut-off frequency vs. collector current for a SiGe HBT [After B. Senapati, 'Modelling and Characterization of SiGe-HBTs for RF Applications', PhD Thesis, Indian Institute of Technology, Kharagpur, 2001]

Table 9.4   Extracted and optimised VBIC model parameters of a SiGe HBT at room temperature [After B. Senapati, 'Modelling and Characterization of SiGe-HBTs for RF Applications', PhD Thesis, Indian Institute of Technology, Kharagpur, 2001.]

Extracted value	Optimised value
IS = 2.5621e-15A, NF = 1.01, NR = 1.03	IS = 4.6221e-15A, NF = 1.0, NR = 1.02
IBEI = 5.5433e-17A, NEI = 1.049	IBEI = 5.5433e-17A, NEI = 1.023
IBEN = 7.4573e-13A, NEN = 2.0	IBEN = 1.4573e-13A, NEN = 2.0
IBCI = 1.543e-16A, NCI = 1.057	IBCI = 2.548e-16A, NCI = 1.035
IBCN = 1.4127e-12A, NCN = 1.55	IBCN = 3.68e-12A, NCN = 1.64
IKF = 5.5e-2A, IKR = 1e-2A	IKF = 2.5e-2A, IKR = 1e-2A
VEF = 40V, VER = 32.5V	VEF = 45V, VER = 39V
RBX = 5.26Ω, RBI = 30.01Ω	RBX = 2.03Ω, RBI = 25Ω
RE = 2.81Ω, RCI = 11.8Ω	RE = 1.51Ω, RCI = 5.2Ω
GAMM = 2e-11, VO = 1, HRCF = 2	GAMM = 2.5e-11, VO = 1, HRCF = 2
AVC1 = 2, AVC2 = 15, RTH = 424 °C/W	AVC1 = 2.5, AVC2 = 20, RTH = 400 °C/W
CJE = 105fF, PE = 0.75, ME = 0.35	CJE = 95fF, PE = 0.76, ME = 0.36
CJC = 79fF, PC = 0.71, MC = 0.38	CJC = 38fF, PC = 0.72, MC = 0.42
TF = 4ps, XTF = 5, ITF = 1.2e-1	TF = 3.7ps, XTF = 5, ITF = 3.7e-1

measurement equipment. A detailed description of the parameter extraction method-ology for all advanced models such as VBIC, HICUM and MEXTRAM for a variety of bipolar processes is available in the respective user manuals.

VBIC and HICUM models have been applied for d.c. modelling of state-of-the-art SiGe and SiGeC HBTs. A comparison is made also with the SGP model [84–89]. The compared effects in SiGe HBTs include Early effect, quasi-saturation, substrate parasitic, avalanche multiplication and self-heating. Extracted d.c. parameters of the SiGe HBTs were implemented in the VBIC and HICUM models using APLAC simulation tool [90].

Figures 9.20 and 9.21 show the forward and reverse Gummel plots of a SiGe HBT, and Figures 9.22 and 9.23 show the corresponding current gains, respectively. Measured data is presented by a solid line, SGP model by open upward triangle, VBIC model by open square and HICUM model by open circle with line. In general, the fitting is good. The decrease of β with increasing dissipated power in the device results in a negative differential output resistance (NDR) owing to the self-heating in the device (see Figure 9.24). The GP model fails to predict this important self-heating effect [91], while VBIC and HICUM capture the shift with an accurate electrother-mal model. It is clear that quasi-saturation [25] and weak avalanche [27, 92] effects are satisfactorily modelled in both HICUM and VBIC. This avalanche modelling is important in view of application in power amplifier circuits in complex modulation

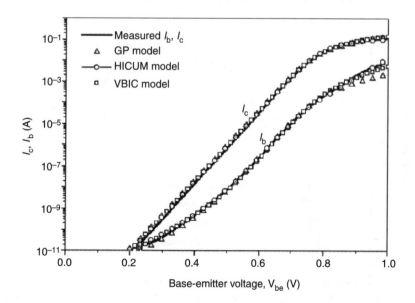

Figure 9.20   Forward Gummel plot of a SiGe HBT [After A. Chakravorty, Compact Modelling of Si-Heterostructure Bipolar Transistors and Active Induc-tor Design for RF Applications, PhD Thesis, IIT Kharagpur, 2005]

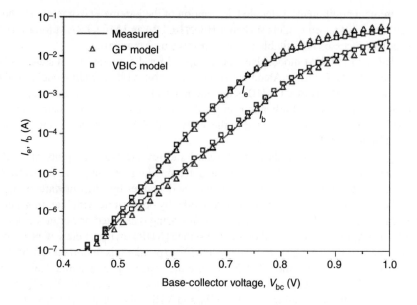

Figure 9.21   Reverse Gummel plot [After A. Chakravorty, Compact Modelling of Si-Heterostructure Bipolar Transistors and Active Inductor Design for RF Applications, PhD Thesis, IIT Kharagpur, 2005]

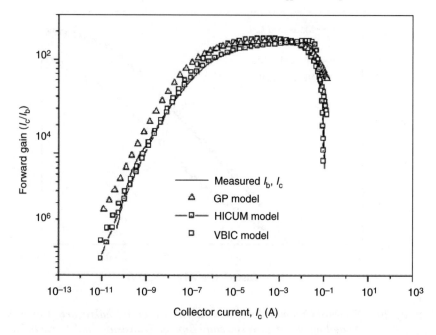

Figure 9.22   Forward d.c. current gain [After A. Chakravorty, Compact Modelling of Si-Heterostructure Bipolar Transistors and Active Inductor Design for RF Applications, PhD Thesis, IIT Kharagpur, 2005]

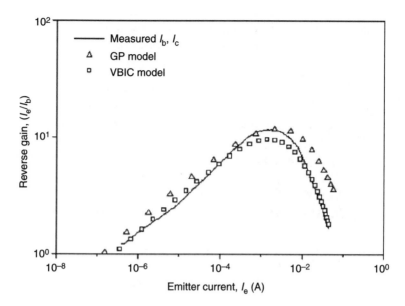

Figure 9.23   Reverse d.c. current gain [After A. Chakravorty, Compact Modelling of Si-Heterostructure Bipolar Transistors and Active Inductor Design for RF Applications, PhD Thesis, IIT Kharagpur, 2005]

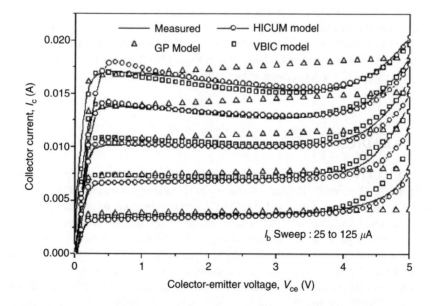

Figure 9.24   Forward output characteristics [After A. Chakravorty, Compact Modelling of Si-Heterostructure Bipolar Transistors and Active Inductor Design for RF Applications, PhD Thesis, IIT Kharagpur, 2005]

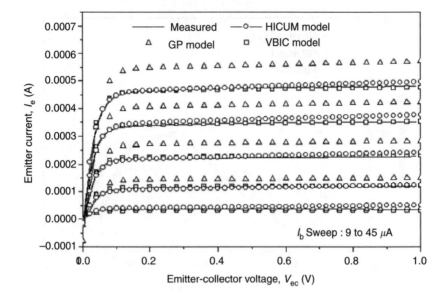

Figure 9.25   Reverse output characteristics of a SiGe HBT [After A. Chakravorty, Compact Modelling of Si-Heterostructure Bipolar Transistors and Active Inductor Design for RF Applications, PhD Thesis, IIT Kharagpur, 2005]

systems [93]. The reverse output characteristics of a SiGe HBT shown in Figure 9.25 shows that the predictions using the VBIC model with the extracted parameters compare favourably with the experimental data for advanced SiGe HBTs. A closer look reveals that in the forward high current region HICUM model performs better, whereas the VBIC model is more suitable for both the forward and reverse regions of SiGe HBTs.

9.8   TRADICA

TRADICA (TRAnsistor DImensioning and CAlculation program) work started in 1986 at Ruhr-University Bochum, Germany [94]. Although it was initially developed as a transistor dimensioning tool, TRADICA can also be used as a model generation and prediction tool. TRADICA contains the geometry dependent formulations that are required to generate model parameters for bipolar transistor models for various processes (such as double-poly self-aligned transistor structures). TRADICA takes process-specific and model-specific parameters as well as geometry description to generate model parameters for transistors of various configurations, which are defined by emitter width and length, number of emitter, base and collector contacts (stripes), and their spatial arrangement. Presently, the program generates a hierarchy of SGP and HICUM models in SPICE format, which can be included in the usual transistor library. Input data for TRADICA are as follows:

1. electrical data, such as sheet resistances, capacitances per unit area and periphery length, including transistor model specific data such as currents per area, transit time and so on;
2. geometry data defined by cross-sections and top-views (design rules, emitter dimensions);
3. physical data such as mobilities, temperature coefficients and electromigration limit;
4. process-related data such as doping concentrations (optional), certain vertical dimensions and process tolerances;
5. operating point information (only for calculating the critical emitter area).

Model parameter determination is based on test structures, mostly transistors, but also process monitors. The program, which connects process and circuit design, offers the following advantages over conventional bipolar transistor modelling approaches:

- The flexibility of the method facilitates quick generation of a large number of transistor configurations once the parameter extraction is finished, making the approach significantly more efficient in terms of parameter extraction and more flexible from a design point of view with respect to circuit optimisation.
- Since the approach is physical and based on process monitors, it facilitates prediction of the influence of process tolerances on device and circuit performance, and eventually statistical design.
- Parameters for transistors not existing on the test chip can also be generated for circuit design at any time, before and during process development, supporting concurrent design. As a result, the number of design cycles and time to market can be significantly reduced.
- Given specific parameters extracted from an arbitrary wafer, TRADICA can shift model parameters for wafers subject to various process variations based on respective process control monitor (PCM) data and analytical prediction equations.

9.9 Scalable modelling: SiGe/SiGeC HBTs

As the complexity of the circuit increases, scalable device models are crucial for the circuit design. An accurate modelling of key parameters is essential for accurate circuit behaviour prediction and circuit design optimisation. From the design point of view, it is desirable to be able to vary the transistor dimension, defined by the emitter width and length together with the number of contacts for base, emitter and collector. Circuit designers need a large range of transistor dimensions for various applications. For example, when designing a low-noise amplifier, the specifications for bandwidth and noise have to be satisfied simultaneously. As the maximum speed and minimum noise in a transistor do generally not occur at the same bias point, one has to properly select the transistor sizes that give the best overall circuit performance.

A geometry scalable compact model requires that all of its equivalent circuit elements are described as a function of layout dimensions [82, 95–97].

To serve the design engineers reliably and effectively through design kits, fast generation of the model parameters for a wide range of transistor sizes is very important in a new technology. It is important to note that geometry effects are not only dependent on the device dimensions but also, for given dimensions, on bias and frequency, which make transistor geometry scaling a difficult task.

Scaling becomes even more complicated for the elements of the substrate and the thermal network due to the large variety of layouts and device designs, for example, junction isolation, deep trench, SOI and metallisation [98,99]. Different sets of scaling equations are needed for these cases. Since a complete description of the scaling equations for bipolar transistors is beyond the scope of this book, the reader is referred to References 2, 63 and 94, and the references therein.

As SiGe technology has become the key technology for the wireless communications market, it is important to understand the scalability of this technology. Standard CMOS models, such as MOS model 11 [100] and BSIM4 [101] give quite reliable scaling results since they describe only lateral current flow in MOS devices. On the other hand, scaling of bipolar transistor model parameters is significantly more involving due to current flow in both vertical and lateral directions. Moreover, the scaling rules depend on the layout configuration, device cross-section and processing data. As a consequence, the geometry scalable bipolar transistor models and procedures are not generally available. Advanced compact models like VBIC, HICUM and MEXTRAM came to play with their intuitive mathematical formulation to model the physical effects such as quasi-saturation, basewidth modulation, weak avalanche, self-heating and so forth, which were not considered in SGP [61, 83, 102, 103].

Work on bipolar transistor scaling so far has generally considered the SGP model, which is inadequate to handle device scaling across a wide range of device sizes, and more sophisticated substitutes are necessary [23]. A scalable, statistical model has been developed by IBM for SiGe HBTs [53, 104], which are components of a commercially available BiCMOS technology for high frequency applications. IBM has implemented a model for their SiGe HBT process [105], in which SGP parameters are scaled from input layout dimensions using simple equations. The SGP model parameters were scaled, and statistics added, using language features built into HSPICE. The resulting fit, for both d.c. and a.c. test conditions have been found to be adequate for most applications [78]. The device scaling and model statistics have also been implemented in the SiGe technology. Extraction of scaled MEXTRAM parameters for high-speed SiGe HBTs based on device physics and general scaling has been reported by Wu *et al.* [49].

Correct scaling of the emitter area and perimeter with drawn dimensions is crucial to developing a scalable model for an npn transistor. The calculations are complicated by photolithographic diffraction effects that result in rounded corners in the image developed on the wafer. In very narrow devices, the rounding of the corners can lead to actual fore-shortening of the emitter length and possibly the width. Figure 9.26 shows an example of a rectangular emitter with corner-rounding and a narrower emitter with

(a)　　　　　　　　　(b)　　　　　　　　(c)

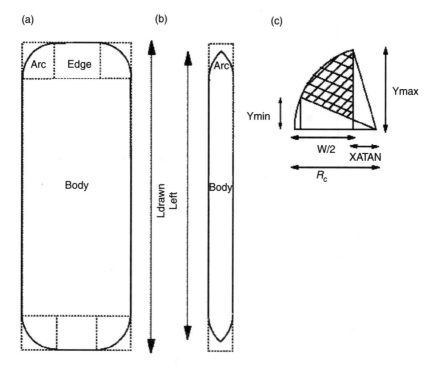

*Figure 9.26　Schematic illustration of corner-rounding and length fore-shortening:
(a) corner-rounding only, (b) length fore-shortening, and (c) arc detail.
(Note: Width, W, and corner-rounding radius, R_C, are in arbitrary
units) [After K. M. Walter et al., IEEE J. Solid-State Circuits, Vol. 33,
1998(1439–1444). ©1998 IEEE]*

length fore-shortening as a result of the photo diffraction. A detail of a rounded corner is also shown.

For analysis, the emitter is divided into three distinct regions: the body, the edge and the arc. The emitter area is calculated as shown in the equations below. These equations use the following parameters: N_{cr} is the corner-rounding radius L is the statistical emitter length $= (L_{\text{DRAWN}} - \Delta L)$ W is the statistical emitter width $= (W_{\text{DRAWN}} - \Delta L)$ For detail, the reader may refer to the reference [78]. Figure 9.27 shows the ratio of the calculated area to the area of a rectangle with the same starting dimensions. It is clearly seen that the corner-rounding always reduces the area, but as the width becomes less than twice the corner-rounding radius, the actual area decreases rapidly. Normally, the reduction is minor for a nominal emitter, but statistically the width can easily become less than the critical value.

The statistical change of emitter dimensions from drawn dimensions (prior to corner-rounding) is calculated by scaling the base current density. The base current density is limited by the finite recombination velocity at the single crystal-to-polySi interface. Therefore, with the proper adjustment for photolithographic biases on the

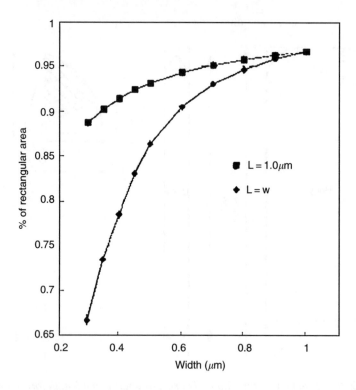

*Figure 9.27   Area reduction due to corner-rounding [After K. M. Walter et al., IEEE
J. Solid-State Circuits, Vol. 33, 1998(1439–1444). ©1998 IEEE]*

image size, the base current density at a given bias will be constant with calculated
perimeter-to-area ratio. This assumes that perimeter effects are minimal because of the
severe limitation imposed by the small recombination velocity at the oxide, making
the current proportional to oxide area only.

Most of the SGP model parameters are then derived from the adjusted areas and
perimeters of the emitter, base and collector areas. Devices with multiple emitter
fingers are handled by modelling a single finger and associated portion of the parasitic capacitances, and then multiplying the result by the number of fingers. Once the
emitter area and perimeter are accurately calculated, the major roadblock to a scalable model is obtaining an analytical expression for base resistance based on device
geometry. The base resistance was calculated following References 4 and 56.

Measured transistor Gummel plots, which show the collector and base current
response to applied base emitter bias, have been compared with the same model
characteristics over a range of emitter widths from 0.3 to 4.0 μm and lengths from
0.5 to 20.0 μm. Figure 9.28 shows the fit for collector current over the linear portion of
the Gummel plot for sub-minimum (0.3×2.5), standard (0.5×2.5) and large (4×5) μm
npn transistors.

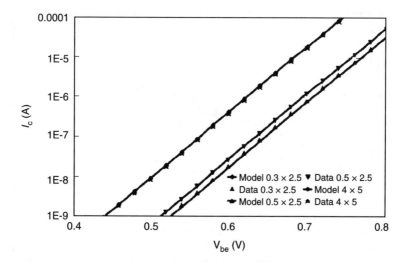

Figure 9.28 I_c *vs.* V_{be} *showing results of emitter scaling [After K. M. Walter et al., IEEE J. Solid-State Circuits, Vol. 33, 1998(1439–1444). ©1998 IEEE]*

Junction capacitances have been determined from low frequency measurements on structures with large area or perimeter and from scattering parameter (s-parameter) measurements on a large set of transistors with varying emitter dimension. Parasitic capacitances associated with various overlaps of electrical junctions have been extracted from the s-parameter measurements, providing very sensitive extraction results. Forward transit time has been fit from analysis of s-parameter data (based on area and perimeter components without parasitics) vs. bias, and fits the model within ±15 per cent. Figure 9.29 illustrates the fit to the f_T vs. I_c curve for three transistors, down to $0.5 \times 1.0 \, \mu$m emitters. Known limitations of the SGP model are evident in the poorer fit at high current density.

Statistical analysis of in-line manufacturing test data shows that the major components of statistical variation in the devices come from (1) emitter size, (2) base pinch resistance (R_{bi}) and (3) vertical profile (for instance, in the case where the emitter doping intersects the germanium profile). As a result of the self-aligned and silicided extrinsic base, total base resistance is small and insensitive to varying layout. Also, tight process control significantly simplifies the development of an accurate statistical model.

Geometry scalable SPICE models for SiGeC HBTs have been proposed by several researchers, and electrical simulation results are compared with physical simulation results [106–108]. Also a simple but accurate scalable modelling strategy for SiGeC HBTs fabricated in a low-cost BiCMOS technology has been reported [109] in VBIC and HICUM, which offers the designer community a wide range of transistor sizes. In this approach, only one full extraction on a basic transistor is performed. To simplify the parameter extraction, emitter length is changed only from a minimum length (0.84 μm) to a maximum by a factor of four (3.36 μm). To get a large emitter area,

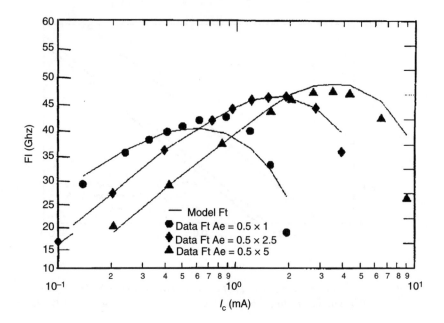

Figure 9.29   Results of scaling on f_T [After K. M. Walter et al., IEEE J. Solid-State Circuits, Vol. 33, 1998(1439–1444). ©1998 IEEE]

several base-emitter structures in parallel on the top of a single collector is used. With this procedure one can change the collector current from minimum to biggest device by a factor of 50. The main advantage of this approach is its simplicity.

Fregonese *et al.* [108] have developed a geometry scalable substrate network for HBTs isolated by deep trench in a BiCMOS technology. The doping and dimensions of the simulated structures are very similar to the 0.25 μm BiCMOS technology from ST Microelectronics [110]. This technology is characterised by a transit frequency of 70 GHz. The physical structure is scalable according to all dimensions and resistivities. The length L, the width W of the N well, the resistivity of the substrate, the substrate contact configuration with respect to its distance d from the deep trench and its contact width W_C were considered in simulation.

The authors have reported a detailed investigation of the substrate impedance as a function of the device dimensions, of the substrate doping and of the substrate contact configuration using a commercial physical 3D simulation tool. The deep trenches are built up with SiO_2 walls filled in with polySi. The substrate thickness is about 300 μm. Owing to its symmetry, only a quarter of the structure was simulated.

The p-n junction and its resistive access RS represent the internal part. This p-n junction can be modelled with capacitor, which is bias dependent. A capacitance, CSUB associated with the material cut-off frequency has been added in parallel to RS. For a high resistivity substrate, however, this capacitance cannot be neglected. The implementation of the scalable substrate network has been added to the HICUM Level 0. The substrate network is connected to the internal collector node. This modified HICUM L0 model is scalable and has been implemented in VERILOG-A. A scalable extraction has been performed on the BiCMOS 0.25 μm technology from ST Microelectronics. In order to evaluate the impact of the substrate network on the electrical characteristics, simulations have been done and compared with real device measurements.

9.10 Role of TCAD

As discussed above, generating the compact model can be difficult and time-consuming because of the amount of required data and number of modelling parameters. Traditionally, TCAD involves process and device experiments only and has direct impact on the process and device development cycle time. With accurate process and device simulation early in the development cycle, TCAD may be used to produce the entire suite of a.c. and d.c. device characteristics and also the predictive compact model. As the technology development progresses, updates to the compact models can be provided to ensure the designers will have minimal design impact from the initial models to final. Thus it is important to understand this new methodology of combining the TCAD process, device and compact modelling strategy in developing state-of-the-art SiGe technologies. Sheridan *et al.* [111] have used the genetic algorithm optimisation techniques for bipolar compact modelling. The technique uses only a minimal set of extracted parameters together with the use of a physically bounded genetic algorithm to efficiently optimise a model parameter set. This significantly reduces the extraction and optimisation effort as well as eliminates inaccuracies resulting from commonly used fitting methods to approximate model equations, allowing designers early access to technology models.

Chakravorty *et al.* [112] have also modelled a low-cost BiCMOS SiGeC HBT using adaptive neuro-fuzzy inference system (ANFIS). Accuracy of the model is checked for all the d.c. and s-parameters in a wide range of bias and frequencies and the average error is found to be less than 4 per cent. Especially in high current and high frequency regions, the ANFIS model is proved to be excellent unlike most of the physics-based equivalent circuit models that fail to track the actual device behaviour. The HBT module used for investigation offers $f_T/f_{max}/BV_{ceo}$ values of 75 GHz/90 GHz/2.4 V. A schematic for this HBT type is shown in Figure 9.30. The measurements are performed on a semi-automatic probe shield system from Suss MicroTec, where d.c. and RF probes are installed simultaneously. Direct current measurements are carried out using a HP4142 d.c. source from Agilent Technologies along with Kelvin probes to avoid the influence of contact resistances. The RF measurements are performed using 50 GHz Infinity Probes from Cascade and PNA3568 from Agilent Technologies. OPEN-SHORT de-embedding procedures are preceded to exclude the pad parasitics. All measurements were done on RF pads. In this way, the substrate current had to be measured from the wafer backside. Measurements are guided with ICCAP from Agilent.

Instead of y-parameters, the s-parameters of the device were used to train and validate the model. Experimental (scattered) as well as the corresponding ANFIS model simulation (line) data are plotted in Figures 9.31–9.34. From Figure 9.31, it is

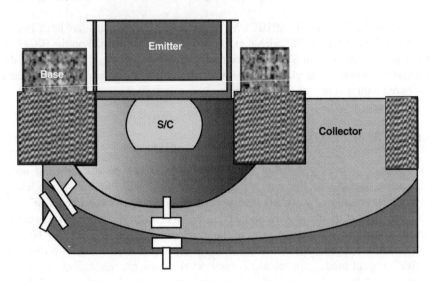

*Figure 9.30   Schematic diagram of a low-cost SiGeC HBT. SIC selectively
implanted collector [After A. Chakravorty, Compact Modelling of Si-
Heterostructure Bipolar Transistors and Active Inductor Design for RF
Applications, PhD Thesis, IIT Kharagpur, 2005]*

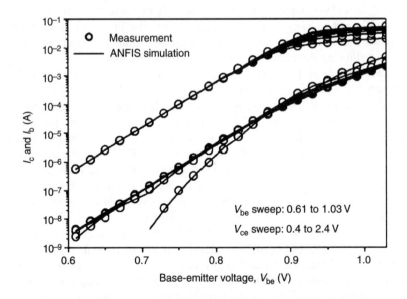

*Figure 9.31   Measured and ANFIS-simulated Gummel plots (collector and base cur-
rent data) of SiGeC HBT [After A. Chakravorty, Compact Modelling
of Si-Heterostructure Bipolar Transistors and Active Inductor Design
for RF Applications, PhD Thesis, IIT Kharagpur, 2005]*

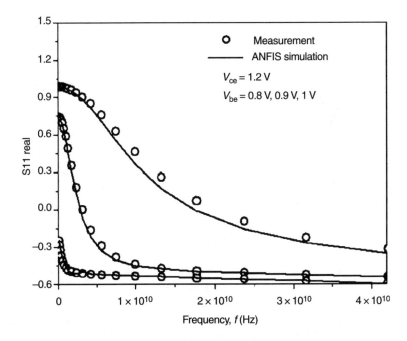

Figure 9.32 *Measured and ANFIS-simulated real (s_{11}) data [After A. Chakravorty, Compact Modelling of Si-Heterostructure Bipolar Transistors and Active Inductor Design for RF Applications, PhD Thesis, IIT Kharagpur, 2005]*

seen that the I_c prediction is accurate, especially in the high current region, where modern RF circuits operate. Figures 9.32 and 9.33 show the real and imaginary part of s_{11} data validation for a range of V_{be} (0.8, 0.9 and 1.0 V) with the proposed model. The mismatch between the measurement and simulation in high frequency region is due to the fact that lesser amount of training data was available in this area. A good agreement of model simulation with the measured s_{21} magnitude data plotted against the frequency, as seen from Figure 9.34, proves the utility of ANFIS model.

9.11   Summary

The aim of this chapter has been to provide a background for bipolar compact modelling. The main improvements of the VBIC model over the SGP model is the addition of Early voltage effect, quasi-saturation, substrate parasitic, avalanche multiplication, self-heating and modelling of the parasitic pnp transistor. A methodology to extract and optimise d.c. and a.c. parameters for the VBIC model is developed and applied to SiGe HBTs. The model has been verified and optimized using a VBIC pseudo-code and an excellent agreement has been found. A comparison of the VBIC and SGP

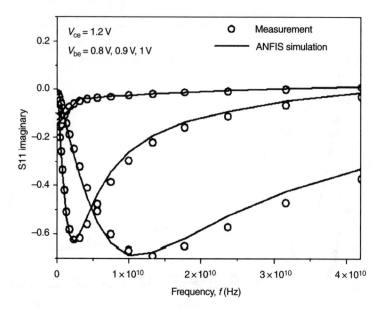

*Figure 9.33   Measured   and   ANFIS-simulated   imaginary   (s_{11})   data   [After
A. Chakravorty, Compact Modelling of Si-Heterostructure Bipolar
Transistors and Active Inductor Design for RF Applications, PhD
Thesis, IIT Kharagpur, 2005]*

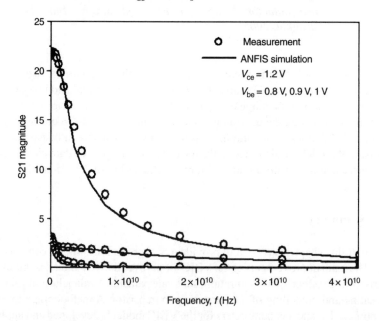

*Figure 9.34   Measured   and   ANFIS-simulated   magnitude   (s_{21})   data   [After
A. Chakravorty, Compact Modelling of Si-Heterostructure Bipolar
Transistors and Active Inductor Design for RF Applications, PhD
Thesis, IIT Kharagpur, 2005]*

model clearly illustrates the weakness of the conventional SPICE Gummel-Poon model when applied to advanced SiGe HBTs.

The MEXTRAM model includes a direct coupling between current and charge, thus leading to coupled d.c. and a.c. behaviour. This has the advantage of a physical representation of the device, which facilitates accurate scaling. The MEXTRAM and VBIC bipolar transistor models are comparable for low and medium collector current densities and frequencies. Parameter conversion from MEXTRAM to VBIC can easily be done for the depletion capacitances and d.c. parameters related to low and medium current levels.

HICUMs major advantages over other bipolar models are its scalability, simple and process-based/related parameter extraction, predictive capability of process and layout variations, and simple numerical formulation. TCAD tool TRADICA has also been discussed. ANFIS automates RF modelling as it is technology independent, neuro-computing based, intelligent and capable of achieving any predetermined accuracy limit required.

References

1 C. K. Maiti, and G. A. Armstrong, *Applications of Silicon-Germanium Heterostructure Devices*. Inst. of Physics Pub., UK, 2001.

2 M. Schroter, H.-M. Rein, W. Rabe, R. Reimann, H.-J. Wassener, and A. Koldehoff, 'Physics and process-based bipolar transistor modelling for integrated circuit design', *IEEE J. Solid-State Circuits*, vol. 34, pp. 1136–1149, 1999.

3 H.-M. Rein, and M. Schroter, 'Experimental determination of the internal base sheet resistance of bipolar transistors under forward-bias conditions', *Solid State Electron.*, vol. 34, pp. 301–308, 1991.

4 M. Schroter, 'Modelling of the low-frequency base resistance of single base contact bipolar transistors', *IEEE Trans. Electron Devices*, vol. 39, pp. 1966–1968, 1992.

5 H.-M. Rein, 'Improving the large-signal models of bipolar transistors by dividing the intrinsic base into two lateral sections', *Electron. Lett.*, vol. 13, pp. 40–41, 1977.

6 M. Pfost, H.-M. Rein, and T. Holzwarth, 'Modelling substrate effects in the design of high-speed Si-bipolar ICs', *IEEE J. Solid-State Circuits*, vol. 31, pp. 1493–1497, 1996.

7 M. Schroter, M. Friedrich, and H.-M. Rein, 'A generalized Integral Charge-Control Relation and its application to compact models for silicon based HBTs', *IEEE Trans. Electron Devices*, vol. 40, pp. 2036–2046, 1993.

8 J. Kim, J.-O. Plouchart, N. Zamdmer, N. Fong, M. Sheronv, Y. Tan, et al., 'Highly manufacturable 40-50 GHz VCOs in a 120 nm system-on-chip SOI technology', in *IEEE IEDM Tech. Dig.*, pp. 367–370, 2003.

9 G. Freeman, M. Meghelli, Y. Kwark, S. Zier, A. Rylyakov, M. A. Sorna, et al., '40-Gb/s Circuit Built from a 120-GHz f_T SiGe Technology', *IEEE J. Solid-State Circuits*, vol. 37, pp. 1106–1114, 2002.

10 J. Ebers, and J. Moll, 'Large-signal behavior of junction transistor', *Proc. IRE*, vol. 42, pp. 1761–1772, 1954.

11 H. K. Gummel, and H. C. Poon, 'An integrated charge control model of bipolar transistors', *Bell Syst. Tech. J.*, vol. 49, pp. 827–852, 1970.

12 H. K. Gummel, 'A charge control relation for bipolar transistors', *Bell Syst. Tech. J.*, vol. 49, pp. 115–120, 1970.

13 J. L. Moll, and I. M. Ross, 'The dependence of transistor parameters on the distribution of base layer resistivity', *Proc. IRE*, vol. 44, pp. 72–78, 1956.

14 C. T. Kirk, 'A theory of transistor cut-off frequency F_T falloff at high current densities', *IRE Trans. Electron Dev.*, vol. ED-9, pp. 164–174, 1962.

15 I. E. Getreu, *Modelling of Bipolar Transistors*. Elsevier, NY, 1978.

16 P. Antognetti, and G. Massobrio, *Semiconductor Device Modelling with SPICE*. McGraw Hill Book Co., NY, 1987.

17 M. Reisch, *High-Frequency Bipolar Transistors*. Springer-Verlag, Berlin, 2003.

18 B. Senapati, and C. K. Maiti, 'Modelling of self-heating effects in SiGe-HBTs', in *Proc. of 11th Intl. Workshop on Physics of Semiconductor Devices, New Delhi*, 2001.

19 X. Cao, J. McMacken, K. Stiles, P. Layman, J. J. Liou, A. Ortiz-Conde, *et al.*, 'Comparison of the new VBIC and conventional Gummel-Poon bipolar transistor models', *IEEE Trans. Electron Devices.*, vol. 47, pp. 427–433, 2000.

20 H. C. D. Graaff and W. J. Kloosterman, 'New formulation of the current and charge relations in bipolar transistor modelling for CACD Purposes', *IEEE Trans. Electron Devices*, vol. 32, pp. 4215–2419, 1985.

21 H.-M. Rein, and M. Schroter, 'A compact physical large-signal model for high-speed bipolar transistors at high current densities - Part II: Two-dimensional model and experimental results', *IEEE Trans. Electron Devices*, vol. 34, pp. 1752–1761, 1987.

22 A. Koldehoff, M. Schroter, and H.-M. Rein, 'A compact bipolar transistor model for very-high-frequency applications with special regard to narrow emitter stripes and high current densities', *Solid State Electron.*, vol. 36, pp. 1035–1048, 1993.

23 C. C. McAndrew, J. A. Seitchik, D. F. Bowers, M. Dunn, M. Foisy, I. Getreu, *et al.*, 'VBIC95, The vertical bipolar inter-company model', *IEEE J. Solid-State Circuits*, vol. 31, pp. 1476–1483, 1996.

24 C. C. McAndrew, J. A. Seitchik, D. F. Bowers, M. Dunn, M. Foisy, I. Getreu, *et al.*, 'VBIC95: An improved vertical IC bipolar transistor model', in *IEEE BCTM Proc.*, pp. 170–177, 1995.

25 G. M. Kull, L. W. Nagel, S.-W. Lee, P. Lloyd, E. J. Prendergast, and H. K. Dirks, 'A unified circuit model for bipolar transistors including quasi-saturation effects', *IEEE Trans. Electron Devices*, vol. 32, pp. 1103–1113, 1985.

26 P. B. Weil, and L. P. McNamee, 'Simulation of excess phase in bipolar transistors', *IEEE Trans. Circuits and Systems*, vol. 25, pp. 114–116, 1978.

27 W. J. Kloosterman, and H. C. de Graaff, 'Avalanche multiplication in a compact bipolar transistor model for circuit simulation', in *IEEE BCTM Proc.*, pp. 103–106, 1988.

28 C. C. McAndrew, and L. W. Nagel, 'SPICE early modelling', in *IEEE BCTM Proc.*, pp. 144–147, 1994.

29 F. Najm, 'VBIC95: An improved bipolar transistor model', *IEEE Circuits and Devices Mag.*, pp. 11–15, 1996.

30 C. Jacoboni, C. Canali, G. Ottaviani, and A. A. Quaranta, 'A review of some charge transport properties of silicon', *Solid State Electron.*, vol. 20, pp. 77–89, 1977.

31 W. J. Kloosterman, and H. C. de Graaff, 'Avalanche multiplication in a compact bipolar transistor model for circuit simulation', *IEEE Trans. Electron Devices*, pp. 1376–1380, 1989.

32 X. Cao, J. McMacken, K. Stiles, P. Layman, J. J. Liou, A. Sun, *et al.*, 'Parameter extraction and optimization for new industry standard VBIC model', in *Intl. Conf. on Advanced Semiconductor Devices and Microsystems*, pp. 107–115, 1998.

33 H. C. De Graaff, and F. M. Klassen, *Compact Transistor Modelling for Circuit Design*. Springer-Verlag, New York, 1990.

34 J. C. J. Paasschens, and W. J. Kloosterman, 'Parameter Extraction for the Bipolar Transistor Model Mextram - Level 504, Nat Lab Unclassified Report, NL-UR 2001/801, Koninklijke Philips Electronics', 2001.

35 J. C. J. Paasschens, and W. J. Kloosterman, 'The Mextram Bipolar Transistor Model – Level 504, Nat Lab Unclassified Report, NL-UR 2000/811, Koninklijke Philips Electronics', 2001.

36 J. C. J. Paasschens, W. J. Kloosterman, and R. vd Toon, 'Model Derivation of Mextram 504, Nat Lab Unclassified Report, NL-UR 2002/806, Koninklijke Philips Electronics', 2002.

37 L. C. N. de Vreede, H. C. de Graaff, K. Mouthaan, M. de Kok, J. L. Tauritz, and R. G. F. Baets, 'Advanced modelling of distorsion effects in bipolar transistors using the Mextram model', *IEEE J. Solid-State Circuits*, vol. 31, pp. 114–121, 1996.

38 G. Niu, J. D. Cressler, and A. J. Joseph, 'Quantifying neutral base recombination and the effects of collector-base junction traps in UHV/CVD SiGe HBTs', *IEEE Trans. Electron Devices*, vol. 45, pp. 2499–2504, 1998.

39 S. L. Salmon, J. D. Cressler, R. C. Jaeger, and D. L. Harame, 'The influence of Ge grading on the bias and temperature characteristics of SiGe HBTs for precision analog circuits', *IEEE Trans. Electron Devices*, vol. 47, pp. 292–298, 2000.

40 W. J. Kloosterman, J. A. M. Geelen, and D. B. M. Klaassen, 'Efficient parameter extraction for the Mextram model', in *IEEE BCTM Proc.*, pp. 70–73, 1995.

41 E. Sonmez, A. Trasser, P. Abele, K. B. Schad, and H. Schumacher, 'Integrated receiver components for low-cost 26 GHz LMDS applications using an 0.8 μm SiGe HBT technology', in *33rd European Microwave Conf.*, pp. 399–402, 2003.

42 P. Deixler, H. G. A. Hulzing, J. J. T. M. Donkers, J. H. Klootwijk, D. Hartskeerl, W. B. de Boer, *et al.*, 'Explorations for high performance SiGe-heterojunction bipolar transistor integration', in *IEEE BCTM Proc.*, pp. 30–33, 2001.

43 F. Yuan, Z. Pei, J. W. Shi, S. T. Chang, and C. W. Liu, 'Mextram modelling of Si/SiGe heterojunction phototransistors', in *IEEE ISDRS Symp. Dig.*, pp. 92–93, 2003.

44 J. C. J. Paasschens, W. J. Kloostennan, and R. J. Havens, 'Modelling two SiGe HBT specific features for circuit simulation', in *IEEE BCTM Proc.*, pp. 38–41, 2001.

45 J. C. J. Paasschens, W. J. Kloosterman, R. J. Havens, and H. C. de Graaff, 'Improved compact modelling of output conductance and cut-off frequency of bipolar transistors', *IEEE J. Solid-State Circuits*, vol. 36, pp. 1390–1398, 2001.

46 H. C. de Graaff, and W. J. Kloosterman, 'Modelling of the collector epilayer of a bipolar transistor in the Mextram model', *IEEE Trans. Electron Devices*, vol. 42, pp. 274–282, 1995.

47 W. J. Kloosterman, J. C. J. Paasschens, and R. J. Havens, 'A comprehensive bipolar avalanche multiplication compact model for circuit simulation', in *IEEE BCTM Proc.*, pp. 172–175, 2000.

48 S. Mijalkovic, and H. C. de Graaff, 'Mextram 504 for SiGe HBT applications', in *Workshop on Semiconductor Advances for Future Electronics*, pp. 115–122, 2001.

49 H. C. Wu, S. Mijalkovic, and J. N. Burghartz, 'Parameter extraction of scalable Mextram model for high-speed SiGe HBTs', in *IEEE BCTM Proc.*, pp. 140–143, 2004.

50 E. Sonmez, W. Durr, P. Abele, K.-B. Schad, and H. Schumacher, 'Parameter extraction of SiGe HBTs for a scalable MEXTRAM model and performance verification by a SiGe HBT MMIC active receive mixer design for 11 GHz', in *Topical Meeting on Silicon Monolithic Integrated Circuits in RF Systems*, pp. 159–162, 2000.

51 U. Erben, and E. Sonmez, 'Fully differential 5 to 6 GHz low noise amplifier using SiGe HBT technnology', *Electron. Lett.*, vol. 40, pp. 39–40, 2004.

52 W. C. Hua, T. Y. Yang, and C. W. Liu, 'The comparison of isolation technolo- gies and device models on SiGe bipolar low noise amplifier', *Appl. Surf. Sci.*, vol. 224, pp. 425–428, 2004.

53 D. Sheridan, M. R. Murty, W. Ansley, and D. Harame, 'IBM Bipolar Model Standardization: VBIC-MEXTRAM-HICUM, Compact Model Counsil (CMC) Meeting', 2002.

54 H. Stubing, and H.-M. Rein, 'A compact physical large signal model for high speed bipolar transistors at high current densities Part I: One- dimensional model', *IEEE Trans. Electron Devices*, vol. ED-34, pp. 1741–1751, 1987. see also correspondence in *IEEE Trans. Electron Devices*, Vol. 35, pp. 1995–1997, 1988.

55 M. Schroter, and H.-M. Rein, 'Transit time of high-speed bipolar transistors in dependence on operating point, technological parameters, and temperature', in *IEEE BCTM Proc.*, pp. 85–88, 1991.

56 M. Schroter and H.-M. Rein, 'Simulation and modelling of the low-frequency base resistance of bipolar transistors in dependence on current and geometry', *IEEE Trans. Electron Devices*, vol. 38, pp. 538–544, 1991.

57 H.-M. Rein, H. Stubing, and M. Schroter, 'Verification of the integral charge-control relation for high-speed bipolar transistors at high current densities', *IEEE Trans. Electron Devices*, vol. 32, pp. 1070–1076, 1985.

58 M. Schroter, and H.-M. Rein, 'Investigation of very fast and high-current transients in digital bipolar circuits by using a new compact model and a device simulator', *IEEE J. Solid-State Circuits*, vol. 30, pp. 551–562, 1995.

59 M. Schroter, Z. Yan, T.-Y Lee, and W. Shi, 'A compact tunnelling current and collector breakdown model', in *IEEE BCTM Proc.*, pp. 203–206, 1998.

60 M. Schroter, and T.-Y. Lee, 'A physics-based minority charge and transit time model for bipolar transistors', *IEEE Trans. Electron Devices*, vol. 46, pp. 288–300, 1999.

61 M. Schroter, and H. Tran, 'Modelling of base-collector junction related effects in heterojunction bipolar transistors', in *Intl NanoTech Meeting Dig.*, pp. 102–107, 2004.

62 S. Wilms, and H.-M. Rein, 'Analytical high-current model for the transit time of SiGe HBTs', in *IEEE BCTM Proc.*, pp. 199–202, 1998.

63 M. Schroter, *HICUM – A scalable physics-based compact bipolar transistor model, ver. 2.1*, 2000.

64 M. Schroter, 'Staying current with HICUM', *IEEE Circuits and Devices Magazine*, vol. 18, pp. 16–25, 2002.

65 M. Schroter, H. Jiang, S. Lehmann, and S. Komarow, 'HICUM/Level0 – A simplified compact bipolar transistor model', in *IEEE BCTM Proc.*, pp. 112–115, 2002.

66 C. C. McAndrew, and L. W. Nagel, 'Early effect modelling in SPICE', *IEEE J. Solid-State Circuits*, vol. 31, pp. 136–138, 1996.

67 S. Mijalkovic, 'Generalized Early factor for compact modelling of bipolar transistors with non-uniform base', *Electron. Lett.*, vol. 39, pp. 1757–1758, 2003.

68 J. A. Seitchik, C. F. Machala, and P. Yang, 'The determination of SPICE Gummel-Poon parameters by a merged optimization-extraction techniques', in *IEEE BCTM Proc.*, pp. 275–278, 1989.

69 B. Senapati, *Modelling and Characterization of SiGe-HBTs for RF Applications*. PhD thesis, Indian Institute of Technology, Kharagpur, 2001.

70 J. S. Yuan, 'Base pushout effect on collector signal delay and Early voltage for Heterojunction Bipolar Transistors', *Solid State Electron.*, vol. 36, pp. 657–660, 1993.

71 Application Note 315, *Practical Applications of the 4145B Semiconductor Parameter Analyzer*, 1987.

72 D. E. Dawson, A. K. Gupta, and M. L. Salib, 'CW measurement of HBT thermal resistance', *IEEE Trans. Electron Devices*, vol. 39, pp. 2235–2239, 1992.

73 N. Bovolon, P. Baureis J. E. Muller, P. Zwicknagl, R. Schultheis, and E. Zanoni, 'A simple method for the thermal resistance measurement of AlGaAs/GaAs

heterojunction bipolar transistors', *IEEE Trans. Electron Devices*, vol. 45, pp. 1846–1848, 1998.

74 S. P. Marsh, 'Direct extraction technique to derive the junction temperature of HBT's under high self-heating bias conditions', *IEEE Electron Devices*, vol. 47, pp. 288–291, 2000.

75 F. X. Sinnesbichler, and G. R. Olbrich, 'Electro-thermal large-signal modelling of SiGe HBTs', in *29th European Microwave Conf.*, pp. 125–128, 1999.

76 J. M. M. Rios, L. M. Lunardi, S. Chandrasekhar, and Y. Miyamoto, 'A self-consistent method for complete small-signal parameter extraction of InP-based heterojunction bipolar transistors (HBT's)', *IEEE Trans. Microwave Th. and Tech.*, vol. 45, pp. 39–45, 1997.

77 D. Costa, W. U. Liu, and J. S. Harris, 'Direct extraction of the AlGaAs/GaAs heterojunction bipolar transistor small-signal equivalent circuit', *IEEE Trans. Electron Devices*, vol. 38, p. 2018, 1991.

78 K. M. Walter, B. Ebersman, D. A. Sunderland, G. D. Berg, G. G. Freeman, R. A. Groves, D. K. Jadus, and D. L. Harame, 'A scaleable, statistical SPICE Gummel-Poon model for SiGe HBTs', *IEEE J. Solid-State Circuits*, vol. 33, pp. 1439–1444, 1998.

79 A. DiVergilio, P. Zampardi, and K. Newton, 'VBIC: A New standard in advanced bipolar modelling', *IBM MicroNews*, vol. 5, pp. 14–17, 1999.

80 B. Senapati, and C. K. Maiti, 'Advanced SPICE modelling of SiGe HBTs using VBIC model', *IEE Proc. Circuits, Devices and Systems*, vol. 149, pp. 129–135, 2002.

81 B. Senapati, and C. K. Maiti, 'Advanced SPICE Modelling of SiGe-HBTs', in *National Seminar on Adv. Transistor Modelling (Invited Paper)*, 2001.

82 T.-Y. Lee, M. Schroter, and M. Racanelli, 'A scalable model generation methodology for bipolar transistors for RF IC design', in *IEEE BCTM Proc.*, pp. 171–174, 2001.

83 D. Berger, D. Celi, M. Schroter, M. Malorny, T. Zimmer, and B. Ardouin, 'HICUM parameter extraction flow for a single transistor geometry', in *IEEE BCTM Proc.*, pp. 116–119, 2002.

84 A. Chakravorty, R. Garg, and C. K. Maiti, 'VBIC-HICUM models applied to advanced BJTs developed in high performance mixed analog/digital 0.5 μm BiCMOS BC050 process', in *Tech. Dig. HOT*, p. 12, 2003.

85 A. Chakravorty, R. Garg, and C. K. Maiti, 'VBIC modelling of high speed bipolar transistors fabricated in a high performance mixed analog/digital 0.5 μm BiCMOS technology', in *Proc. NCIIT*, pp. 94–96, 2003.

86 A. Chakravorty, B. Senapati, G. Dalapati, R. Garg, C. K. Maiti, G. A. Armstrong, *et al.*, 'HICUM modelling of SiGe-HBTs fabricated in wafer bonded SOI substrates', in *Proc. CODEC*, 2004.

87 A. Chakravorty, B. Senapati, G. Dalapati, R. Garg, C. K. Maiti, G. A. Armstrong, *et al.*, 'VBIC Modelling of SiGe-HBTs Fabricated in Wafer Bonded SOI Substrates', in *Proc. IWPSD*, pp. 591–593, 2003.

88 B. Senapati, R. F. Scholz, D. Knoll, B. Heinemann, and A. Chakravorty, 'Application of the VBIC model for SiGe:C heterojuction transistors', in *Proc. Mixed Design of Integrated Circuits and Systems*, pp. 94–96, 2004.

89 A. Chakravorty, R. Garg, and C. K. Maiti, 'Comparison of state-of-the-art bipolar compact models for SiGe-HBTs', *Appl. Surf. Sci.*, vol. 224, pp. 354–360, 2004.

90 APLAC Solutions Corporation, 'APLAC Manual, ver. 7.91', 2003.

91 M. Reisch, 'Self-heating in BJT circuit parameter extraction', *Solid State Electron.*, vol. 35, pp. 677–679, 1992.

92 G. Niu, J. D. Cressler, U. Gogineni, and D. L. Harame, 'Collector-base junction avalanche multiplication effects in advanced UHV/CVD SiGe HBTs', *IEEE Electron Devices Lett.*, vol. 19, pp. 288–290, 1998.

93 U. Erben, M. Wahl, A. Schuppen, and H. Schumacher, 'Class-A SiGe HBT Power Amplifiers at C-Band Frequencies', *IEEE Microwave and Guided Lett.*, vol. 5, pp. 435–436, 1995.

94 M. Schroter, 'TRADICA User Manual', 2002.

95 S. P. Voinigescu, M. C. Maliepaard, J. L. Showell, G. E. Babcock, D. Marchesan, M. Schroter, *et al.*, 'A scalable high-frequency noise model for bipolar transistors with application to optimal transistor sizing for low-noise amplifier design', *IEEE J. Solid-State Circuits*, vol. 32, pp. 1430–1439, 1997.

96 H.-M. Rein, 'A simple method for separation of the internal and external (peripheral) currents of bipolar transistors', *Solid State Electron.*, vol. 27, pp. 625–632, 1984.

97 M. Schroter and D. J. Walkey, 'Physical modelling of lateral scaling in bipolar transistor', *IEEE J. Solid-State Circuits*, vol. 31, pp. 1484–1492, 1996.

98 D. J. Walkey, T. J. Smy, D. Marchesan, H. Tran, C. Reimer, T. C. Kleckner, *et al.*, 'Extraction and modelling of thermal behaviour in trench isolated bipolar structures', in *IEEE BCTM Proc.*, pp. 97–100, 1999.

99 D. J. Walkey, T. J. Smy, R. G. Dickson, J. S. Brodsky, D. T. Zweidinger, and R. M. Fox, 'Equivalent circuit modelling of static substrate thermal coupling using VCVS representation', *IEEE J. Solid-State Circuits*, vol. 37, pp. 1198–1206, 2002.

100 R. M. D. A. Velghe, D. B. M. Klaassen, and F. M. Klaassen, 'MOS Model 9'. Unclassified report NL-UR 003/94, Philips Electronics N.V., 1994.

101 Y. Cheng, M. Chan, K. Hui, M. Jeng, Z. Liu, J. Huang, J. Chen, R. Tu, P.K. Ko, and C. Hu, 'BSIM3v3 Manual'. University of California/Berkeley, 1996.

102 M. Schroter, 'Compact bipolar transistor modelling – Issues and possible solutions', in *Intl NanoTech Meeting Dig.*, pp. 282–285, 2003.

103 M. Schroter, H. Tran, and W. Kraus, 'Germanium profile design options for LEC-SiGe HBTs', *Solid State Electron.*, vol. 48, pp. 1133–1146, 2004.

104 M. Ramana Murty, K. M. Newton, S. L. Sweeney, D. C. Sheridan, and D. L. Harame, 'Implementation of a scalable statistical VBIC model for large signal and intermodulation distortion analysis of SiGe HBTs', in *IEEE MTT-S Digest*, pp. 2165–2168, 2002.

105 D. J. Harame, J. H. Comfort, J. D. Cressler, E. F. Crabbe, J. Y. C. Sun, B. S. Meyerson, and T. Tice, 'Si/SiGe epitaxial-base transistors-part I: materials, physics, and circuits', *IEEE Trans. Electron Devices*, vol. 42, pp. 455–467, 1995.

106 R. F. Scholz, A. Chakravorty, D. Knoll, C. K. Maiti, 'Simple Emitter and Collector Scaling Approach with VBIC for Low-cost SiGe:C HBT's', in *Tech. Dig. Workshop on Compact Modelling for RF/Microwave Applications*, 2003.

107 R. F. Scholz, B. Senapati, A. Chakravorty, D. Knoll, and C. K. Maiti, 'Simple emitter and collector scaling approach with VBIC for low-cost SiGe:C HBTs', in *IC-CAP European User Meeting*, 2003.

108 S. Fregonese, D. Celi, T. Zimmer, C. Maneux, and P. Y. Sulima, 'A scalable substrate network for compact modelling of deep trench insulated HBT', *Soild State Electron.*, vol. 49, pp. 1623–1631, 2005.

109 D. Knoll, K. E. Ehwald, B. Heinemann, A. Fox, K. Blum, H. Rucker, *et al.*, 'A flexible, low-cost, high performance SiGe:C BiCMOS process with a one-mask HBT module', in *IEEE IEDM Tech. Dig.*, pp. 783–786, 2002.

110 H. Baudry, B. Szelag, F. Deleglise, M. Laurens, J. Mourier, F. Saguin, *et al.*, 'BiCMOS7RF: a highly-manufacturable 0.25-μm BiCMOS RF-applications-dedicated technology using non selective SiGe:C epitaxy', in *IEEE BCTM Proc.*, pp. 207–210, 2003.

111 D. C. Sheridan, J. B. Johnson, and J. Watts, 'Use of genetic algorithm optimisation techniques for bipolar compact modelling, *Workshop on Compact Modelling for RF/Microwave Applications*', 2003.

112 A. Chakravorty, R. F. Scholz, B. Senapati, D. Knoll, A. Fox, R. Garg, *et al.*, 'Accurate Modelling of SiGe:C-HBTs using Adaptive Neuro-Fuzzy Inference System', *Mat. Sci. Semicond. Processing*, vol. 8, pp. 307–311, 2005.

Chapter 10

Design and simulation of high-speed devices

One of the main differences between Si VLSI and RF electronics is the choice of semiconductor materials and transistor types. Until recently, CMOS has been the standard device and Si is the only semiconductor used in VLSI. In RF electronics, on the other hand, a wide variety of different semiconductor materials (i.e., Si, SiGe, GaAs, InP and wide bandgap materials) and various transistor types, such as bipolar junction transistor, heterojunction bipolar transistor, metal–semiconductor field-effect transistor (MESFET), high electron mobility transistor (HEMT) and metal–oxide–semiconductor field-effect transistor (MOSFET) have found their applications.

Bipolar junction transistors (BJTs) have the benefits of the silicon technology, for example, very high integration and low-cost production, but are limited to lower frequencies. Important steps forward to faster silicon-based devices were the invention of the polySi emitter transistor and now the SiGe HBTs, which are competitive in terms of speed to the III–V devices. HBTs and HEMTs are among the most advanced semiconductor devices. They both benefit from the use of heterojunctions formed by different materials. The HBTs make use of wide bandgap emitter and narrow bandgap base. In an npn transistor this favours the electron injection from the emitter to the base, and restricts hole injection from the base the emitter. This advantage can be maintained even if the base is highly doped to get a low base resistance and the emitter is lightly doped. Microwave, millimetre-wave and high-speed digital HBT ICs are used for microwave power and low power wireless communications applications between 0.9 and 100 GHz.

HBTs based on AlGaAs/GaAs, InGaAs/InP and Si/SiGe have shown superior electrical properties compared to Si BJTs. Much of this improvement is due to the high-base doping achievable through the use of wide band-gap emitters as well as superior material properties of the state-of-the-art HBTs due to bandgap engineering techniques and epitaxial growth developments. For example, by varying Al content in AlGaAs/GaAs HBT a graded bandgap can be obtained. The resulting accelerating field decreases the time that the electrons needed to transport across the base and thus

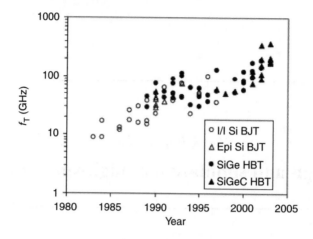

Figure 10.1   The trend of cut-off frequency f_T over the past two decades, categorised by conventional implanted-base Si BJTs (I/I Si BJT), epitaxial-base Si BJTs (Epi Si BJT), SiGe HBTs (SiGe HBT), and SiGe HBTs with carbon-doped base (SiGeC HBT) [After J.-S. Rieh et al., Proc. IEEE, Vol. 93, 2005(1522–1538). © 2005 IEEE]

increases f_T. Furthermore, using MBE, it is possible to grow extremely thin layers with a thickness of only a few nanometres and sharp interfaces between adjacent layers.

SiGe HBTs with $f_T > 300$ GHz are becoming more promising each day and threatening the GaAs market due to competitive material properties, yield, cost and integration potentials [1, 2]. SiGe HBTs have started replacing III–V devices for several applications, such as low noise amplifiers and frequency dividers up to 100 GHz [3], and are being considered essential for 40 Gb/s optical communication systems. Transit frequencies, f_T of 350 GHz [2], maximum oscillation frequencies, f_{max} of 285 GHz, and ring oscillator delays of 4.2 ps [4] have been reported. Figure 10.1 shows the current status of peak-f_T of SiGe HBTs over the last couple of years. The devices are fully compatible with the existing state-of-the-art 0.13 μm CMOS technology [4, 5].

AlGaAs/GaAs HBTs are commercially available today and commonly used in RF power amplifiers due to higher power densities and higher f_T and f_{max} without the limitation of photolithography. This means better device matchings when compared to III–V FETs. Although brittle and expensive, InP based HBTs show even higher speed performance due to superior material properties of the InP material. InP HBTs with 370 GHz f_T and 459 GHz f_{max} have been fabricated [6]. Rodwell *et al.* [7] have demonstrated an InP-based HBT with f_{max} of 1.08 THz fabricated with a transferred substrate method. Also, UIUC group has reported a single HBT with 506 GHz f_T. All these devices especially the InP-based HBT show aggressive characteristic frequencies when compared to traditional homojunction Si BJTs, where 100 GHz f_{max} and 84 GHz f_T are available from experimental devices [8]. Cut-off frequencies f_T of about

150 GHz and maximum oscillation frequencies f_{max} of more than 250 GHz [9, 10] have been reported for HBTs on GaAs. Transfer substrate InAlAs/InGaAs HBTs on GaAs with 250 GHz [11] and record InP-based HBTs with 800 GHz were demonstrated [12] but they are still lacking level of integration ($\sim$1000 transistors per chip) compared to the GaAs-based HBTs.

It is important to find the potentials and limitations of these devices and analyse them under common figure of merit (FOM) definitions such as current–voltage characteristics, stability, power gain analysis, cut-off and maximum oscillation frequencies and noise figure as well as to make a meaningful comparison which is necessary for a technology choice especially in RF circuit and system level applications such as power amplifier, low noise amplifier circuits and transceiver/receiver systems. A common definition of FOMs is necessary to be able to choose the right device or material type. Schwierz and Liou have reported a comprehensive study for the definition of FOMs for RF heterojunction bipolar transistors [13]. The most relevant figures of merit of transistors for RF/microwave applications are [14]:

- peak f_T and f_{max} values which determine the ultimate circuit speed;
- peak f_T current density J_{c,F_T} which determines the power dissipation of a high-speed circuit at a given data rate;
- $I_{c,F_T}/C_{bc}(C_{gd})$, also known as the intrinsic slew-rate of the transistor, is defined as the ratio of the peak f_T current and the output capacitance;
- BV_{cbo}/BV_{dg} is the breakdown voltage of the transistor which limits the maximum output swing achievable in the technology;
- turn-on/threshold voltage: V_{be}/V_t limits the minimum value of the power supply voltage and, usually, power dissipation of a circuit and favours CMOS over bipolar devices and InP HBTs over SiGe HBTs;
- minimum feature size affects power dissipation and integration levels and favours Si CMOS and SiGe BiCMOS over III–V technologies owing to the more mature processing techniques;
- thermal resistance is closely related to the semiconductor material and to the minimum feature size and affects indirectly the integration levels and the transistor speed that can be achieved under reliable operating conditions.

In the following, a comparison of three state-of-the-art heterojunction bipolar transistors, the AlGaAs/GaAs HBT, the Si/SiGe HBT and the InGaAs/InP HBT, is presented. The trade-offs between the devices are examined, focusing on their intrinsic properties and electrical parameters, and on how these properties and parameters relate to the reported and potential device performance. Heterostructures, which are key elements for some advanced RF transistors, are described, and an overview of the different transistor types and their figures of merit is given. A comprehensive overview of the complex relationship between IC technologies, devices and RF circuit performance in terms of key RF figures of merits may be found in Reference 15. It covers CMOS, BiCMOS and bipolar technologies and also the trends in RF technologies.

10.1   SiGe vs. GaAs HBTs

The SiGe HBTs are double heterojunction bipolar transistors (DHBTs) as the SiGe material is used as a narrow bandgap material in the p-type base. The emitter and the collector are silicon and have larger bandgap. The AlGaAs/GaAs and InGaP/GaAs HBTs benefit from a single heterojunction formed between the AlGaAs wide bandgap emitter and the GaAs p-type base. InP/InGaAs and InAlAs/InGaAs grown on InP substrate give double heterojunction devices as both emitter and collector regions include wide bandgap materials. HBTs based on III–V semiconductors are currently the fastest devices.

During the last decade, there have been a lot of debates and discussions on the competition between SiGe and GaAs HBTs for high-frequency RF applications. An early review by Cressler [16] gives an excellent description of the advantages of SiGe HBT over Si bipolar for high-frequency analog applications. It is clear that reported data show a large overlap in the performance of GaAs and SiGe HBTs [17].

Some of the key physical properties of Si and GaAs relevant to the electrical design of SiGe and GaAs HBTs are listed in Table 10.1 [18]. It shows that the dielectric constant, ϵ, for Si and for GaAs are about the same. Therefore, the depletion layer capacitance of a p-n junction in Si or in GaAs is a function of the junction depletion layer width only. The electron diffusion coefficient D, which is related to mobility, is much larger in GaAs than in Si. For fields less than about 3×10^4 V/cm the velocity in GaAs is larger than that in Si. However, for fields larger than about 3×10^4 V/cm, the opposite is true. Compared with field-effect devices bipolar transistors employ vertical current transport, which offer better utilization of wafer area and thus lead to higher power density. The bipolar approach also offers higher linearity at higher power levels, superior power-added-efficiency and low frequency noise as necessary for RF circuits and systems.

Just like Si bipolar transistor, SiGe HBTs use polySi emitter. The emitter poly is usually doped with As to a concentration of $>10 \times 10^{21}$ cm^{-3}. There is no additional

Table 10.1   Some properties of Si and GaAs at 300 K

Material parameter	Si	GaAs
Dielectric constant	11.9	13.1
Energy gap (eV)	1.12	1.424
Electron drift mobility (cm^2/Vs)	1500	8500
	~200	700
	at 2×10^{18} doping	at 2×10^{19} doping
Intrinsic carrier concentration (cm^3)	1.45×10^{10}	~1.79×10^6
Minority carrier lifetime (s)	2.5×10^{-3}	~10^{-3}
Thermal conductivity (W/cm-K)	1.5	0.46
Semi-insulating substrate	SOI	Ion bombardment

emitter bandgap engineering. The base region usually has an average doping concentration of $>10 \times 10^{18}$ cm^{-3}. In some designs, a very thin i-layer is sandwiched between the emitter poly and the base layer to increase e-b junction breakdown voltage and to reduce e-b junction capacitance. However, due to out-diffusion from the emitter and base layers, the effect of this i-layer, while not negligible, is not very large. The net is that the e-b junction capacitance is about that of a n$^+$p-junction with the p-region doped at $>10 \times 10^{18}$ cm^{-3}. In GaAs HBT, the emitter is engineered using AlGaAs to have an energy gap much larger than that of the GaAs base layer. It allows a much more lightly doped emitter, typically about $>5 \times 10^{17}$ cm^{-3} than the base, which is typically doped to about 4×10^{19} cm^{-3}. Thus, for the same emitter size, the e-b junction capacitance of GaAs HBT is much smaller than that of SiGe HBT.

In Chapter 6, the working principle of HBTs has been discussed. It should be noted that bandgap grading is available to both SiGe and GaAs HBTs, although it is applied much more widely to SiGe than to GaAs. For a linearly graded base bandgap the current gain, (β), Early voltage, (V_A) and the base transit time, (τ_b), when compared to Si counterpart are given by Equations (6.6), (6.7) and (6.9), respectively. The collector doping concentration, N_c must be large enough to avoid base-pushout or Kirk effect. As a general rule, to avoid Kirk effect, the collector current density should be less than qvN$_c$, where v is the electron velocity. Under normal operation, the electric field in the collector is in the range $3-10 \times 10^4$ V/cm. At these fields, v is approximately the same for Si and GaAs. Therefore, for well designed SiGe and GaAs HBTs operating at the same collector current density, the b-c junction capacitances are about the same. However, the collector-substrate capacitance of a SiGe HBT is larger than that of GaAs HBT unless an SOI substrate is used.

Since the bandgap of silicon is smaller than that of GaAs, the diode turn-on voltage of a p-n diode in Si is smaller than that in GaAs. This voltage difference can be estimated from the forward current density of a diode. From the material properties shown in Table 10.1, one can therefore deduce that at 300 K, the diode turn-on voltage for a GaAs diode is about 0.28 V larger than that of a Si diode. The larger diode turn-on voltage implies that a circuit built with GaAs HBTs must use a larger power supply voltage than one built with SiGe HBTs. For an ECL circuit, this difference is about 20 per cent. Thus, GaAs HBT circuits inherently dissipate more power than SiGe HBT circuits.

At low current densities where Kirk effect is negligible, the cut-off frequency of a bipolar transistor is given by Equation (6.11), where the first term represents the delay time due to charging of the emitter capacitance, the collector capacitance and the parasitic capacitances, connected to the base lead. As discussed before, for the same size emitter, emitter capacitance is smaller for GaAs HBT than for SiGe HBT, and collector capacitance is about the same for the two devices if SOI is used for SiGe HBT. If SOI is not used, then collector capacitance is much larger for SiGe HBT. Therefore, for the same design rules and collector current densities, the first term in Equation (6.11) is smaller for GaAs HBT than for SiGe HBT.

For devices without base bandgap grading, τ_b is inversely proportional to carrier mobility. Using the mobility values in Table 10.1, τ_b is a factor of three lower for a

GaAs base layer than for a Si base layer of the same width. From Equation (6.9), a SiGe HBT will need a base bandgap grading of $\Delta E_g \sim 150$ meV to achieve the same τ_b as a GaAs HBT with no base bandgap grading. With base bandgap grading in both devices, τ_b will be lower for GaAs HBT than for a SiGe HBT. On the other hand, τ_c should be about the same for SiGe HBT and GaAs HBT if both devices are optimised for the same collector current densities. It has been shown that at low collector current density f_T is larger for GaAs HBT (and larger yet for GaInAs), consistent with the discussion above [19–22]. At higher collector current, f_T is limited by base and collector transit times as indicated in Equation (6.9) and by the Kirk effect, which dominates at very large current densities, so that f_T falls off rapidly with increasing collector current. GaAs HBTs are usually designed to operate at collector current density of less than 0.5 mA/μm^2, where Kirk effect is minimal, to avoid local over-heating. SiGe HBTs without SOI do not have local over-heating problems, and hence are usually designed to operate at much larger collector current density, where Kirk effect is important but is minimised by collector profile engineering. However, if an SOI substrate is used, then SiGe HBTs will be limited to even lower current densities than GaAs HBTs because of the very small thermal conductivity of SiO$_2$.

The maximum oscillation frequency, f_{max}, is related to f_T by Equation (6.12), where R_b is the base resistance and C_{bc}, is the collector-base capacitance. For comparable advanced device structures, emitter sizes and layouts, R_b should be somewhat smaller for GaAs HBT than for SiGe HBT, and C_{bc} should be about the same for both devices. Therefore, f_{max} is potentially larger for GaAs HBT than for SiGe HBT. However, SiGe HBTs without SOI can be designed to operate at much larger current densities, with smaller emitters and accompanying parasitics, and hence could have larger f_{max} than GaAs HBTs.

10.2   Device simulation

Heterostructure bipolar transistors and high electron mobility transistors (HEMTs) are among the most advanced semiconductor devices. In this section, the simulation of SiGe, III–V GaAs and InP based heterostructure semiconductor devices is considered. A common feature is the lack of a rigorous approach to III–V and IV–IV group semiconductor materials modelling. As an example, modelling of AlGaAs, InGaAs, or even InAlAs and InGaP is restricted to slight modifications of the GaAs material properties. Critical issues concerning simulation of heterostructures, such as interface modelling at heterojunctions and insulator surfaces as well as hydrodynamic and high field effects modelling, carrier energy relaxation, impact ionisation, gate current modelling and self-heating effects are mostly not considered. In addition, quantum mechanical effects are often neglected or accounted for only by simple models for quantum corrections [23, 24], as solving the Schrodinger or the Wigner equation is extremely expensive in terms of computational resources.

To enable predictive simulation of semiconductor devices better carrier transport models are required. The drift-diffusion (DD) transport model is the simplest and most popular model used for device simulation [25]. However, with downscaling the feature sizes, non-local effects become more pronounced and may be accounted for by using an energy-transport (ET) or hydrodynamic (HD) transport model.

During the past two decades Monte Carlo (MC) methods for solving the time-dependent Boltzmann equation have been developed [26, 27] and applied for device simulation [28–30]. Reduction of the demand on computational resources is still an issue for MC simulation. Also, the MC algorithms encounter serious difficulties when applied to advanced semiconductor devices.

The critical modelling issues essential for heterostructure device simulation are addressed in this chapter and special attention is focused on the description of the anisotropic majority/minority electron mobility in strained-SiGe grown on Si. A direct approach to obtain scattering parameters (s-parameters) and other derived figures of merit of SiGe HBTs by means of small-signal a.c. analysis is also discussed. Results from two-dimensional hydrodynamic simulations of SiGe HBTs are presented and compared with experimental data. Several examples are chosen from the work of Palankovski *et al.* [31–35] to demonstrate technologically important issues which can be addressed and solved by device simulation. In the following, a review on the available state-of-the-art device simulators and a discussion on the choice of current transport models to be used are presented. Critical modelling issues, such as bandgap narrowing, anisotropic electron minority mobility in strained-SiGe, carrier transport through heterointerfaces, carrier generation/recombination and lattice self-heating are also addressed. For more details on device simulation and simulators, and material parameters the reader may refer to Reference 36.

10.2.1 Device simulators

With the downscaling of device dimensions, device and circuit simulations need to be carried out by state-of-the-art tools, accounting for physical effects on a microscopic scale. To enable predictive simulation of semiconductor devices, proper models describing carrier transport are also required. The continuously increasing computational power of computer systems now allows the use of technology computer aided design (TCAD) tools on a very large scale.

Currently, for the silicon industry, where process, device and interconnect simulation tools form a virtual workbench from material analysis to chip design are available, III–V simulation is mainly focused on device and circuit aspects. A common feature is the lack of a rigorous approach to III–V group semiconductor materials modelling. As an example, modelling of AlGaAs, InGaAs, or even InAlAs and InGaP is restricted to slight modifications of the GaAs material properties. For heterojunction devices, for example, SiGe HBTs, owing to the extensive number of process steps involved, device simulation is focused on process control and inverse modelling, for example, of geometry. Several commercial device simulators, company-developed simulators and university-developed simulators claim the capability to handle heterostructure devices. These simulators differ considerably in scope (one-, quasi-two, two-, quasi-three, or three-dimensional), in the choice of carrier transport model (drift-diffusion, energy-transport or MC statistical solution of the Boltzmann equation) and in the capability of including electrothermal effects.

The quality of the physical models can be questioned as the model parameters for SiGe are often simply inherited from parameters for silicon. Critical issues concerning simulation of heterostructures, such as interface modelling at heterojunctions and at silicon/polySi interfaces are frequently not considered. Hydrodynamic and high-field effects, such as carrier energy relaxation, impact ionisation and self-heating effects are often ignored.

The two-dimensional device simulator PISCES and its various derivatives in TCAD products from commercial vendors incorporates modelling capabilities for SiGe based devices. PISCES also incorporates more limited modelling capabilities for GaAs and InP-based devices. One of its many modifications G-PISCES from Gateway Modelling [37] has been extended by the inclusion of a full set of III–V models. Examples of MESFETs, HEMTs and HBTs for several material systems, for example, InAlAs/InGaAs, AlGaAs/InGaAs, AlGaAs/GaAs and InGaP/GaAs HBTs are demonstrated. A disadvantage of this simulator is the lack of appropriate ET or HD transport models, necessary to model high-field effects, which are available in the original version of PISCES.

The device simulator MEDICI from Synopsys which is also based on PISCES, offers simulation features for SiGe/Si HBTs and AlGaAs/InGaAs/GaAs HEMTs. Advantages of this simulator are hydrodynamic simulation capabilities and rigorous approach to generation/recombination processes. In addition, it includes module treating anisotropic material properties. However, it has been successfully used for the simulation of AlGaAs/GaAs HBTs [38].

At the quantum level, among others, one-dimensional Schrodinger-Poisson solver NEMO [39] based on non-equilibrium Green's functions, is offered for sub-0.1 μm SiGe structure simulation. Gateway Modelling offers a one-dimensional Schrodinger-Poisson solver POSES [40] for charge analysis in HEMT devices for process control. In the program SIMBA [41] a link between a one-dimensional Schrodinger solver and a two-dimensional Poisson solver provides drift-diffusion transport simulation of GaN HEMTs [42]. The three-dimensional version of SIMBA has been used for thermal optimisation of GaN HEMT layouts [43].

The two- and three-dimensional device simulator DESSIS from ISE has demonstrated a rigorous approach to semiconductor physics modelling and includes extensive interface trap modelling to and III–V materials [44]. Interface tunnelling is included in a thermionic field emission model. The density-gradient method is used to model quantum effects in heterostructure devices [44, 45]. Quasi-two-dimensional approaches using simplified one-dimensional current equation are demonstrated by BIPOLE3 from BIPSIM which additionally features good models for polySi. A full hydrodynamic transport model has been coupled with a Schrodinger solver at the University of Leeds [46, 47]. These simulations have been verified against MC simulations for devices with gate lengths down to 50 nm [46].

The quasi-two-dimensional Silvaco simulator FastBlaze has capabilities of simulation of heterostructure devices. The two- and three-dimensional simulator ATLAS [48] from Silvaco can also be adapted for the simulation of AlGaAs/GaAs and AlGaAs/InGaAs/GaAs heterostructures and PHEMTs. Some optimisation results for SiGe HBTs were reported with another university-developed simulator, SCORPIO

Table 10.2 Comparison of device simulators. DD: drift-diffusion, ET: energy-transport, HD: hydrodynamic, TFE: thermionic field emission, TE: thermionic emission [After V. Palankovski and S. Selberherr, Proc. Symp. on Diagnostics and Yield: Advanced Silicon Devices and Technologies for ULSI Era, 2003(1–11)]

Simulator	Dimension	Model	Features
SCORPIO	1D	DD	SiGe HBT
POSES	1D		Schrodinger-Poisson
NEMO	1D		Schrodinger-Poisson
BIPOLE3	quasi-2D	DD	polySi
ATLAS	2D	DD, ET	TE, heterojunction model
FastBlaze	quasi-2D	HD	
APSYS	2D	HD	Optical, interfaces
G-PISCES	2D	DD	III–V models
PISCES	2D	DD,ET	polySi, harmonic balance
MEDICI	2D	DD,HD	Anisotropic properties, TFE
FIELDAY	2D, 3D	DD	Electrothermal
MINIMOS-NT	2D, 3D	DD,HD	
DESSIS	2D, 3D	DD,HD	Trap modelling, TFE model

[49]. Some important features of several heterostructure device simulators currently available are summarised in Table 10.2.

A software interface between the device model and the compact large-signal model within the Microwave Design System (MDS and ADS) has been offered by Agilent. FastBlaze can be combined with the Advanced Design System (ADS) and has an interface with the microwave circuit simulator. An extraction with subsequent multi-tone excitation calculations has been presented in Reference 50.

10.3 Material parameters

Understanding the performance of SiGe HBTs is to a great extent determined by ones knowledge of the strained-$Si_{1-x}Ge_x$ layer properties. The incorporation of Ge significantly changes the properties of the base region and the base-emitter and base-collector junctions in a $Si/Si_{1-x}Ge_x/Si$ heterostructure. Various material properties essential for device simulation and their dependence on the composition for coherently strained-$Si_{1-x}Ge_x$ alloy on Si (100) substrate must be known. Commercially available simulation software such as Silvaco-ATLAS, MEDICI and university software such as PISCES-2ET [48, 51, 52] have been used for the simulation of SiGe HBTs [25].

Most of the available simulators handle strained-SiGe materials with a very conservative approach, especially in the carrier mobility models. Also, the reported models for SiGe material parameters are scattered over a vast body of literature.

It has been reported that for adequate modelling, a proper selection of model options are also needed [53, 54]. More accurate strained layer mobility models for majority and minority carrier are required to take into account the different mobilities along parallel (in plane) and perpendicular (out-of-plane) to the growth plane of the strained-SiGe layer. A MC model for mobility in SiGe has been developed and validated with the available transport data [55].

In this section, focus is placed on developing accurate SiGe material parameter models for robust numerical simulation for SiGe HBTs that fit into the TCAD software framework currently used for the design of heterostructure technologies. The dependence of various material parameters on Ge mole fraction (x) is identified. The models developed are based on the computed and experimental data available in the literature. The models are based on those incorporated in Silvaco ATLAS two-dimensional DD device simulator [48]. The models developed should account for the effect of the heterojunctions and non-uniform Ge and doping densities. Important material parameters which are needed for device simulation include:

- energy bandgap and bandgap narrowing;
- valence and conduction band edge offset due to the composition change and electron affinity;
- effective mass for electron and hole for the determination of effective densities of states for the conduction and valence bands;
- dielectric constant;
- electron and hole mobilities and their dependence on composition, doping density, electric field and temperature;
- saturation velocity;
- minority carrier lifetime and corresponding coefficients for various recombination mechanisms including Auger and radiative;
- coefficients for impact ionisation.

Heterointerface modelling is a key issue for devices which include abrupt junctions. Thermionic emission and field emission effects critically determine the current transport parallel and perpendicular to the heterointerfaces. All important physical effects, such as bandgap narrowing, anisotropic electron minority mobility in strained-SiGe, Shockley–Read–Hall recombination, surface and Auger recombination, and impact ionisation are taken into account. III–V materials and SiGe are known to have reduced heat conductivity in comparison to silicon [56].

Self-heating effects are accounted for by solving the lattice heat flow equation self-consistently with the energy transport equations. Advanced device simulation allows precise physics-based extraction of small-signal parameters [57,58]. A model for carrier-temperature-dependent energy relaxation times [59] has been developed as well as model for lattice-temperature-dependent saturation velocities [60].

The model parameters are checked against several independent technologies to obtain concise set used for all simulations. Reviewing simulation of HBTs and submicron heterojunction field-effect transistors with gate-lengths below 100 nm, solutions of energy transport equations are necessary to account for non-local effects, such as velocity overshoot. Advanced device simulation allows precise physics-based extraction of small-signal parameters [57, 58].

10.3.1 Bandgap and bandgap narrowing

The band edge offsets caused by the composition change are critical in determining the heterostructure device characteristics. In order to determine the band alignment for a given $Si_{1-x}Ge_x/Si$ heterointerface, the bandgap of the constituent layer as well as the band edge discontinuity, ΔE_c or ΔE_v, must be known. The first estimate of the ΔE_v was reported by van de Walle and Martin [61]. People and Bean [62] have shown the band alignment of coherently strained-$Si_{1-x}Ge_x/Si$ heterostructure on $Si_{1-y}Ge_y$ substrate. Both type-I and type-II band alignments are possible depending on the type of strain. In SiGe the strain-dependent change of the bandgap due to Ge content has been reported [63]. The temperature-dependent bandgaps of the constituents, E_g^{Si} and E_g^{Si} are calculated by the commonly used model of Varshni [64]

$$E_g = E_{g,0} - \frac{\alpha \cdot T_L^2}{\beta + T_L} \tag{10.1}$$

where $E_{g,0}$ is the bandgap at $T_L = 0\,K$. The parameter values are summarised in Table 10.3. The dependence on the material composition x is then introduced by

$$E_g^{SiGe} = E_g^{Si} \cdot (1-x) + E_g^{Ge} \cdot x + C_g \cdot (1-x)x \tag{10.2}$$

where x is the Ge mole fraction and $C_g = -0.4$ eV. This one-valley bandgap fit can be applied to the case of the strained-$Si_{1-x}Ge_x$ grown on Si as shown in Figure 10.2. Depending on the strain, the bandgap can become smaller than that of pure Ge [25]. In the unstrained case, however, an X-to-L gap transition is observed at about $x = 0.85$ which has to be accounted by the model as well. The stress-dependent change of the bandgap is an effect which must be separated from dopant-dependent bandgap narrowing (BGN) which depends on the semiconductor material composition, the doping concentration and the lattice temperature [65].

The importance of BGN in heavily-doped semiconductors is well known. In the last 20 years, BGN in silicon has been studied by various research groups [66, 67].

Table 10.3 *Parameter values for modelling the bandgap energy [After V. Palankovski and S. Selberherr, Proc. Symp. on Diagnostics and Yield: Advanced Silicon Devices and Technologies for ULSI Era, 2003(1–11)]*

Material	$E_{g,0}$ (eV)	α (eV/K)	β (K)
Si	1.1695	4.73×10^{-4}	636
Ge	0.7437	4.774×10^{-4}	235

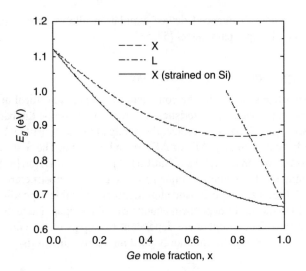

Figure 10.2   Material composition dependence of the L and X-bandgaps in $Si_{1-x}Ge_x$ at 300 K [After V. Palankovski and S. Selberherr, Proc. Symp. on Diagnostics and Yield: Advanced Silicon Devices and Technologies for ULSI Era, 2003(1–11)]

However, there is a lack of experimental information on the band structure of many technologically significant materials including SiGe and on the heavy doping effects in them. In these devices, with base doping levels close to 10^{20} cm^{-3}, BGN is expected to have a significant influence on the electrical characteristics.

The first empirical expression for the apparent BGN in Si derived from the measurement of collector current of an npn transistor (for p-type Si), was reported by Slotboom and de Graaff [66]. Lanyon and Tuft [68] theoretically predicted BGN in Si using the concept of stored electrostatic energy of majority–minority carrier pairs. del Alamo *et al.* [69] presented an empirical expression for the apparent BGN in n-type Si. The expression yielded a value of about 35 mV less than that obtained from the expression of Slotboom and de Graaff. Swirhun *et al.* [70] reported that the apparent BGN is smaller in n-type than in p-type Si which is in agreement with the Slotboom-de Graaff expression.

Klaassen [67] modified the Slotboom and de Graaff model [71] to accurately describe the apparent BGN for both n- and p-type Si as a function of impurity concentration and is given by

$$\Delta E_g^{bgn}(N) = V_1 \left[\ln\left(\frac{N}{N_{\text{ref}}}\right) + \sqrt{\left[\ln\left(\frac{N}{N_{\text{ref}}}\right)\right]^2 + C} \right] \qquad (10.3)$$

Equation (10.3) is frequently used in device simulators and the parameters used by different authors are given in Table 10.4. Temperature dependence of the apparent BGN has been studied by Swirhun *et al.* [72] and it was shown that apparent BGN is

Table 10.4   *Parameter values used in simulation for apparent bandgap narrowing*

Parameter	Ref. [66] (p-type)	Ref. [69] (n-type)	Ref. [70] (p-type)	Ref. [67] (n- and p-type)
V_1(mV)	9.0	9.35	9.0	6.92
N_{ref}(cm^{-3})	1.0×10^{17}	7.0×10^{17}	1.0×10^{17}	1.3×10^{17}
C	0.5	0	0.5	0.5

independent of temperature. This is in agreement with the results of Slotboom [73] and Wagner and del Alamo [74].

In strained-Si$_{1-x}$Ge$_x$, the BGN as a function of Ge mole fraction (for <0.3) of the p-type Si$_{1-x}$Ge$_x$ has been calculated by Jain *et al.* [75]. However, the predicted values are not consistent with the experimental results reported using optical techniques [76–78], which suggest that the BGN and Fermi level E_f (measured from the valence band edge) in p-type Si$_{1-x}$Ge$_x$ layer should be significantly greater than those in p-type Si. Several authors [79, 80] have shown that the strain changes the effective density of states mass (m^*_{DOS}) of the hole. Poortmans *et al.* [81] have recalculated the BGN for strained layers taking into account the change in effective mass.

The apparent BGN in bipolar transistors with epitaxial-Si, epitaxial-SiGe and ion implanted bases has been measured from the temperature dependence of the collector current density [82–84]. An analytical expression for the apparent BGN for strained-Si$_{1-x}$Ge$_x$ layers, based on the reported experimental data and consistent with the Klaassen model [67] is as follows:

$$\Delta E_g^{bgn}(N) = 0.00692 \left[\ln \left(\frac{N}{1.3 \times 10^{17}} \right) + \sqrt{\left[\ln \left(\frac{N}{1.3 \times 10^{17}} \right) \right]^2 + 0.5} \right]$$

(10.4)

In Figure 10.3, BGN vs. material composition in boron-doped Si$_{1-x}$Ge$_x$ is compared to the model developed by Jain *et al.* [75]. The decrease of the BGN with increase of the Ge fraction has been experimentally observed [85, 86]. Their approach explains this effect by the decreased density of states in the valence band and increase of the relative permittivity in the strained-SiGe alloy.

In general, the valence band discontinuity for Si$_{1-x}$Ge$_x$/Si grown on (001)-Si$_{1-y}$Ge$_y$ can be calculated from [87]

$$\Delta E_v(x) = (0.74 - 0.53y)x$$

(10.5)

where y is the Ge concentration in the substrate. Valence band discontinuity of Si/Si$_{1-x}$Ge$_x$/Si heterostructures has been evaluated experimentally by several workers using X-ray photoemission spectroscopy (XPS) [88], admittance spectroscopy (AS) [89, 90], capacitance-voltage [91], current–voltage (I–V) [92] and from HBT

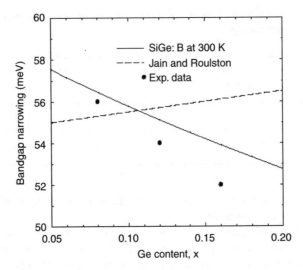

Figure 10.3   *Doping-dependent bandgap narrowing vs. Ge content in p-SiGe compared to experimental data [After V. Palankovski and S. Selberherr, Proc. Symp. on Diagnostics and Yield: Advanced Silicon Devices and Technologies for ULSI Era, 2003(1–11)]*

devices [93, 94]. It is observed that the experimental results are in good agreement with the theoretical predictions.

The band edge offsets are determined from the change of the electron affinity and the bandgap. By designating of the shift of the conduction band edge relative to the relaxed-Si [95], the electron affinity as function of the mole fraction is given by

$$\chi(x) = \chi_{Si} + 0.196x - 0.396x^2 \tag{10.6}$$

where $\chi_{Si} = 4.05$ is the electron affinity of silicon. However, this formula is valid for a Ge mole fraction less than 0.3.

The total bandgap difference of the strained-$Si_{1-x}Ge_x$ layer grown on Si(001) substrate has been computed and compared with the experimental data by many research groups. The best fit of the curve can be expressed as [96]

$$\Delta E_g(x) = 0.96x - 0.43x^2 + 0.17x^3 \tag{10.7}$$

The temperature dependence of the bandgap with Ge mole fraction (x) for a strained-$Si_{1-x}Ge_x$ layer grown on a (001) Si substrate is given by

$$E_g(x, T) = E_g(0) - \frac{\alpha(x)T^2}{T + \beta(x)} - \Delta E_g(x) \tag{10.8}$$

where $E_g(0) = 1.17$ eV stands for Si bandgap at $T = 0\,K$. The composition dependence of temperature parameters, $\alpha(x)$ and $\beta(x)$ can be expressed as $\alpha(x) = 4.73 \times 10^{-4}(1-x) + 4.77 \times 10^{-4}x$ and $\beta(x) = 636(1-x) + 235x$.

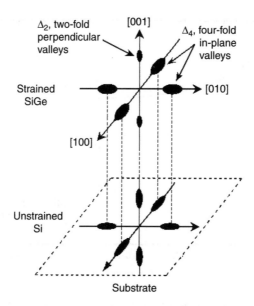

Δ_2, two-fold perpendicular valleys

[001]

Δ_4, four-fold in-plane valleys

Strained SiGe

[010]

[100]

Unstrained Si

Substrate

Figure 10.4   *The constant energy surface of strained-SiGe grown on (001) Si substrate. Note that the size change of each valley in a constant energy surface indicates a shift up (smaller) or down (larger) in energy*

10.3.2   Conduction band density of states

The calculation of the effective density of states is based on the density of state effective mass, which in turn depends on the Ge mole fraction. Moreover, one needs to consider for electron the effective masses at different valleys in the conduction band in order to construct the overall effective masses for use in computing the densities of states. The conduction band edge of Si consist of a set of six ellipsoidal energy shells located at equivalent positions in k-space ⟨001⟩ direction as shown in Figure 10.4.

In strained-$Si_{1-x}Ge_x$ alloys, with increasing Ge content, the energies of the X and L conduction band minima approach each other. Furthermore, the six-fold degeneracy of the conduction band in the X valley will be split into an energy-lowered four-fold quadruplet and the energy raised two-fold pair [87, 97]. Herring and Vogt [98] have reported that due to strain, although there is no apparent change in the shape of the energy surface, the strain causes a shift in energy as shown in Figure 10.4.

The effective mass of electrons in the X and L valley of strained-$Si_{1-x}Ge_x$ for Ge mole fraction < 0.3 is similar to that of Si. The temperature dependence of transverse and longitudinal effective mass in the X valley are generally assumed to be the same as Si [99]. The conduction band density of states for strained-$Si_{1-x}Ge_x$ as a function of Ge fraction is shown in Figure 10.5. The density of states decreases slowly as the Ge mole fraction increases, due to the reduction in the population at the band minima.

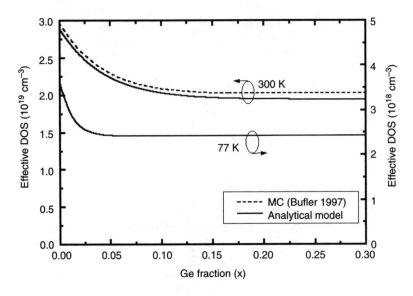

Figure 10.5   The conduction band density of states for strained-$Si_{1-x}Ge_x$ as a function of Ge fraction for 300 and 77 K. Monte Carlo result at 300 K is also shown

10.3.3   Valence band density of states

The presence of strain alters the valence band structure as well. The most obvious change is that it lifts the degeneracy of the four-fold states so that an energy gap separates the heavy and light hole bands. The change in the valence band effective mass due to the presence of a biaxial stress is expected to be same as for the uniaxial stress case. This is due to the fact that the biaxial stress can be decomposed into a hydrostatic term and an uniaxial term [87]. The hydrostatic stress term simply shifts all of the energy levels of the valence bands equally, not affecting the effective mass. On the other hand, the uniaxial stress splits the heavy and light hole bands and changes the valence band structure [100], as shown in Figure 10.6.

Depending on the direction of the strain, the light hole band will either move down (compressive strain) or up in energy (tensile strain) relative to the heavy hole band. The addition of strain causes a drastic change in mass for the heavy holes, especially at low k values where it becomes comparable to the light hole mass. Hasegawa [101] has shown that the highest energy state among the valence bands can be either $\langle 3/2, 3/2 \rangle$ or the $\langle 3/2, 1/2 \rangle$ state, and thus the usual meaning of the 'heavy' and 'light' hole bands is lost. Hasegawa has also shown that the transverse and longitudinal effective masses are not equal. The strain or stress effects on the valence band structure have been explained on the basis of the coupling interactions between the heavy and light hole bands [100], and the lift of the degeneracy in bulk-Si heavy and light hole bands and the changes of the effective masses in the presence of the uniaxial stress have been experimentally confirmed [102]. The coupling interactions of the spin-orbit band with

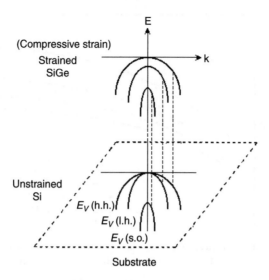

Figure 10.6   *Shift of the top of each valley in the valence band as a function of strain energy*

the heavy and light hole bands can be significant under large stress or strain and thus cannot be neglected. A strong reduction of the effective mass has been reported by many authors [79, 80, 103, 104]. The valence band density of states effective mass as a function of temperature is shown in Figure 10.7.

In contrast to the conduction band effective mass, that for the valence band is expected to be more temperature sensitive. Theoretical calculations of Madarasz [105] for Si show that the earlier work of Barber [106] underestimated the temperature dependence. The temperature dependence of density of state (DOS) effective mass of holes for Si shows a good fit to the expression

$$\frac{m^*_{dv}(T)}{m_o} = 0.545 + 3.41 \times 10^{-3}T - 5.66 \times 0^{-6}T^2 + 3.53 \times 10^{-9}T^3 \quad (10.9)$$

where T is the absolute temperature and m_o is the electron rest mass. Valence band density of state effective mass of strained-$Si_{1-x}Ge_x$ depends both on temperature and Ge mole fraction as

$$m^*_{dv}(x, T) = \frac{m^*_{dv}(T)}{m^*_{dv}(0)} m^*_{dv}(x) \quad (10.10)$$

where $m^*_{dv}(0)$ is the DOS effective mass of Si. Finally, the temperature and Ge mole-fraction-dependent valence band density of states of strained-$Si_{1-x}Ge_x$ can be expressed as

$$N_v(x, T) = 2.54 \times 10^{19} [m^*_{dv}(x, T)]^{3/2} \left(\frac{T}{300}\right)^{3/2} \quad (10.11)$$

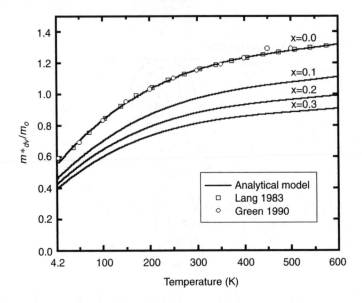

Figure 10.7   The valence band density of states effective mass as a function of temperature. Models for Si are also shown

The valence band density of states for strained-$Si_{1-x}Ge_x$ as a function of Ge fraction is shown in Figure 10.8. It is seen that the density of states slowly decrease as the Ge composition is increased due to reduction in the effective mass.

10.3.4 $Si_{1-x}Ge_x$ mobility

The hole and electron transport for strained and unstrained SiGe have been modelled by many workers [79, 107–110]. The enhanced low temperature mobilities have been observed for both holes and electrons [111]. MC simulations of electron mobility in heavily doped SiGe at room temperature indicate μ_n will be almost 50 per cent higher than for Si due to the smaller effective mass in SiGe [55, 112, 113].

In addition to phonon, impurity and alloy scattering mechanisms, strain plays a major role in determining carrier mobility. The drift hole mobility for unstrained and coherently strained-$Si_{1-x}Ge_x$ ($x < 0.3$) alloy with doping levels 10^{15}–10^{19} cm^{-3} were analytically obtained by Manku *et al.* [114]. These mobilities were calculated by taking into account the acoustic, optical, alloy and ionised-impurity scattering and were compared with experimental result for boron doping concentration of 2×10^{19} cm^{-3}. The results showed an increase in the mobility relative to that of Si.

However, in the above, the majority and minority carrier mobilities were assumed to be equal. Bufler *et al.* [115] have reported that minority carrier mobilities in unstrained and strained-$Si_{1-x}Ge_x$ are different as in the case of Si. The electron and hole mobilities for strained-$Si_{1-x}Ge_x$ were calculated by taking into account the minority and majority carrier transport parameters.

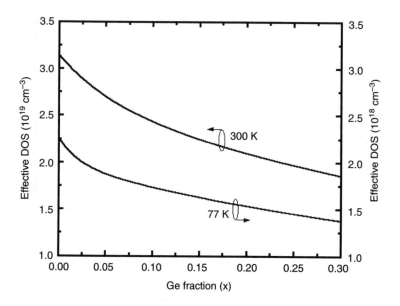

Figure 10.8   *The valence band density of states for strained-Si$_{1-x}$Ge$_x$ as a function of Ge fraction for 300 and 77 K*

Table 10.5   *Hole mobility parameters for strained-SiGe*

N_{ref} (cm^{-3})	α	β	γ	δ
2.35×10^{17}	-0.57	-2.23	-2.546	0.90

MC schemes generally require a huge computing resource for obtaining the mobility parameter as the free flight of the electron under an accelerated electric field, subject to the scattering events is repeated until a stable value is reached. For device simulation purpose, however, a simple and less involved method for calculating the hole mobility in SiGe is desirable and is given by

$$\mu_p^{s-SiGe} = \mu_{min} T_n^{\alpha} + \frac{\mu_o T_n^{\beta}}{1 + (N/N_{ref})^{\delta} T_n^{\gamma}} \tag{10.12}$$

where $T_n = T/300$, and μ_{min} and μ_o are given as [116]

$$\mu_{min} = 44 - 20x + 850x^2; \quad \mu_o = 400 + 29x + 4737x^2 \tag{10.13}$$

The temperature dependence of parameters α, β and γ are assumed to be the same as those of Si, and are shown in Table 10.5. The doping dependence of hole mobility

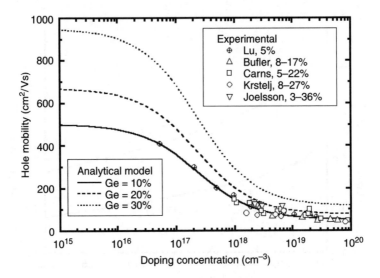

Figure 10.9   *The hole mobility for strained-$Si_{1-x}Ge_x$ as function of Ge compo-*
sition for different doping level. The experimental electron mobility
data points are taken from Lu et al. [117], Bufler et al. [118], Carns
et al. [116], Krstelj et al. [119], and Joelsson et al. [120]

of strained-SiGe is shown in Figure 10.9. The experimental hole mobility data from
several reports are included for comparison purposes.

As the minority carrier mobility is of considerable importance for bipolar transistors, an analytical low field mobility model which distinguishes between majority
and minority electron mobilities has also been developed [65] using MC simulation
data for electrons in Si. MC simulation which accounts for alloy scattering and the
splitting of the anisotropic conduction band valleys due to strain [121] in combination with an accurate ionised impurity scattering model [122] allows one to obtain
results for SiGe for the complete range of donor and acceptor concentrations and Ge
contents. Commonly the same functional form is used to fit the doping dependence
of the in-plane mobility component for $x = 0$ and $x = 1$ (Si and strained-Ge on Si).
The material composition dependence is modelled by [31]

$$\frac{1}{\mu(x)} = \frac{1-x}{\mu^{Si}} + \frac{x}{\mu^{Ge}} + \frac{(1-x) \cdot x}{C_\mu} \tag{10.14}$$

where C_μ equals to $140 \, \text{cm}^2/\text{Vs}$ and $110 \, \text{cm}^2/\text{Vs}$ for doping levels below and above
C_{mid}, respectively. Figure 10.10 shows the in-plane minority electron mobility in
$Si_{1-x}Ge_x$ as function of x at 300 K for different acceptor doping concentrations.
The model parameters used for SiGe at 300 K are summarised in Table 10.6. The
component of the mobility perpendicular to the surface is then obtained by multiplication factor given by the ratio of the two mobility components. The good agreement

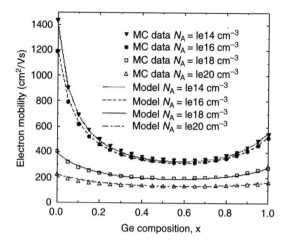

Figure 10.10 *Minority electron mobility in* $Si_{1-x}Ge_x$ *as function of x for in-plane direction: the model gives good agreement with Monte Carlo simulation data [After V. Palankovski and S. Selberherr, Proc. Symp. on Diagnostics and Yield: Advanced Silicon Devices and Technologies for ULSI Era, 2003(1–11)]*

Table 10.6 *Parameter values for the majority/minority electron mobility at 300 K [After V. Palankovski and S. Selberherr, Proc. Symp. on Diagnostics and Yield: Advanced Silicon Devices and Technologies for ULSI Era, 2003(1–11)]*

Parameter	Si	Ge (on Si)	Unit
μ_n^L	1430	560	cm^2/Vs
μ_{mid}^{maj}	44	80	cm^2/Vs
μ_{hi}^{maj}	58	59	cm^2/Vs
μ_{mid}^{min}	141	124	cm^2/Vs
μ_{hi}^{maj}	218	158	cm^2/Vs
α	0.65	0.65	—
β	2.0	2.0	—
C_{mid}	1.12×10^{17}	4.0×10^{17}	cm^{-3}
C_{hi}^{maj}	1.18×10^{20}	4.9×10^{18}	cm^{-3}
C_{hi}^{min}	4.35×10^{19}	5.4×10^{19}	cm^{-3}

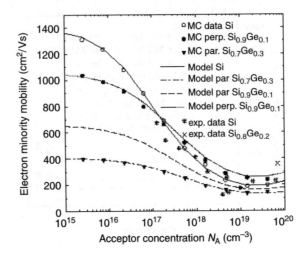

Figure 10.11   *Minority electron mobility in Si$_{1-x}$Ge$_x$ as function of N$_A$ and x: the model gives good agreement with measurements and Monte Carlo simulation data both for in-plane and perpendicular to the surface directions [After V. Palankovski and S. Selberherr, Proc. Symp. on Diagnostics and Yield: Advanced Silicon Devices and Technologies for ULSI Era, 2003(1–11)]*

of the model with the measured and the MC simulation data, both for in-plane and perpendicular to the surface directions, is illustrated in Figure 10.11.

10.3.5   Saturation velocity

In Si, the saturation velocity, v$_{sat}$, depends strongly on temperature [123, 124]. The expressions for temperature-dependent saturation velocity for electrons and holes respectively, are given by [125]

$$v_{\text{sat}n}(T) = 1.45 \times 10^7 \sqrt{\tan h\left(\frac{155}{T}\right)}$$

$$(10.15)$$

$$v_{\text{sat}p}(T) = 9.05 \times 10^6 \sqrt{\tan h\left(\frac{312}{T}\right)}$$

and represent fits to the experimental data [123, 125].

In strained-Si$_{1-x}$Ge$_x$, the saturation velocity depends on the Ge mole fraction. Since the alloy scattering is an important parameter in SiGe alloys, it affects the velocity-field characteristics for carrier transport. Several workers [115, 126, 127] have shown that the saturation velocity decreases as the Ge mole fraction is increased. Bufler *et al.* [115] have investigated the Ge mole fraction dependence of electron and hole saturation velocity, v$_{\text{sat}n,\text{sat}p}(x)$, using MC simulation. The best fit of these results

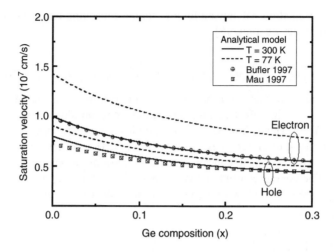

Figure 10.12 *The electron and hole saturation velocity for strained-*$\text{Si}_{1-x}\text{Ge}_x$* as a function of Ge composition at 300 and 77 K. The MC electron saturation velocity results are taken from Bufler et al. [115] and hole saturation velocity from H. Mau [128]*

to an analytical expression is given as

$$V_{\text{sat}n,\text{sat}p}(x, T) = v_{\text{sat}n,\text{sat}p}(T)\left[\frac{\alpha}{\alpha + x(1 - x)}\right] \tag{10.16}$$

where $V_{\text{sat}n,\text{sat}p}(T)$ is saturation velocity for Si as in Equation (10.15) and constant $\alpha = 0.255$. The composition and temperature-dependent electron and hole saturation velocities for strained-$\text{Si}_{1-x}\text{Ge}_x$ are shown in Figure 10.12.

10.3.6 Generation and recombination

The dominant recombination processes in bulk-Si are Shockley–Read–Hall (SRH) and Auger recombination. Radiative recombination is negligible since Si is an indirect bandgap semiconductor, and recombination involving excitons and shallow-level traps is only important at a low temperature. Misfit dislocations in SiGe layers are undesirable since the formation of efficient generation recombination (GR) centre degrades device performance and introduces noise. Misfit dislocation generation can be caused by improper growth condition or strained layer thickness exceeding the critical thickness [129–132]. Pseudomorphic or commensurate growth lowers the energy of the interfacial atoms at the expense of stored strain energy within the coherently strained layer and the $\text{Si}_{1-x}\text{Ge}_x$ epilayer undergoes a biaxial in-plane compressive strain due to the large lattice constant of $\text{Si}_{1-x}\text{Ge}_x$ compared to that of Si. This biaxial strain is composed of a uniaxial extension in the growth direction and a hydrostatic strain in all directions [133].

SRH recombination is dominant for low doping concentrations and is determined from the following equation [18]:

$$R_{SRH} = \frac{np - n_{ie}^2}{\tau_p[n - n_{ie}\exp(E_t - E_i/kT)] + \tau_n[p - n_{ie}\exp(E_t - E_i/kT)]} \quad (10.17)$$

where E_t is the energy level for the recombination centres, E_i is the intrinsic Fermi energy and n_{ie} is the effective intrinsic carrier concentration including bandgap narrowing effects. The default trap energy level is the intrinsic level, E_i.

The minority-carrier lifetimes resulting from SRH recombination is inversely proportional to the dopant concentration and depends strongly on temperature [134–136]. For doping concentrations up to 10^{19} cm^{-3} an empirical fit to experimental data gives the following temperature dependence minority-carrier lifetime [134, 137] as

$$\tau_{n,p} = \frac{\tau_o(T/300)^\gamma}{1 + (N/N_{SRH})} \quad (10.18)$$

for both electrons and holes respectively, and N is the local (total) impurity concentration, τ_o is the minority carrier lifetime in lightly doped Si and N_{SRH} is the reference doping. A good fit to experimental data [134, 138] is achieved by setting the parameters as given in Table 10.7. However, as τ_o is very much process dependent, values shown may not be accurate in all cases.

King et al. [94] have investigated the dependence of the minority carrier recombination lifetime on oxygen concentration in CVD grown strained-SiGe layer. Ghani et al. [139] have investigated the effect of oxygen concentration on the recombination lifetime of minority electrons. The lifetime value was found to reduce from 0.1 μs for an oxygen concentration of $<10^{18}$ cm^{-3} down to a few ps for oxygen concentrations in excess of $<10^{20}$ cm^{-3}. Minority carrier lifetime, obtained from a contactless photocarrier lifetime measurements in MBE-grown Si/SiGe heterostructures by Higashi et al. [140], was found to degrade from 50 to 100 μs for SiGe layer to $\approx$100 ns for a fully processed SiGe HBT devices. Generation lifetimes in the strained layers have also been measured in a strained layer MOS capacitor by a modified Zerbst technique [141]. Studies on the determination of minority carrier lifetime in SiGe-based HBTs have been reported and it has been shown that the base recombination lifetime increases in SiGe HBTs [142]. Since strained-SiGe is very similar to

Table 10.7   Recombination parameters for electron and hole in Si

Parameter	Electron	Hole
τ_o (s)	1×10^{-7}	1×10^{-7}
N_{SRH} (cm^{-3})	5×10^{16}	5×10^{16}
γ	-1.5	-1.5

Si in band structure for low Ge mole fraction (below $x = 0.20$), Si recombination model is generally assumed for SiGe [133] although the minority carrier lifetime in SiGe are believed to be somewhat shorter than Si minority carrier lifetime.

Auger recombination become dominant for total dopant concentration beyond 10^{18} cm^{-3} [143]. The model of the Auger recombination process is given by

$$R_{Au} = (C_{cn}n + C_{cp}p)(np - n_{ie}^2) \tag{10.19}$$

The Auger coefficients C_{cn} and C_{cp} are very weakly temperature dependent and are represented using simple power laws. Selberherr [144] has reported a fit to the data of Dziewor and Schmidt [143] for Si as

$$C_{cn} = 1.26 \times 10^{-31}T^{0.14}$$
$$C_{cp} = 3.55 \times 10^{-32}T^{0.18} \tag{10.20}$$

10.3.7 Impact ionisation

Impact ionisation is the third generation–recombination mechanism that should be considered if the carrier energies are high enough (>100 kV/cm) to cause interband transition. The generation rate of electron-hole pairs due to impact ionisation for Si has been modelled as [145]

$$G = \alpha_n \frac{J_n}{q} + \alpha_p \frac{J_p}{q} \tag{10.21}$$

where α_n and α_p are the electron and hole ionisation rates that are functions of electric field (E) and is given by

$$\alpha_{n,p} = \alpha_{n,p}^{\infty} \exp\left[-\left(\frac{E_{n,p}^{crit}}{E}\right)^{\beta_{n,p}}\right] \tag{10.22}$$

In both electronic and optoelectronic devices, the impact ionisation-induced breakdown is of great importance. Electron-initiated impact ionisation rate (α_n) and hole-initiated impact ionisation rate (α_p) limit the high power performance of electronic devices. The ratio of electron and hole impact ionisation coefficient is especially important in low noise applications. The impact ionisation process depends largely on the band-structure and bandgap. In strained-SiGe, as Ge content is increased the bandgap decreases, and it is expected that the impact ionisation will play a major role at high electric field.

However, as the Ge composition is increased, as discussed before, the alloy scattering increases up to a Ge mole fraction of 50 per cent. The increase in alloy scattering randomises the carrier distribution and suppresses the energy that the carriers can gain from the field. This, in turn, tends to suppress the impact ionization. Thus, there is a competition between the effect of decreasing bandgap and increasing alloy scattering. However, once the Ge content exceeds 50 per cent, the effect of alloy scattering weakens.

The experimental impact ionisation coefficient ratio (α_n/α_p) has been determined using $Si_{1-x}Ge_x$ photodiodes for the entire composition range $0.08 < x < 1$. The measured value of the ratio varies from 3.3 at $x = 0.08$ to 0.3 at $x = 1$ [146]. Yeom *et al.* [147] have calculated the impact ionisation coefficient ratio (α_n/α_p) for SiGe alloys and reported similar results.

10.4   Simulation of III–V devices

Modelling of advanced III–V semiconductor devices appears to be a less rigorous approach than for silicon devices as the material properties for modelling of AlGaAs, InGaAs or even InAlAs, and InGaP are restricted to slight modifications of GaAs material properties. Also, the database for novel materials, such as GaN or GaSb, which have entered the III–V material systems with impressive device results, is still relatively limited. Modelling of stress-induced changes of the physical properties of strained material layers and consideration of piezoelectric effects is a subject of ongoing research [36].

The two-dimensional device simulator MINIMOS-NT has been extended to deal with different complex materials and structures, such as binary and ternary semiconductor III–V alloys with arbitrary material composition profiles. It is well known that GaAs-HBTs with InGaP emitter material can be improved with respect to reliability if the emitter material covers the complete p-doped base layer forming outside the active emitter, the so-called InGaP ledge. Palankovski *et al.* [148] have reported the influence of the ledge thickness and of the surface charges on the device performance and its impact on reliability.

Palankovski *et al.* [149] have addressed these critical modelling issues for III–V devices based on GaAs, AlAs, InAs, InP and GaP, their ternary alloys and non-ideal dielectrics in the 2D device simulator MINIMOS-NT [150]. Various important physical effects, such as BGN [151], surface recombination and self-heating, are taken into account. A new universal low field mobility model based on MC simulations, distinguishing between majority and minority electron mobilities has also been implemented in MINIMOS-NT [65]. This model is significant as underestimation of electron mobility in the GaAs base of HBTs can result in significant overestimation of the extracted base resistance. The model parameters have been verified against several independent HEMT, HBT technologies and submicron HFETs with gate-lengths down to 100 nm used for millimetre-wave devices. In simulation, solutions of energy transport equations were used to account for non-local effects, such as velocity overshoot. A new model for carrier temperature-dependent energy relaxation times [59] was developed as well as a model for lattice temperature-dependent saturation velocities [60]. Interested readers may refer to an excellent review on modelling of current transport in ultra-scaled field-effect transistors by Sverdlov *et al.* [152].

For HFET devices the Schottky contact model determines the possibility to calculate realistic gate currents, based on thermionic effects. Thermionic emission (TE) and field emission effects critically determine the current transport parallel and perpendicular to the heterointerfaces. Carrier tunnelling must be included to describe the current transport from the channel to the contacts in HEMT devices depending on the alloying of the ohmic contacts.

Another critical issue for recessed HFETs and for HBTs is the description of the semiconductor/insulator interface. Fermi-level pinning prevails especially for typical barrier materials such as AlGaAs or InAlAs, for ledge materials such as InGaP and insulators such as SiN. Sophisticated thermionic-field emission interface models [153] in conjunction with the hydrodynamic transport model are needed in simulation. When assessing a complete device description, the inclusion of holes and hot hole effects is necessary.

Trap modelling, for example, the inclusion of DX centres and especially dynamic trap modelling are the most challenging issues for III–V devices, since the carrier lifetimes cannot be generalised without deep insight into process technology, for example, from 1/f noise measurements. Yang *et al.* [154] have stressed the importance of Auger recombination for InGaAs-base HBTs relative to SRH recombination.

III–V materials are known to have a reduced heat conductivity in comparison to Si [56]. To account for self-heating effects the lattice heat flow equation is solved self-consistently with the energy transport equations (system of six partial differential equations). The thermal conductivity and the specific heat are expressed as functions of the lattice temperature and in the case of semiconductor alloys of the material composition [33]. Figures 10.13 and 10.14 present comparisons between experimental data [155–158] and simulations for the thermal conductivity in the temperature range 300–800 K. The parameter values used are summarised in Tables 10.8 and 10.9. In Figures 10.15 and 10.16, comparisons between experimental data from [156–161] and the results obtained with the model are shown for the thermal conductivity at 300 K.

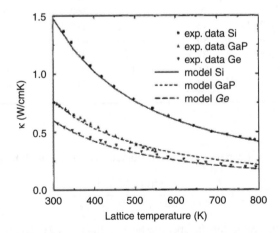

Figure 10.13   Temperature dependence of the thermal conductivity. Comparison between experimental data and the model for Si, Ge, and GaP [After V. Palankovski et al., IEEE Trans. Electron Devices, Vol. 48, 2001(1264–1269). © 2001 IEEE]

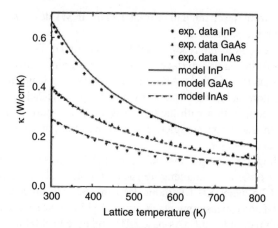

Figure 10.14   Temperature dependence of the thermal conductivity. Comparison between experimental data and the model for InP, GaAs, and InAs [After V. Palankovski et al., IEEE Trans. Electron Devices, Vol. 48, 2001(1264–1269). © 2001 IEEE]

Table 10.8   Parameter values for mass density, thermal conductivity, and specific heat of several important semiconductors

Material	ρ (g/cm^3)	κ_{300} (W/K m)	α	C_{300} (J/K kg)	C_L (J/K kg)	β
Si	2.33	148	−1.65	711	255	1.85
Ge	5.327	60	−1.25	360	130	1.3
GaAs	5.32	46	−1.25	322	50	1.6
AlAs	3.76	80	−1.37	441	50	1.2
InAs	5.667	27.3	−1.1	394	50	1.95
InP	4.81	68	−1.4	410	50	2.05
GaP	4.138	77	−1.4	519	50	12.6

The specific heat capacity of the considered material is modelled with

$$C_L(T_L) = C_{300} + C_L \cdot \frac{(T_L/300\,\mathrm{K})^\beta - 1}{(T_L/300\,\mathrm{K})^\beta + (C_L/C_{300})} \tag{10.23}$$

where C_{300} is the value for the specific heat at 300 K [162]. The model is used for the basic semiconductor materials. In Figures 10.17 and 10.18, comparisons between experimental data and the results obtained with the model for the specific heat are shown. Note the excellent agreement it gives in a wide temperature range (0–800 K). The parameter values used are summarised in Table 10.8.

Table 10.9   Parameter values for thermal conductivity

Material	C (W/K m)
SiGe	2.8
AlGaAs	3.3
InGaAs	1.4
InAlAs	3.3
InAsP	3.3
GaAsP	1.4
InGaP	1.4

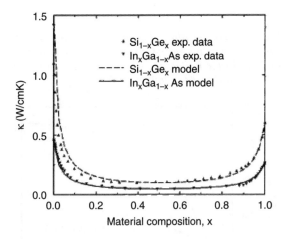

Figure 10.15   *Material composition dependence of the thermal conductivity. Comparison between experimental data and the model for SiGe and InGaAs [After V. Palankovski et al., IEEE Trans. Electron Devices, Vol. 48, 2001(1264–1269). © 2001 IEEE]*

The advance of device simulation further allows a precise physics-based small-signal extraction [57,163]. Measured bias dependent s-parameters serve as a valuable source of information when compared to bias dependent s-parameters simulated from a device simulator, for example, from MINIMOS-NT. By simulating in the frequency domain, important small-signal figures of merit, such as the cut-off frequency and the maximum oscillation frequency can be efficiently extracted [164]. On the other hand, non-linear periodic steady-state analysis can be performed in the time domain to obtain large-signal figure of merit parameters, such as distortion, power, frequency and noise as well in the context of coupled device and circuit simulation [165].

Figure 10.19 shows the simulated and measured output characteristics of a pseudomorphic $Al_{0.2}Ga_{0.8}As/In_{0.2}Ga_{0.8}As/GaAs$ high-power HEMT with gate-length

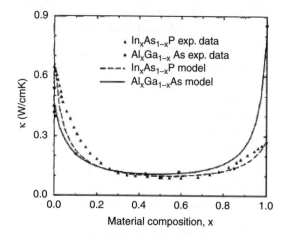

Figure 10.16   *Material composition dependence of the thermal conductivity. Comparison between experimental data and the model for InAsP and AlGaAs [After V. Palankovski et al., IEEE Trans. Electron Devices, Vol. 48, 2001(1264–1269). © 2001 IEEE]*

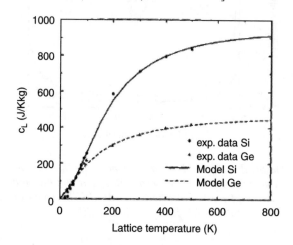

Figure 10.17   *Temperature dependence of the specific heat. Comparison between experimental data and the model for Si and Ge [After V. Palankovski et al., IEEE Trans. Electron Devices, Vol. 48, 2001(1264–1269). © 2001 IEEE]*

210 nm at substrate temperature of 300 K. The device is designed for K_a-band applications. Self-heating effects are accounted for as well as impact ionisation. Special care was put to match the output conductance for $V_{ds} > 5$ to characterise the device in the region, which defines the clipping of a voltage swing on an applied load-line and thus linearity.

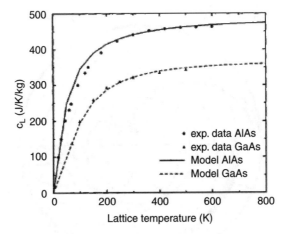

Figure 10.18 *Temperature dependence of the specific heat. Comparison between experimental data and the model for GaAs and AlAs [After V. Palankovski et al., IEEE Trans. Electron Devices, Vol. 48, 2001 (1264–1269). © 2001 IEEE]*

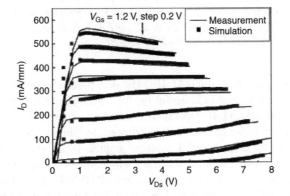

Figure 10.19 *Output characteristics of a pseudomorphic $Al_{0.2}Ga_{0.8}As/In_{0.2}Ga_{0.8}As/GaAs$ high-power HEMT with gate-length 210 nm at substrate temperature of 300 K [After V. Palankovski et al., IEEE J. Solid-State Circuits, Vol. 36, 2001(1365–1370). © 2001 IEEE]*

Figure 10.20 demonstrates a comparison of the simulated and measured s-parameters of a gate length of 140 nm pseudomorphic $Al_{0.2}Ga_{0.8}As/In_{0.2}Ga_{0.8}As/GaAs$ HEMT at 373 K from 0.5 to 50 GHz using 0.5 GHz steps. No fitting is applied, just the parasitic elements extracted from the measurements are used. The overall agreement is considered good. The discrepancies found for s_{12} between simulation and measurement are due to both a systematic error in the determination of s_{12} and the simulation itself.

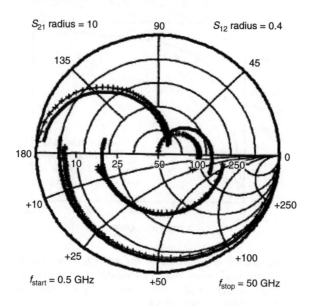

Figure 10.20   Simulated (−) and measured (+) s-parameters of a $L_g = 140$ nm HEMT at $T_L = 373$ K from 0.5 to 50 GHz using 0.5 GHz steps [After V. Palankovski et al., IEEE J. Solid-State Circuits, Vol. 36, 2001(1365–1370). © 2001 IEEE]

For high speed InAlAs/InGaAs HEMTs, the precise evaluation of low voltage or low power capabilities is useful for the development for high speed optical data transmission beyond 40 Gbit/s. The comparison of several lattice matched and metamorphic technologies further allowed to obtain consistent simulation parameters also for this material system. Figure 10.21 shows simulation and measurements for two different substrate temperatures for a composite channel $In_{0.52}Al_{0.48}As/In_{0.66}Ga_{0.34}As/In_{0.53}Ga_{0.47}As/InP$ HEMT for gate length of 150 nm.

The III–V HBTs are considered essential for high-power amplifiers at 3 V power supply, as they offer high current amplification and power-added efficiency (PAE). Figure 10.22 shows the simulated forward Gummel plot of such $3 \times 30 \, \mu m^2$ AlGaAs/GaAs HBT with an InGaP ledge compared to experimental data. A good agreement at room temperature and the simulated Gummel plot at 373 K demonstrates the suitability of MINIMOS-NT to reproduce the thermal device behaviour correctly. It is often problematic to achieve realistic results in simulation of output HBT characteristics, especially for high-power devices as the power dissipation increases with collector-to-emitter voltage, gradually elevating the junction temperature above the ambient temperature [166]. This leads to gradually decreasing collector currents at constant applied base current or, respectively, gradually increasing at constant base-to-emitter voltage.

Figure 10.23 shows the simulated output device characteristics compared to measurements for constant base voltage in the range 1.4–1.45 V using a 0.01 V step.

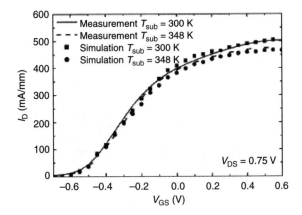

Figure 10.21   Transfer characteristics of a composite channel InAlAs/InGaAs/InP HEMT with gate length of 150 nm for two different temperatures [After V. Palankovski et al., IEEE J. Solid-State Circuits, Vol. 36, 2001(1365–1370). © 2001 IEEE]

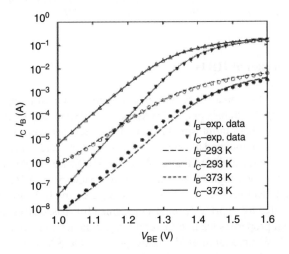

Figure 10.22   Forward Gummel plots at $V_{bc} = 0$ V for AlGaAs/GaAs HBT: Comparison with measurement data at $T_L = 293$ K and $T_L = 373$ K [After V. Palankovski et al., IEEE J. Solid-State Circuits, Vol. 36, 2001(1365–1370). © 2001 IEEE]

Note the significant disagreement between simulation without self-heating (SH) and the measured data and the good agreement when self-heating is included in the simulation. The lattice temperature can reach a temperature as much as 400 K. Such a lattice temperature rise significantly changes the material properties of the device and hence its electrical characteristics [167].

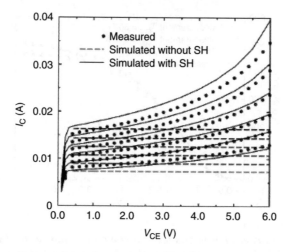

Figure 10.23   *HBT output characteristics: simulation of with and without self-heating compared to measurement data at constant V_{be} stepped from 1.4 to 1.45 V [After V. Palankovski et al., IEEE J. Solid-State Circuits, Vol. 36, 2001(1365–1370). © 2001 IEEE]*

10.5   Comparison of HBTs

Esame *et al.* [168] have made a comprehensive comparison of three state-of-the-art heterojunction bipolar transistors; AlGaAs/GaAs, Si/SiGe and InGaAs/InP HBTs. Direct current and bias point simulations of the devices were performed using Agilent's ADS design tool and a comparison is given for a wide range of FOM specifications. It has been concluded that GaAs based HBTs are suitable for high-power applications due to their high-breakdown voltages, SiGe-based HBTs are promising for low noise applications due to their low noise figures and InP will be the choice, if very high-data rates are of primary importance since InP-based HBT transistors have superior material properties leading to Terahertz frequency operation.

For device I–V curve families V_{ce} is swept from 0 to 5 V and I_b from 10 to 100 mA. As seen from Figure 10.24, for 70 μA base current the AlGaAs/GaAs device's collector current is 6 mA, while it is 11 mA for the Si/SiGe device and 4 mA for the InGaAs/InP HBT at their optimal V_{ce} for class A operation. Since the breakdown voltage phenomena were not modelled for the AlGaAs/GaAs and InGaAs/InP devices, no collector-base breakdown can be seen in Figure 10.24(a) and (c). However, for the Si/SiGe device, this breakdown can be seen at around 5 V of collector-emitter voltage (Figure 10.24b). This low BV_{cbo} is the main problem for the SiGe-based heterojunction transistors and limits their use in some application areas such as base stations of cellular networks, where supply voltages exceed 10 V. A solution may be to design a double heterojunction bipolar transistor, which provides high-breakdown voltages owing to the use of a wide-gap material in the collectors.

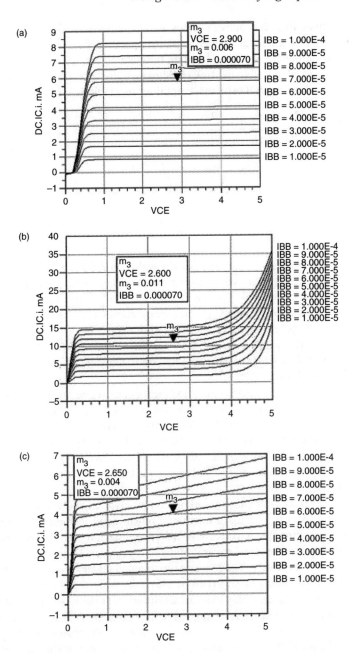

Figure 10.24 I_c *vs.* V_{ce} *curves of (a) AlGaAs/GaAs HBT, (b) Si/SiGe HBT, and (c) InGaAs/InP HBT [After O. Esame et al., Microelectronics J., Vol. 35, 2004(901–908)]*

RF transistors, like other active devices are unconditionally stable at any operating frequency above a critical frequency, f_k. At operating frequencies below f_k, the transistor is conditionally stable and certain termination conditions can cause oscillations. As discussed in Reference 169, stability measure '$B > 0$' and stability factor '$k > 1$' are the necessary and sufficient conditions for stability. k and B are defined as

$$k = \frac{1 - |s_{11}|^2 - |s_{22}|^2 + |\Delta|^2}{2(s_{12}s_{21})} \tag{10.24}$$

$$B = 1 + |s_{11}|^2 - |s_{22}|^2 - |\Delta|^2 \tag{10.25}$$

where

$$|\Delta| = |s_{11}s_{22} - s_{12}s_{21}| \tag{10.26}$$

If $k > 1$, the transistor is unconditionally stable, and for $k < 1$, it works in the region of conditional stability where unintended oscillations may occur. These sets of expressions reduce to '$k > 1$' since in this case B is positive for all the devices. In Figure 10.25, stability factor k vs. frequency curves for $70\,\mu$A base current is given and '$k = 1$' points are marked. The AlGaAs/GaAs device is unconditionally stable over 12.3 GHz while f_k is 31.7 GHz for the Si/SiGe HBT and 56.5 GHz for InGaAs/InP HBT. Collector voltages are set to optimum values for class A operation in the stability analysis.

For f_T calculation, V_{ce} is set to optimal value for class A operation for each device, and select different base current values over a wide range of frequency spectrum. For all devices, I_b is swept from 10 to 100 μA with 10 μA steps. The f_T values and also the short current gain can be read from Figure 10.26. For $I_b = 100\,\mu$A, the AlGaAs/GaAs HBT shows an f_T of 95.7 GHz while SiGe HBT shows 278 GHz and the InGaAs/InP HBT 302.6 GHz. However, β at $I_b = 100\,\mu$A for the above devices are 25, 51 and 40, respectively.

There are several power gain definitions commonly used to characterise RF transistors. If a transistor is to achieve maximum power gain, then power matching is required. For operating frequencies above f_k, power matching is obtained when both the input and output of the transistor are conjugately impedance-matched to the signal source and load respectively. The power gain obtained under these matching conditions is the maximum available gain (MAG). However, when the transistor is operated at frequencies below f_k, auxiliary external admittances at the input and output have to be connected to the transistor to suppress its tendency to oscillate. The maximum oscillation frequency is defined as the frequency at which the transistor still provides a power gain MAG if the stability factor k > 1. It is equal to the maximum stable gain, $|s_{21}|/|s_{12}|$, if k < 1 and the maximum unilateral gain if $s_{12} = 0$. For $I_b = 100\,\mu$A, the AlGaAs/GaAs HBT shows an f_{max} of 151.7 GHz while SiGe HBT shows 134.6 GHz and the InGaAs/InP HBT 416.6 GHz. However, the current gains at $I_b = 100\,\mu$A are 25, 51 and 40, respectively. The corresponding graphs for MAG are illustrated in Figure 10.27.

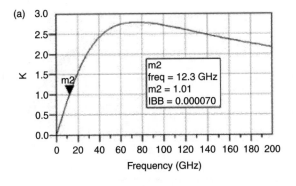

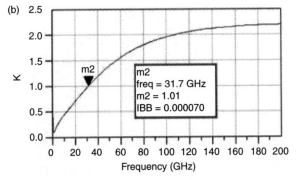

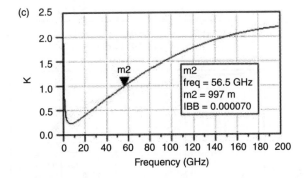

Figure 10.25 *Stability factor k vs. frequency curves for (a) AlGaAs/GaAs HBT,*
(b) Si/SiGe HBT, and (c) InGaAs/InP HBT [After O. Esame et al.,
Microelectronics J., Vol. 35, 2004(901–908)]

Si/SiGe HBTs show better noise performance than their AlGaAs/GaAs and
InGaAs/InP counterparts. NF_{min} is 5 dB for the GaAs based device at its optimum bias
for class A operation. It is 1.5 dB for the SiGe HBT and 2.735 dB for the InP HBT for a
wide range of V_{ce} values. This advantage makes SiGe attractive for applications such
as low noise amplifiers (LNAs), where minimum noise figure is the critical FOM.

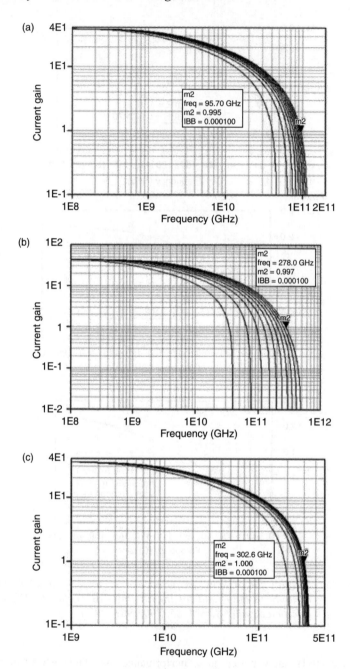

Figure 10.26　Short-circuit current gain vs. frequency curves for (a) AlGaAs/GaAs HBT, (b) Si/SiGe HBT, and (c) InGaAs/InP HBT [After O. Esame et al., Microelectronics J., Vol. 35, 2004(901–908)]

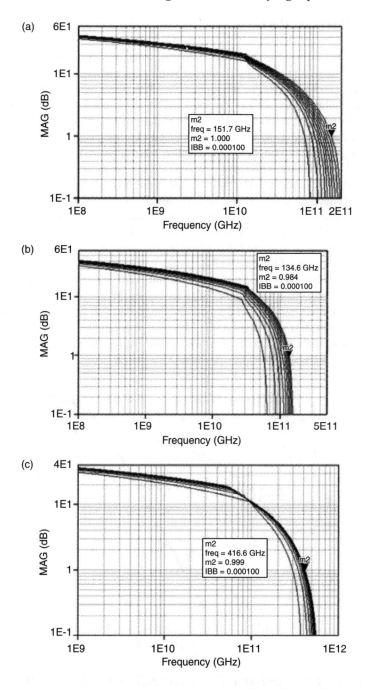

Figure 10.27   *MAG vs. frequency curves for (a) AlGaAs/GaAs HBT, (b) Si/SiGe*
HBT, and (c) InGaAs/InP HBT [After O. Esame et al.,
Microelectronics J., Vol. 35, 2004(901–908)]

10.6   RFMOS vs. HBTs

There has been tremendous progress in consumer RF products with the availability
of SiGe BiCMOS and CMOS technologies with good passive elements. Therefore,
there is an increasing trend to migrate applications from SiGe BiCMOS to a pure
CMOS technology, especially in the sub-10 GHz range. Exploring the design space
to create acceptable circuits with MOSFETs will need device models and design kits
that are accurate over a wide range of frequency and bias conditions. Since HBTs
have been used extensively in RF applications, accurate high-frequency models and
design tools are already available.

Most high volume consumer wireless applications still operate below 10 GHz
in carrier frequency, and therefore high f_T can be potentially traded for significant
savings in power. A similar argument can also hold for SiGe HBTs, where f_T can
exceed 350 GHz [170]. As CMOS technology scales into the sub-100 nm nodes,
RF performance can approach that of SiGe, for example, a $f_T = 243$ GHz and
$f_{max} = 208$ GHz for n-MOSFET [171].

A comparison of speed and power performance of the state-of-the-art CMOS and
SiGe RF transistors has also been reported by Jagannathan *et al.* [172]. Figure 10.28
shows the Gummel ($V_c = V_b = 0$ V) and subthreshold I–V ($V_{ds} = 1$ V) of the HBT
and n-FET respectively. Current in the FET at high gate bias is limited by velocity
saturation and mobility reduction, while a combination of emitter resistance and

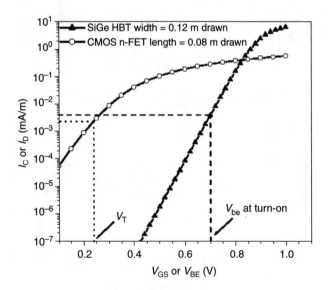

Figure 10.28   *Direct current I–V of n-MOSFET and SiGe HBT. Device turn-
on voltages are indicated [After B. Jagannathan et al., Proc.
Silicon Monolithic Integrated Circuits in RF Systems Meeting,
2004(115–118). © 2004 IEEE]*

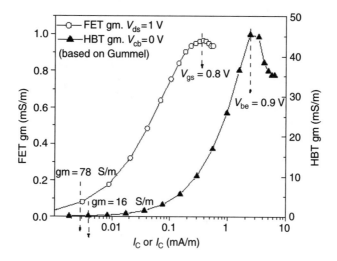

Figure 10.29 g_m *comparison of n-MOSFET and HBT.* g_m *at device 'turn-on' and gate (base) bias at peak* g_m *is indicated [After B. Jagannathan et al., Proc. Silicon Monolithic Integrated Circuits in RF Systems Meeting, 2004(115–118). © 2004 IEEE]*

base-widening mechanisms limit it in the HBT. Figure 10.29 shows the corresponding comparison of g_m as a function of current. g_m at V_t (turn-on V_{be}) for the two devices and V_{gs} (V_{be}) at peak g_m are marked on the figure as well. g_m for the FET is superior to that of the HBT at low bias current.

Figure 10.30 compares f_T of the two devices for two supply voltages. Under maximum supply conditions, peak f_T of the FET is ~150 GHz compared to ~225 GHz in the bipolar device. Note that bias points for peak f_T corresponds to that for peak g_m in the device. At similar f_T, the current in the FET is lower compared to the bipolar. In spite of the large g_m difference, current at peak f_T for the FET is ~4x lower compared to the HBT, an indication of the much smaller input capacitance in the FET. Figure 10.31 shows f_T as the function of power consumed. The lower current operation of the FET leads to the lower power consumption for similar values of f_T. A substantial advantage in power for the FETs, however, results because the supply voltage can be lowered much more aggressively. Figure 10.32 compares Mason's gain (U) of HBT and CMOS at 15 GHz as a function of both V_{gs} (V_{be}) and V_{ds} (V_{ce}). For V_{gs} (V_{be}) variation, the drain (collector) is held at a constant 1 V, and for V_{ds} (V_{ce}) variation the gate (base) bias corresponds to peak f_T. Despite the f_T differences, U of similar magnitude can be obtained in both devices.

CMOS technology (90 nm node) has been introduced in communications [173], suggesting that the advanced CMOS process technologies can deliver necessary analog/RF performance which may replace SiGe HBTs for the majority of communications applications below 5 GHz range. Kuhn *et al.* [174] have compared RF CMOS performance from a 90 nm derivative communications process technology to

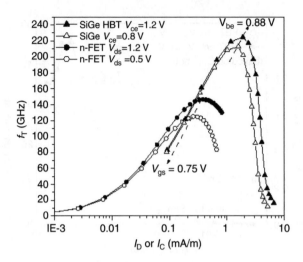

Figure 10.30 f_T vs. current comparison of n-MOSFET and SiGe HBT for various supply voltages [After B. Jagannathan et al., Proc. Silicon Monolithic Integrated Circuits in RF Systems Meeting, 2004(115–118). © 2004 IEEE]

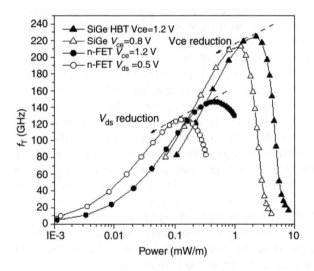

Figure 10.31 f_T vs. power comparison of n-MOSFET and SiGe HBT for various supply voltages [After B. Jagannathan et al., Proc. Silicon Monolithic Integrated Circuits in RF Systems Meeting, 2004(115–118). © 2004 IEEE]

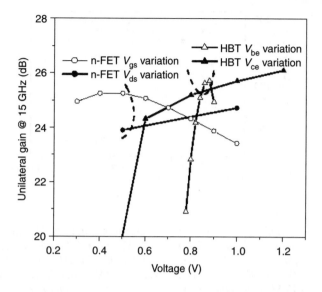

Figure 10.32   *U gain vs. $V_{gs}(V_{be})$ and $V_{ds}(V_{ce})$ comparison of n-MOSFETs and SiGe HBT [After B. Jagannathan et al., Proc. Silicon Monolithic Integrated Circuits in RF Systems Meeting, 2004(115–118). © 2004 IEEE]*

SiGe HBT performance. NMOS performances at $f_T/f_{max} = 209/248$ GHz (70 nm) and $f_T/f_{max} = 166/277$ GHz (80 nm) with F_{min} at 0.3 dB (2 GHz) and 0.6 dB (10 GHz) have also been reported.

Small-signal RF performance is strongly influenced both by output conductance and gate resistance. Both parameters increase significantly as the gate length approaches the minimum length for a technology, and thus devices slightly greater in length than the minimum will frequently deliver higher f_{max} as shown in Figure 10.33. A peak f_T of 166 GHz and the 277 GHz peak f_{max} have been reported from an NMOS with dimensions 80 nm × 2.5 μm × 6 device. At the end of 2004, a combination $f_T/f_{max} = 209$ GHz/248 GHz and a peak $f_{max} = 277$ GHz corresponds to the highest reported in the literature for an NMOS device, and comparable to the best published HBT data (see Figure 10.34).

Broadband noise performance has been considered to be a significant limitation of CMOS in comparison to SiGe HBTs. Figure 10.35 compares F_{min} vs. frequency to CMOS and HBT devices from the literature. For operating point comparison, Figure 10.36 compares F_{min} versus bias current for different technologies, where values of F_{min} are among the lowest reported in CMOS and are comparable to published SiGe HBT data.

Thus in future RF CMOS could be a preferred technology for a majority of communication products that combine analog/mixed-signal/RF circuits with digital logic. The NMOS transistors use a high volume digital CMOS process and thus can be produced more cheaply than SiGe HBTs.

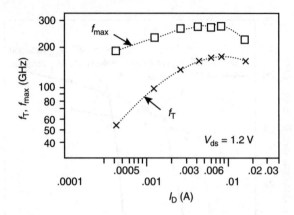

Figure 10.33 f_T and f_{max} for an $80\,nm \times 2.5\,\mu m \times 6$ device. The measurement is made by biasing the device at $V_{ds} = 1.2\,V$ and then stepping V_{gs} through a series of values. Peak f_T is 166 GHz and peak f_{max} is 277 GHz at $V_{gs} = 0.8\,V$ and $V_{ds} = 1.2\,V$ [After K. Kuhn et al., IEEE VLSI Tech. Symp. Dig., 2004(224–225). © 2004 IEEE]

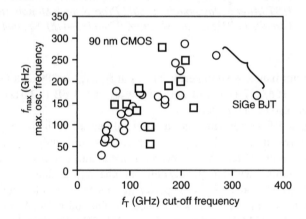

Figure 10.34 f_T and f_{max} are compared with the reported values for CMOS devices [After K. Kuhn et al., IEEE VLSI Tech. Symp. Dig., 2004(224–225). © 2004 IEEE]

10.7 Summary

The main purpose of this chapter was to discuss the physical models necessary for simulation of advanced heterostructure devices, such as HBTs or HEMTs. Important material parameters of strained-SiGe have been considered. Comprehensive analytical models for material parameter of strained-SiGe films have been presented. The model equations account for valence band discontinuity, heavy doping effects,

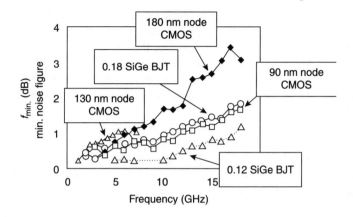

Figure 10.35   F$_{min}$ *comparison vs. frequency of RF CMOS against a variety of reported RF CMOS and HBT devices [After K. Kuhn et al., IEEE VLSI Tech. Symp. Dig., 2004(224–225). © 2004 IEEE]*

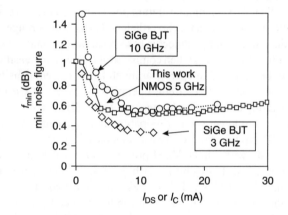

Figure 10.36   F$_{min}$ *comparison vs. bias of RF CMOS against a variety of reported RF CMOS and HBT devices [After K. Kuhn et al., IEEE VLSI Tech. Symp. Dig., 2004(224–225). © 2004 IEEE]*

valence and conduction band effective densities of states and ionised doping concentration. Minority and majority electron and hole mobilities in the whole range from low to very high electric field have also been discussed in detail. The models are based on computed and experimental data available in the literature. SiGe material parameter models developed can be incorporated in commercially available simulators for simulation of SiGe HBTs. New models for physical properties with respect to material composition and strain conditions owing to lattice mismatch have to be developed.

For advanced heterostructure simulations, anisotropic effective carrier masses, density of states and carrier mobilities should be considered. A built-in feature for automatic estimation of the strain condition based on empirical relation for the growth-condition-dependent critical layer thickness may also be developed. Possible simulation applications in the future could be the simulation of leakage currents at low bias in GaAs-based HBTs, breakdown simulation in III–V devices and thermal investigations not only in the device but also in the interconnects. The accuracy of the device simulation for all devices has to be verified not only against d.c. but also against RF measurements.

The big challenge remaining for III–V device simulation is improved understanding of processed semiconductors after different manufacturing steps, for example, before and after etching. Although a variety of simulators have successfully demonstrated some level of agreement with measurements, the understanding of changes of transport and interface parameters remains the ultimate goal of process control.

A comprehensive study has been made towards the definition of FOMs for RF heterojunction bipolar transistors. From a power perspective, SiGe-, GaAs- and InP-based devices show slightly different d.c. power consumption and output power values at 10 GHz. InGaAs/InP HBT has the least power consumption since its output current is smaller, while SiGe HBT maintains output power nearly twice its counterparts and it is relatively more power efficient. Nevertheless, GaAs is the strongest candidate for power applications due to large collector-base breakdown voltage. GaAs HBTs are inherently faster than SiGe HBTs when compared at low current densities. SiGe HBTs without SOI can be designed to operate at much larger current densities than GaAs HBTs. Comparing a GaAs HBT circuit operating at say 0.5 mA/μm^2 with a SiGe HBT circuit operating at say 2.0 mA/μm^2, the SiGe HBT circuit could have better performance and much lower power dissipation, but larger substrate-coupling noise than the GaAs HBT circuit. Thus, the real advantage of SiGe HBTs, without SOI, lies in their compatibility with silicon VLSI processes and scalability to higher current densities. SiGe-based HBTs exhibit lower breakdown voltage since the bandgap of silicon is less than that of GaAs. This makes it more difficult to use SiGe for high-power applications. However, this problem can be reduced by use of a double HBT, since a collector-base heterojunction with wide band-gap collector gives rise to higher breakdown voltage.

The characteristic frequencies f_T and f_{max} depend on material properties, device model parameters, device geometry and bias. Optimised SiGe based HBTs show better (>200 GHz) characteristic frequency values when compared with traditional GaAs-based HBTs. Nevertheless, speed-related FOMs are comparable for the two material systems. InP-based HBT is the most promising due to superior properties of InP material system. It is well known that f_T is more important for digital circuits, while for analog applications, f_{max} is more significant. Manufacturers of RF transistors often strive for $f_T \sim f_{max}$ so that the devices are useful for different applications. Another significant difference between AlGaAs/GaAs HBT, Si/SiGe HBT and InGaAs/InP HBT is their noise performance. The noise figure of SiGe-based HBTs is lower than III–V HBTs at high frequencies, making SiGe a better choice for low-noise applications.

References

1 B. Jagannathan, M. Khater, F. Pagette, J.-S. Rieh, D. Angell, H. Chen, *et al.*, 'Self-aligned SiGe NPN transistor with 285 GHz f_{max} and 207 GHz f_t in a manufacturable technology', *IEEE Electron Devices Lett.*, vol. 23, pp. 258–260, 2002.

2 J.-S. Rieh, B. Jagannathan, H. Chen, K. T. Schonenberg, D. Angell, A. Chinthakindi, *et al.*, 'SiGe HBTs with cut-off frequency of 350 GHz', in *IEEE IEDM Tech. Dig.*, pp. 771–774, 2002.

3 J. Bock, H. Schafer, H. Knapp, D. Zoschg, K. Aufinger, M. Wurzer, *et al.*, 'Sub 5 ps SiGe bipolar technology', in *IEEE IEDM Tech. Dig.*, pp. 763–766, 2002.

4 B. Jagannathan, M. Meghelli, A. V. Rylyakov, R. A. Groves, A. K. Chinthakindi, C. M. Schnabel, *et al.*, 'A 4.2-ps ECL ring-oscillator in 285-GHz f_{max} SiGe technology', *IEEE Electron Devices Lett.*, vol. 23, pp. 541–543, 2002.

5 T. Hashimoto, Y. Nonaka, T. Saito, K. Sasahara, T. Tominari, K. Sakai, *et al.*, 'Integration of 0.13-μm CMOS and high performance self-aligned SiGe HBT featuring low base resistance', in *IEEE IEDM Tech. Dig.*, pp. 779–782, 2002.

6 Z. Griffith, Y. Kim, M. Dahlstrom, A. C. Gossard, and M. J. W. Rodwell, 'InGaAs-InP metamorphic DHBTs grown on GaAs with lattice-matched device performance and f_τ, $f_{max} > 268$ GHz', *IEEE Electron Devices Lett.*, vol. 25, pp. 675–677, 2004.

7 M. J. W. Rodwell, M. Urteaga, T. Mathew, D. Scott, D. Mensa, Q. Lee, *et al.*, 'Submicron scaling of HBTs', *IEEE Trans. Electron Devices*, vol. 48, pp. 2606–2624, 2001.

8 E. Ohue, Y. Kiyota, T. Onai, M. Tanabe, and K. Washio, '100-GHz f_T Si homo-junction bipolar technology', in *IEEE VLSI Tech. Symp. Dig.*, pp. 106–107, 1996.

9 T. Niwa, Y. Amamiya, M. Mamada, and H. Shimawaki, 'High-β AlGaAs/InGaAs HBTs with reduced emitter resistance for low-power-consumption high-speed ICs', in *Proc. Intl. Symp. on Compound Semiconductors*, pp. 309–312, 1999.

10 T. Oka, K. Hirata, K. Ouchi, H. Uchiyama, T. Taniguchi, K. Mochizuki, *et al.*, 'Advanced performance of small-scaled InGaP/GaAs HBTs with over 150 GHz and over 250 GHz', in *IEEE IEDM Tech. Dig.*, pp. 653–656, 1998.

11 D. Mensa, Q. Lee, J. Guthrie, S. Jaganathan, and M. Rodwell, 'Transferred-substrate HBTs with 250 GHz current-gain cut-off frequency', in *IEEE IEDM Tech. Dig.*, pp. 657–660, 1998.

12 Q. Lee, S. Martin, D. Mensa, R. Smith, J. Guthrie, and M. Rodwell, 'Submicron transferred-substrate heterojunction bipolar transistors', *IEEE Electron Devices Lett.*, vol. 20, pp. 396–398, 1999.

13 F. Schwierz, and J. J. Liou, 'Semiconductor devices for RF applications: evolution and current status', *Microelectron. Reliab.*, vol. 41, pp. 145–168, 2001.

14 S. P. Voinigescu, D. S. McPherson, F. Pera, S. Szilagyi, M. Tazlauanu, and H. Tran, 'A comparison of silicon and III–V technology performance and building block implementations for 10 and 40 Gb/s optical networking ICs', *Int. J. High Speed Electron. Syst.*, vol. 13, pp. 27–58, 2003.

15 H. S. Bennett, R. Brederlow, J. C. Costa, P. E. Cottrell, W. M. Huang, A. A. Immorlica, Jr., *et al.*, 'Device and technology evolution for Si-based RF integrated circuits', *IEEE Trans. Electron Devices*, vol. 52, pp. 1235–1258, 2005.

16 J. D. Cressler, 'Re-engineering silicon: Si-Ge heterojunction bipolar technology', *IEEE Spectrum*, vol. 32, pp. 49–55, 1995.

17 T. H. Ning, 'Trade-offs between SiGe and GaAs bipolar ICs', in *Proc. Int. Conf. Solid-State and Integrated Circuit Technology*, pp. 434–438, 1995.

18 S. M. Sze, *Physics of Semiconductor Devices*. John Wiley & Sons, New York, 2nd ed., 1981.

19 D. L. Harame, J. M. C. Stork, B. S. Meyerson, K. Y.-J. Hsu, J. Cotte, K. A. Jenkins, *et al.*, 'Optimization of SiGe HBT technology for high speed analog and mixed-signal applications', in *IEEE IEDM Tech. Dig.*, pp. 71–74, 1993.

20 E. F. Crabbe, B. Meyerson, J. Stork, and D. Harame, 'Vertical profile optimization of very high frequency epitaxial Si- and SiGe-base bipolar transistors', in *IEEE IEDM Tech. Dig.*, pp. 83–86, 1993.

21 T. Uchino, T. Shiba, T. Kikuchi, Y. Tamaki, A. Watanabe, Y. Kiyota, *et al.*, '15-ps ECL/74-GHz f_T Si bipolar technology', in *IEEE IEDM Tech. Dig.*, pp. 67–70, 1993.

22 J. F. Jensen, M. Hafizi, W. E. Stanchina, R. A. Metzger, and D. B. Rensch, '39.5-GHz static frequency divider implemented in AlInAs/GaInAs HBT technology', in *GaAs IC Symp. Tech. Dig.*, pp. 101–104, 1992.

23 W. Hansch, T. Vogelsang, R. Kircher, and M. Orlowski, 'Carrier transport near the Si SiO$_2$ interface of MOSFET', *Solid State Electron.*, vol. 32, pp. 839–849, 1989.

24 K. Dragosits, V. Palankovski, and S. Selberherr, 'Two-dimensional modeling of quantum mechanical effects in ultra-short CMOS devices', in *Advances in Simulation, Systems Theory and Systems Engineering*, (*N. Mastorakis, V. Kluev, and D. Koruga, Eds.*), pp. 113–116, WSEAS Press, 2002.

25 C. K. Maiti, and G. A. Armstrong, *Applications of Silicon-Germanium Heterostructure Devices*. Inst. of Physics Pub., UK, 2001.

26 C. Jacoboni, and P. Lugli, *The Monte Carlo Method for Semiconductor Device Simulation*. Springer-Verlag, New York, 1989.

27 K. Hess, Ed., *Monte Carlo Device Simulation: Full Band and Beyond*. Kluwer, Boston, MA, 1991.

28 S. Laux, and M. Fischetti, 'The DAMOCLES Monte Carlo device simulation program', in *Computational Electronics*, (*K. Hess, J. Leburton, and U. Ravaioli, Eds.*), pp. 87–92, Kluwer, Boston, 1991.

29 H. Kosina, and S. Selberherr, 'A hybrid device simulator that combines Monte Carlo and drift-diffusion analysis', *IEEE Trans. Computer-Aided Design*, vol. 13, pp. 201–210, 1994.

30 W. Engl, A. Emunds, B. Meinerzhagen, H. Peifer, and T. Thoma, 'Bridging the gap between the hydrodynamic and the Monte Carlo model- An attempt', in *Proc. VLSI Process/Device Modeling Workshop*, pp. 32–33, 1989.

31 V. Palankovski, and S. Selberherr, 'Critical modeling issues of SiGe semiconductor devices', in *Proc. Symp. on Diagnostics and Yield: Advanced Silicon Devices and Technologies for ULSI Era*, pp. 1–11, 2003.

32 V. Palankovski, and S. Selberherr, 'Challenges in modeling of high-speed electron devices', in *Proc. IWPSD-2003*, pp. 45–50, 2003.

33 V. Palankovski, R. Schultheis, and S. Selberherr, 'Simulation of power heterojunction bipolar transistors on gallium arsenide', *IEEE Trans. Electron Devices*, vol. 48, pp. 1264–1269, 2001.

34 V. Palankovski, S. Wagner, and S. Selberherr, 'Numerical analysis of compound semiconductor RF devices', in *GaAs IC Symp.*, pp. 107–110, 2003.

35 V. Palankovski, and S. Selberherr, 'Analysis of high speed heterostructure devices', in *Proc. MIEL*, pp. 115–122, 2004.

36 V. Palankovski, and R. Quay, *Analysis and Simulation of Heterostructure Devices*. Springer, New York, 2004.

37 R. Anholt, *'GATES', in Electrical and Thermal Characterization of MESFET's, HEMT's, and HBTs*. Artech House, Boston, 1995.

38 J. J. Liou, *Principles and Analysis of AlGaAs/GaAs Heterojunction Bipolar Transistors*. Artech House, Boston, 1996.

39 NEMO, 'http://www.cfdrc.com/nemo/', 1999.

40 R. Anholt, *'POSES', in Electrical and Thermal Characterization of MESFET's, HEMT's, and HBTs*. Artech House, Boston, 1995.

41 W. Klix, and R. Stenzel, 'Dreidimensionale Numerische Simulation Elektronischer Bauelemente,' http://www.htwdresden.de/klix/simba/welcome.html, 2005.

42 R. Stenzel, C. Pigorsch, W. Klix, A. Vescan, and H Leier, 'Simulation of AlGaN/GaN-HFETs includme spontaneous and piezoelectric polarization charges', in *Proc. Intl. Symp. on Compound Semiconductors*, pp. 511–514, 1999.

43 R. Dietrich, A. Wieszt, A. Vescan, H. Leier, R. Stenzel, and W. Klix, 'Power handling limits and degradation of large area AlGaN/GaN RF-HEMTs', *Solid State Electron.*, vol. 47, pp. 123–125, 2003.

44 E. Lyumkis, R. Mickevicius, O. Penzin, B. Polsky, and K. E. Sayed, 'Numerical analysis of electron tunneling through hetero-interfaces and Schottky barriers in heterostructure devices', in *GaAs IC Symp.*, pp. 129–132, 2000.

45 E. Lyumkis, R. Mickevicius, O. Penzin, B. Polsky, K. E. Sayed, A. Wettstein, *et al.*, 'Simulations of quautum transport in HEMT using density gradient model', in *GaAs IC Symp. Tech. Dig.*, pp. 233–236, 2002.

46 C. Morton, and C. Snowden, 'Comparison of Quasi-2D and ensemble Monte Carlo simulations for deep submicton HEMTs', in *IEEE MTT-S Proc.*, pp. 153–156, 1998.

47 C. Morton, C. Snowden, and M. Howes, 'HEMT physical model for MMIC CAD', in *Proc. European Microwave Conf.*, pp. 199–204, 1995.

48 Silvaco International, *Silvaco-ATLAS User's Manual*, 2004.

49 D. M. Richey, J. D. Cressler, and A. J. Joseph, 'Scaling issues and Ge profile optimization in advanced UHV/CVD SiGe HBT's', *IEEE Trans. Electron Devices*, vol. 44, pp. 431–440, 1997.

50 R. Johnson, Z. Kachwalla, and C. Snowden, 'Multi-tone microwave simulation of HEMTs using a physics-based electro-thermal CAD model', in *Proc. European Microwave Conf.*, pp. 107–110, 1999.

51 Avant Corporation, *Medici - Semiconductor Simulation in 2D*, 1997.

52 Stanford University, *PISCES-2ET 2-D Device Simulator*, 1994.

53 S. Amon, B. Ferk, D. Vrtacnik, D. Resnik, D. Krizaj, and S. Sokolic, '2D numerical modeling of SiGe structures', in *Mediterranean Electrotechnical Conf.*, vol. 1, pp. 326–329, 1998.

54 C. K. Maiti, and G. A. Armstrong, 'Simulation of silicon germanium HBTs using ATLAS', *Simulation Standard*, pp. 6–8, 1997.

55 B. Pejcinovic, L. E. Kay, T. W. Tang, and D. H. Navon, 'Numerical simulation and comparison of Si BJT's and $Si_{1-x}Ge_x$ HBT's', *IEEE Trans. Electron Devices*, vol. 36, pp. 2129–2137, 1989.

56 V. Palankovski, and S. Selberherr, 'Thermal models for semiconductor device simulation', in *Proc. European Conf. on High Temperature Electronics*, pp. 25–28, 1999.

57 R. Quay, R. Reuter, V. Palankovski, and S. Selberherr, 'S-parameter simulation of RF-HEMTs', in *Proc. EDMO*, pp. 13–18, 1998.

58 S. Wagner, V. Palankovski, T. Grasser, R. Schultheis, and S. Selberherr, 'Small-signal analysis and direct S-parameter extraction', in *Proc. EDMO*, pp. 50–55, 2002.

59 B. Gonzalez, V. Palankovski, H. Kosina, A. Hernandez, and S. Selberherr, 'An energy relaxation time model for device simulation', *Solid State Electron.*, vol. 43, pp. 1791–1795, 1999.

60 R. Quay, C. Moglestue, V. Palankovski, and S. Selberherr, 'A temperature dependent model for the saturation velocity in semiconductor materials', *Materials Sci. in Semicond. Processing*, vol. 3, pp. 149–155, 2000.

61 C. G. van de Walle, and R. M. Martin, 'Theoretical study of Si/Ge interfaces', *J. Vac. Sci. Technol. B*, vol. 3, pp. 1256–1259, 1985.

62 R. People, and J. C. Bean, 'Band alignments of coherently strained Ge_xSi_{1-x}/Si heterostructures on $\langle 001 \rangle > Ge_ySi_{1-y}$ substrates', *Appl. Phys. Lett.*, vol. 48, pp. 538–540, 1986.

63 J. Eberhardt, and E. Kasper, 'Bandgap narrowing in strained SiGe on the basis of electrical measurements on Si/SiGe/Si hetero bipolar transistors', *Materials Sci. and Eng.*, vol. B89, pp. 93–96, 2002.

64 Y. Varshni, 'Temperature dependence of the energy gap in semiconductors', *Physica*, vol. 34, pp. 149–154, 1967.

65 V. Palankovski, G. Kaiblinger-Grujin, and S. Selberherr, 'Implications of dopant-dependent low-field mobility and band gap narrowing on the bipolar device performance', *J. Phys. IV*, vol. 8, pp. 91–94, 1998.

66 J. W. Slotboom, and H. C. de Graff, 'Measurement of bandgap narrowing in Si bipolar transistor', *Solid State Electron.*, vol. 19, pp. 857–862, 1976.

67 D. B. M. Klaassen, J. M. Slotboom, and H. C. De Graaff, 'Unified apparent bandgap narrowing in n- and p-type silicon', *Solid State Electron.*, vol. 35, pp. 125–129, 1992.

68 H. P. D. Lanyon, and R. A. Tuft, 'Bandgap narrowing in moderately to heavily doped silicon', *IEEE Trans. Electron Devices*, vol. 26, pp. 1014–1018, 1979.

69 J. A. del Alamo, and R. M. Swanson, 'Measurement of steady-state minority-carrier transport parameters in heavily doped n-type silicon', *IEEE Trans. Electron Devices*, vol. 34, pp. 1580–1589, 1987.

70 S. E. Swirhun, Y. H. Kwark, and R. M. Swanson, 'Measurement of electron life-time, electron mobility and bandgap narrowing in heavily doped p-type silicon', in *IEEE IEDM Tech. Dig.*, pp. 24–27, 1986.

71 J. W. Slotboom, and H. C. deGraff, 'Bandgap narrowing in silicon bipolar transistor', *IEEE Trans. Electron Devices*, vol. 24, pp. 1123–1125, 1977.

72 S. E. Swirhun, D. E. Kane, and R. M. Swanson, 'Temperature dependence of minority electron mobility and band-gap narrowing in p^+-Si', in *IEEE IEDM Tech. Dig.*, pp. 298–301, 1988.

73 J. W. Slotboom, 'The pn-product in silicon', *Solid State Electron.*, vol. 20, pp. 279–283, 1977.

74 J. Wagner, and J. A. del Alamo, 'Band gap narrowing in heavily doped silicon: a comparison of optical and electrical data', *J. Appl. Phys.*, vol. 63, pp. 425–429, 1988.

75 S. C. Jain, and D. J. Roulston, 'A simple expression for band gap narrowing (BGN) in heavily doped Si, Ge, GaAs and Ge_xSi_{1-x} strained layers', *Solid State Electron.*, vol. 34, pp. 453–465, 1991.

76 B. Y. Tsaur, C. K. Chen, and S. A. Manno, 'Long-wavelength Ge_xSi_{1-x}/Si heterojunction infrared detector and 400 X 400- element imager arrays', *IEEE Electron Device Lett.*, vol. 12, pp. 293–296, 1991.

77 T. L. Lin, A. Ksendzov, S. M. Dejewsky, E. W. Jones, R. W. Fathaure, T. N. Krabach, *et al.*, 'SiGe/Si heterojunction internal photoemission long-wavelength infrared detectors fabricated by molecular beam epitaxy', *IEEE Electron Devices Lett.*, vol. 38, pp. 1141–1144, 1991.

78 P. M. Garone, V. Venkataraman, and J. C. Strum, 'Hole mobility enhancement in MOS-gated Ge_xSi_{1-x}/Si heterostructure inversion layers', *IEEE Electron Device Lett.*, vol. 13, pp. 56–58, 1992.

79 T. Manku, and A. Nathan, 'Effective mass for strained p-type $Si_{1-x}Ge_x$', *J. Appl. Phys.*, vol. 69, pp. 8414–8416, 1991.

80 M. V. Fischetti, and S. E. Laux, 'Band structure, deformation potentials, and carrier mobility in strained Si, Ge, and SiGe alloys', *J. Appl. Phys.*, vol. 88, pp. 2234–2252, 1996.

81 J. Poortmans, S. C. Jain, D. H. J. Totterdell, M. Caymax, J. F. Nus, R. P. Mertens, *et al.*, 'Theoretical calculation and experimental evidence of the real and apparent bandgap narrowing due to heavy doping in p-type Si and strained $Si_{1-x}Ge_x$ layers', *Solid State Electron.*, vol. 36, pp. 1763–1771, 1993.

82 B. L. Tron, M. D. R. Hashim, P. Ashburn, M. Mouis, A. Chantre, and G. Vincint, 'Determination of bandgap narrowing and parasitic energy barriers in SiGe HBT's integrated in a bipolar technology', *IEEE Trans. Electron Devices*, vol. 44, pp. 715–722, 1997.

83 P. Ashburn, H. Boussetta, M. D. R. Hashim, A. Chantre, M. Mouis, G. J. Parker, *et al.*, 'Electrical determination bandgap narrowing in bipolar transistors with epitaxial Si, epitaxial $Si_{1-x}Ge_x$, and ion implanted bases', *IEEE Trans. Electron Devices*, vol. 43, pp. 774–782, 1996.

84 I. M. Anteney, G. Lippert, P. Ashburn, H. J. Osten, B. Heinemann, G. J. Parker, *et al.*, 'Characterization of the effectiveness of carbon incorporation in SiGe for the elimination of parasitic energy barriers in SiGe HBTs', *IEEE Electron Device Lett.*, vol. 20, pp. 116–118, 1998.

85 Z. Matutinovic-Krstelj, V. Venkataraman, E. Prinz, J. Sturm, and C. W. Magee, 'A comprehensive study of lateral and vertical current transport in $Si/Si_{1-x}Ge_x/Si$ HBT's', in *IEEE IEDM Tech. Dig.*, pp. 87–90, 1993.

86 M. Libezny, S. Jain, J. Poortmans, M. Caymax, J. Nijs, R. Mertens, *et al.*, 'Photoluminescence determination of the Fermi energy in heavily doped strained $Si_{1-x}Ge_x$ layers', *Appl. Phys. Lett.*, vol. 64, pp. 1953–1955, 1994.

87 R. People, 'Indirect band gap of coherently strained Ge_xSi_{1-x} bulk alloys on $\langle 100 \rangle$ silicon substrates', *Phys. Rev.*, vol. 32, pp. 1405–1408, 1985.

88 W.-X. Ni, J. Knall, and G. V. Hansoon, 'New method to study band offsets applied to strained $Si/Si_{1-x}Ge_x$ (100) heterojunction interfaces', *Phys. Rev. B*, vol. 36, pp. 7744–7747, 1987.

89 S. Takagi, J. L. Hoyt, K. Rim, J. J. Welser, J. F. Gibbons, G. van de Wall, *et al.*, 'Evaluation of the valence band discontinuity of $Si/Si_{1-x}Ge_x/Si$ heterostructures by application of admittance spectroscopy to MOS capacitors theoretical study of Si/Ge interfaces', *IEEE Trans. Electron Devices*, vol. 45, pp. 494–501, 1998.

90 K. Nauka, T. I. Kamins, J. E. Turner, C. A. King, J. L. Hoyt, and J. F. Gibbons, 'Admittance spectroscopy measurements of band offsets in $Si/Si_{1-x}Ge_x/Si$ heterostructures', *Appl. Phys. Lett.*, vol. 60, pp. 195–197, 1992.

91 K. Rim, S. Takagi, J. J. Welser, J. L. Hoyt, and F. Gibbons, 'Capacitance-voltage characteristics of p-Si/SiGeC MOS capacitors', *Mat. Res. Soc. Symp. Proc.*, vol. 379, pp. 327–332, 1995.

92 C. H. Gan, J. A. del Alamo, B. R. Bennett, B. S. Meyerson, E. F. Crabbe, C. G. Sodini, *et al.*, '$Si/Si_{1-x}Ge_x$ valence band discontinuity measurements using a semiconductor-insulator-semiconductor (SIS) heterostructure', *IEEE Trans. Electron Devices*, vol. 41, pp. 2430–2439, 1994.

93 S. C. Jain, J. Poortmans, S. S. Iyer, J. J. Loferski, J. Nijs, R. Mertens, *et al.*, 'Electrical and optical bandgaps of Ge_xSi_{1-x} strained layers', *IEEE Trans. Electron Devices*, vol. 40, pp. 2338–2343, 1993.

94 C. A. King, J. L. Hoyt, and J. F. Gibbons, 'Bandgap and transport properties of $Si_{1-x}Ge_x$ by analysis of nearly ideal $Si/Si_{1-x}Ge_x/Si$ heterojunction bipolar transistor', *IEEE Trans. Electron Devices*, vol. 36, pp. 2093–2104, 1989.

95 C. Jungemann, S. Keith, and B. Meinerzhagen, 'Full-band Monte Carlo device simulation of a Si/SiGe-HBT with a realistic Ge profile', *IEICE Trans. Electron.*, vol. E83, pp. 1228–1234, 2000.

96 J. C. Bean, 'Silicon-based semiconductor heterostructures: Column IV bandgap engineering', *Proc. IEEE*, vol. 80, pp. 571–587, 1992.

97 M. M. Rieger, and P. Vogl, 'Electronic-band parameters in strained $Si_{1-x}Ge_x$ Alloys on $Si_{1-y}Ge_y$ substrates', *Phys. Rev. B*, vol. 48, pp. 14276–14287, 1993.

98 C. Herring, and E. Vogt, 'Transport and deformation-potential theory for many-valley semiconductor with anisotropic scattering', *Phys. Rev.*, vol. 101, pp. 944–961, 1956.

99 M. A. Green, 'Intrinsic concentration, effective densities of states, and effective mass in Si', *J. Appl. Phys.*, vol. 67, pp. 2944–2954, 1990.

100 G. E. Pikus, and G. L. Bir, 'Effect of determination on the hole energy spectrum of germanium and silicon', *Soviet Phys. Solid state*, vol. 1, p. 1502, 1960.

101 H. Hasegawa, 'Theory of cyclotron resonance in strained silicon crystals', *Phys. Rev.*, vol. 129, pp. 1029–1040, 1963.

102 J. C. Hensel, and G. Feher, 'Cyclotron resonance experiments in uniaxially stress silicon', *Phys. Rev.*, vol. 129, pp. 1041–1062, 1963.

103 Y. Fu, Q. Chen, and M. Willander, 'Resonant tunneling of holes in Si/Ge_xSi_{1-x}', *J. Appl. Phys.*, vol. 70, pp. 7468–7473, 1991.

104 S. K. Chun, and K. L. Wang, 'Effective mass and mobility of holes in strained $Si_{1-x}Ge_x$ layers on (001) $Si_{1-y}Ge_y$ substrate', *IEEE Trans. Electron Devices*, vol. 39, pp. 2153–2164, 1992.

105 F. L. Madarasz, J. E. Lang, and P. M. Hemeger, 'Effective masses for non-parabolic bands in p-type silicon', *J. Appl. Phys.*, vol. 52, pp. 4646–4648, 1981.

106 H. D. Barber, 'Effective mass and intrinsic concentration in silicon', *Solid State Electron.*, vol. 10, pp. 1039–1051, 1967.

107 J. M. Hinckley, and J. Singh, 'Hole transport theory in pseudomorphic $Si_{1-x}Ge_x$ alloys grown on Si(001) substrates', *Phys. Rev. B*, vol. 41, pp. 2912–2926, 1990.

108 K. Takeda, A. Taguchi, and M. Sakata, 'Valence-band parameters and hole mobility of Ge-Si alloys - theory', *J. Phys. C: Solid State Phys.*, vol. 16, p. 2237, 1983.

109 S. Krishnamurthy, A. Sher, and A. B. Chen, 'Generalized brooks formula, and the electron mobility in $Si_{1-x}Ge_x$ alloys', *Appl. Phys. Lett.*, vol. 47, pp. 160–162, 1985.

110 T. Manku, and A. Nathan, 'Electron drift mobility model for devices based on unstrained and coherently strained $Si_{1-x}Ge_x$ grown on $\langle 001 \rangle$ silicon substrate', *IEEE Trans. Electron Devices*, vol. 39, pp. 2082–2089, 1992.

111 H. Jorke, and H. J. Herzog, 'Mobility enhancement in modulation doped Si-Si$_{1-x}$Ge$_x$ superlattice grown by molecular beam epitaxy', in *Proc. Silicon Molecular Beam Epitaxy, Electrochem. Soc.*, p. 352, 1985.

112 C. Smith, and A. D. Welbourn, 'Prospects for a hetero-structure bipolar transistor using a silicon-germanium alloy', in *IEEE BCTM Proc.*, pp. 57–64, 1987.

113 L. E. Key, and T. W. Tang, 'Monte Carlo calculation of strained and unstrained electron mobilities in Si$_{1-x}$Ge$_x$ using an improved ionized-impurity model', *J. Appl. Phys.*, vol. 70, pp. 1483–1488, 1991.

114 T. Manku, J. M. McGregor, A. Nathan, D. J. Roulston, J. P. Noel, and D. C. Houghton, 'Drift hole mobility in strained and unstrained doped Si$_{1-x}$Ge$_x$ alloys', *IEEE Trans. Electron Devices*, vol. 40, pp. 1990–1996, 1990.

115 F. M. Bufler, P. Graf, B. Meinerzhagen, D. Adeline, M. M. Rieger, H. Kabbel, *et al.*, 'Low- and high-field electron-transport parameters for untrained and strained Si$_{1-x}$Ge$_x$', *IEEE Electron Device Lett.*, vol. 18, pp. 264–266, 1997.

116 T. K. Carns, S. K. Chun, M. O. Tanner, K. L. Wang, T. I. Kamins, J. E. Turner, *et al.*, 'Hole mobility measurements in heavily doped Si$_{1-x}$Ge$_x$ strained layers', *IEEE Trans. Electron Devices*, vol. 51, pp. 1273–1281, 1994.

117 Q. Lu, M. R. Sardela, Jr., T. R. Bramblett, and J. E. Greene, 'B-doped fully strained Si$_{1-x}$Ge$_x$ layers grown on Si(001) by gas-source molecular beam epitaxy from Si$_2$H$_6$, Ge$_2$H$_6$, and B$_2$H$_6$: charge transport properties', *J. Appl. Phys.*, vol. 80, pp. 4458–4466, 1996.

118 F. M. Bufler, P. Graf, and B. Meinerzhagen, 'Hole transport investigation in unstrained and strained SiGe', *J. Vac. Sci. Technol. B*, vol. 16, pp. 1667–1669, 1998.

119 Z. Matutinovic-Krstelj, V. Venkataraman, E. J. Prinz, J. C. Sturm, and C. W. Magee, 'Base resistance and effective bandgap reduction in n-p-n Si/Si$_{1-x}$Ge$_x$/Si HBTs with heavy base doping', *IEEE Trans. Electron Devices*, vol. 43, pp. 457–466, 1996.

120 K. B. Joelsson, Y. Fu, W. X. Ni, and G. V. Hansson, 'Hall factor and drift mobility for hole transport in strained Si$_{1-x}$Ge$_x$ alloys', *J. Appl. Phys.*, vol. 81, pp. 1264–1269, 1997.

121 S. Smirnov, H. Kosina, and S. Selberherr, 'Investigation of the electron mobility in strained Si$_{1-x}$Ge$_x$ at high Ge composition', in *Proc. SISPAD*, pp. 29–32, 2002.

122 H. Kosina, and G. Kaiblinger-Grujin, 'Ionized-impurity scattering of majority electrons in silicon', *Solid State Electron.*, vol. 42, pp. 331–338, 1998.

123 C. Canali, G. Majni, R. Minder, and G. Ottaviani, 'Electron and hole drift velocity measurements in silicon and their empirical relation to electric field and temperature', *IEEE Trans. Electron Devices*, vol. 22, pp. 1045–1047, 1975.

124 C. Jacoboni, C. Canali, G. Ottaviani, and A. A. Quaranta, 'A review of some charge transport properties of silicon', *Solid State Electron.*, vol. 20, pp. 77–89, 1977.

125 N. Ahmad, and V. K. Arora, 'Velocity-field profile on n-silicon: A theoretical analysis', *IEEE Trans. Electron Devices*, vol. 33, pp. 1075–1077, 1986.

126 S. H. Li, J. M. Hinckley, J. Singh, and P. K. Bhattacharya, 'Carrier velocity-field characteristics and alloy scattering potential in $Si_{1-x}Ge_x/Si$', *Appl. Phys. Lett.*, vol. 63, pp. 1393–1395, 1993.

127 M. Ershov, and V. Ryzhii, 'High -field electron transport in SiGe alloy', *J. Appl. Phys.*, vol. 33, pp. 1365–1371, 1994.

128 H. Mau, *Anpassung und Implementation des Energietransportmodells zur vergleichenden Simulation mit dem Drift-Diffusions-Modell an SiGe Hetero-bipolartransistoren*. PhD thesis, Technische Universit"at Ilmenau, Institut f"ur Festk"orperelektronik, 1997.

129 J. C. Bean, L. C. Feldman, A. T. Fiory, S. Nakahara, and I. K. Robinson, 'Ge_xSi_{1-x}/Si strained layer superlattice growth by molecular beam epitaxy', *J. Vac. Sci. Technol. A*, vol. 2, pp. 436–440, 1984.

130 J. W. Matthews, and A. E. Blakeslee, 'Defects in epitaxial multilayers – I. Misfit dislocations', *J. Crystal Growth*, vol. 27, pp. 118–125, 1974.

131 D. C. Houghton, D. D. Perovie, J.-M. Baribeau, and G. C. Weatherly, 'Misfit strain relaxations in Ge_xSi_{1-x}/Si heterostructures: The structural stability of buried strained layers and strained-layer superlattices', *J. Appl. Phys.*, vol. 67, pp. 1850–1862, 1990.

132 E. Kasper, 'Growth and properties of Si/SiGe superlattices', *Surface Sci.*, vol. 174, pp. 630–639, 1986.

133 S. C. Jain, *Germanium-Silicon Strained Layers and Heterostructures*. Academic Press Inc., New York, 1994.

134 J. G. Fossum, and D. S. Lee, 'A physical model for the dependence of carrier lifetime on doping density in nondegenerate silicon', *Solid State Electron.*, vol. 25, pp. 741–747, 1982.

135 J. G. Fossum, R. P. Mertens. D. S. Lee, and J. F. Nijs, 'Carrier recombination and lifetime in highly doped silicon', *Solid State Electron.*, vol. 26, pp. 569–576, 1983.

136 B. Zetterlund, and A. J. Steckl, 'Low temperature recombination life time in Si MOSFET's', in *IEEE IEDM Tech. Dig.*, pp. 284–288, 1980.

137 M. S. Tyagi, and R. V. Overstraeten, 'Minority carrier recombination in heavily-doped silicon', *Solid State Electron.*, vol. 26, pp. 577–597, 1983.

138 A. Schenk, 'A model for the field and temperature dependence of Shockley-Read-Hall lifetimes in silicon', *Solid State Electron.*, vol. 35, pp. 1585–1596, 1992.

139 T. Ghani, J. L. Hoyt, D. B. Nobel, J. F. Gibbons, J. E. Turner, and T. I. Kamins, 'Effect of oxygen on minority carrier lifetime and recombination currents in $Si_{1-x}Ge_x$ heterostructure devices', *Appl. Phys. Lett.*, vol. 58, pp. 1371–1374, 1991.

140 G. S. Higashi, J. C. Bean, C. Buescher, R. Yadvish, and H. Temkin, 'Improved minority-carrier lifetimes in Si/SiGe heterojunction bipolar transistors grown by molecularbeam epitaxy', *Appl. Phys. Lett.*, vol. 56, pp. 2560–2568, 1990.

141 P. V. Schwartz, and J. C. Strum, 'Microsecond carrier lifetimes in strained silicon-germanium alloys grown by rapid thermal chemical vapor deposition', *Appl. Phys. Lett.*, vol. 57, pp. 2004–2006, 1990.

142 C. A. King, J. L. Hoyt, C. M. Gronet, J. F. Gibbons, M. P. Scott, and J. Turner, 'Band gap and transport properties of $Si_{1-x}Ge_x$ by analysis of nearly ideal $Si/Si_{1-x}Ge_x$ heterojunction bipolar transistor produced by limited reaction processing', *IEEE Trans. Electron Devices*, vol. 10, pp. 52–54, 1989.

143 J. Dziewor, and W. Schmid, 'Auger coefficient for highly doped and highly excited silicon', *Appl. Phys. Lett.*, vol. 31, pp. 346–348, 1977.

144 S. Selberherr, 'Low temperature MOS device modeling', in *Low Temperature Electronics and High Temperature Superconductors, Electrochem. Soc.*, vol. 88-9, pp. 70–86, 1987.

145 S. Selberherr, *Analysis and Simulation of Semiconductor Devices*. Springer Verlag, Vienna, 1984.

146 J. Lee, A. L. Gutierrez-Aitken, S. H. Li, and P. K. Bhattacharya, 'Responsivity and impact ionization coefficients of $Si_{1-x}Ge_x$ photodiodes', *IEEE Trans. Electron Devices*, vol. 43, pp. 977–981, 1996.

147 K. Yeom, J. M. Hinckley, and J. Singh, 'Calculation of electron and hole impact ionization coefficients in SiGe alloys', *J. Appl. Phys.*, vol. 80, pp. 6773–6782, 1996.

148 V. Palankovski, S. Selberherr, R. Quay, and R. Schultheis, 'Analysis of HBT behavior after strong electrothermal stress', in *Proc. SISPAD*, pp. 245–248, 2000.

149 V. Palankovski, R. Quay, and S. Selberherr, 'Industrial application of heterostructure device simulation', *IEEE J. Solid State Circuits*, vol. 36, pp. 1365–1370, 2001.

150 T. Binder, K. Dragosits, T. Grasser, R. Klima, M. Knaipp, H. Kosina, *et al.*, 'MINIMOS-NT User's Guide', in *Institut fur Mikroelektronik, Technische Universitat, Wien*, 1998.

151 V. Palankovski, G. Kaiblinger-Grujin, H. Kosina, and S. Selberherr, 'A Dopant-Dependent Band Gap Narrowing Model Application for Bipolar Device Simulation', in *Simulation of Semiconductor Processes and Devices, (K. D. Meyer and S. Biesemans Eds.)*, pp. 105–108, Springer, New York, 1998.

152 V. Sverdlov, H. Kosina, and S. Selberherr, 'Modeling current transport in ultra-scaled field-effect transistors', *Microelectron. Reliab.*, vol. 47, pp. 11–19, 2007.

153 D. Schroeder, *Modelling of Interface Carrier Transport for Device Simulation*. Springer-Verlag, New York, 1994.

154 K. Yang, J. C. Cowles, J. R. East, and G. I. Haddad, 'Theoretical and experimental d.c. characterization of InGaAs-based abrupt emitter HBTs', *IEEE Trans. Electron Devices*, vol. 42, pp. 1047–1058, 1995.

155 J. A. King, *Material Handbook for Hybrid Microelectronics*. Artech House, Norwood, MA, 1988.

156 A. Katz, *Indium Phosphide and Related Materials*. Artech House, Norwood, MA, 1992.

157 P. D. Maycock, 'Thermal conductivity of silicon, germanium, III–V compounds and III–V alloys', *Solid State Electron.*, vol. 10, pp. 161–168, 1967.

158 M. Landolt, and J. Bornstein, *Numerical Data, and Functional Relationships in Science and Technology*. Springer-Verlag, New York, 1987.

159 S. Adachi, *Physical Properties of III–V Semiconductor Compounds*. Wiley, New York, 1992.

160 S. Adachi, *Properties of Aluminum Gallium Arsenide*. IEE-INSPEC, UK, 1993.

161 P. Bhattacharya, *Properties of Lattice-Matched and Strained Indium Gallium Arsenide*. IEE-INSPEC, UK, 1993.

162 S. Tiwari, *Compound Semiconductor Device Physics*. Academic, New York, 1992.

163 R. Quay, H. Massler, W. Kellner, T. Grasser, V. Palankovski, and S. Selberherr, 'Simulation of gallium-arsenide based high electron mobility transistors', in *Proc. SISPAD*, pp. 74–77, 2000.

164 S. Wagner, V. Palankovski, T. Grasser, G. Rohrer, and S. Selberherr, 'A direct extraction feature for scattering parameters of SiGe-HBTs', in *Intl. SiGe Technology and Device Meeting*, pp. 83–84, 2003.

165 Y. Hu, and K. Mayaram, 'Periodic steady-state analysis for coupled device and circuit simulation', in *Proc. SISPAD*, pp. 90–94, 2000.

166 W. Liu, *Fundamentals of III–V Devices: HBT's, MESFET's, and HFETs/HEMTs*. Wiley, New York, 1999.

167 V. Palankovski, S. Selberherr, and R. Schultheis, 'Simulation of heterojunction bipolar transistors on gallium-arsenide', in *Proc. SISPAD*, pp. 227–230, 1999.

168 O. Esame, Y. Gurbuz, I. Tekin, and A. Bozkurt, 'Performance comparison of state-of-the-art heterojunction bipolar devices (HBT) based on AlGaAs/GaAs, Si/SiGe and InGaAs/InP', *Microelectronics J.*, vol. 35, pp. 901–908, 2004.

169 G. Gonzales, *Microwave Transistor Amplifiers*. Prentice-Hall, Englewood Cliffs, NJ, 1984.

170 J.-S. Rieh, B. Jagannathan, H. Chen, K. Schonenberg, S.-J. Jeng, M. Khater, *et al.*, 'Performance and design considerations for high speed SiGe HBTs of $f_T/f_{max} = 375$ GHz/210 GHz', in *Int. Conf. Indium Phosphide and Related Materials*, pp. 374–377, 2003.

171 N. Zamdmer, J. Kim, R. Trzeinski, J.-O. Plouchart, S. Narasimha, M. Khare, *et al.*, 'A 243-GHz F_t and 208-GHz F_{max} 90-nm SOI CMOS SoC technology with low-power millimeter-wave digital and RF circuit capability', in *IEEE VLSI Tech. Symp. Dig.*, pp. 98–99, 2004.

172 B. Jagannathan, D. Greenberg, D. I. Sanderson, J-S. Rieh, J. Pekarik, J. O. Plouchart, *et al.*, 'Speed and power performance comparison of state-of-the-art CMOS and SiGe RF transistors', in *Proc. Silicon Monolithic Integrated Circuits in RF Systems Meeting*, pp. 115–118, 2004.

173 K. Kuhn, M. Agostinelli, S. Ahmed, S. Chambers, S. Cea, S. Christensen, *et al.*, 'A 90 nm communication technology featuring SiGe HBT transistors, RF CMOS, precision R-L-C RF elements and 1 μm^2 6-T SRAM Cell', in *IEEE IEDM Tech. Dig.*, pp. 73–76, 2002.

174 K. Kuhn, R. Basco, D. Becher, M. Hattendorf, P. Packan, I. Post, *et al.*, 'A comparison of state-of-the-art NMOS and SiGe HBT devices for analog/mixed-signal/RF circuit applications', in *IEEE VLSI Tech. Symp. Dig.*, pp. 224–225, 2004.

Chapter 11
Passive components

Passive components are indispensable in the design and development of microchips for high-frequency applications. Inductors in particular are used frequently in radio frequency integrated circuits (RFICs) such as low noise amplifiers (LNAs) and oscillators. Currently analog/mixed-signal/RF product designs incorporating both CMOS and bipolar active elements have emerged as a potential technology for communication applications. Radio frequency (RF) system-on-chip (SoC) designs require high-performance passive components – particularly on – chip inductors and capacitors. As chip geometries get smaller and operating speeds increase, the physical effects of parasitic capacitance, resistance and inductance become increasingly significant. For RF and wireless applications, the key requirements include high Q passive elements, superior active devices, robust frequency domain tools and accurate parasitic extraction. The technology and tool set must support building monolithic RF building blocks with an eye towards integrating the entire radio on a chip. Important requirements can be more easily understood by examining the features and technology requirements of some key wireless building blocks (see in Table 11.1).

While all of these components can be realised in Si CMOS technology, their specific designs necessitate special consideration owing to the requirement of high quality factor Q at relatively high frequencies. Inductance, in particular, can affect the timing as much as 30 per cent for designs operating above 500 MHz. For SoC design, an integrated approach to inductance analysis enables designers to accurately model all the parasitic effects – resistance, capacitance and inductance. The main problem that designers of multi-million gate designs face during full chip analysis is the large amount of data as detailed modelling of every capacitive and inductive interaction in a design, which in an average one million-gate design results in more than a billion capacitors and inductors. In order to perform accurate inductance analysis, high-performance extraction and circuit simulation tools are required.

Unlike older BiCMOS designs, modern designs are adopting approaches where CMOS digital cells from well-characterised libraries are being mixed with specialized analog BiCMOS (or bipolar) modules to deliver simultaneously optimised digital and

Table 11.1   Features and requirements of basic RF wireless building blocks. (After D. Harame et al., Proc. 4th IEEE Intl. Caracas Conf. Devices, Circuits and Systems, 2002 (D052-1–D052-17).)

Circuit	Main feature	Technology requirements
LNA	Low noise figure Gain per stage High linearity	Low n_{Fmin}, low HBT R_b High Early voltage High f_T/f_{max}
Mixer	High linearity High port isolation Carrier leakage	High Early voltage High f_T/f_{max} Isolation technologies Small mismatch
VCO	Low phase noise Farther from carrier Closer to carrier Tuning range (freq/V) Low gain variation	$\leq$ high-Q passives $\leq$ low 1/f noise HBT Varactor tuning range(C/V)
Synthesizer	Low phase noise Low reference spurs Power consumption Chip area	Same as VCO Small FET asymmetry Small gate length for low Supply operation, reduced area
PA	Power added efficiency Gain per stage Robustness	Low loss matching Interconnect circuits (high-Q inductor, thick metal) Max available gain HBT High breakdown HBT

analog elements [1]. A key difference between digital and mixed-signal processes is the presence of passive elements. The optimisation for analog/mixed-signal/RF is more difficult, owing to conflicting digital and analog needs. The first problem is lack of commonality between digital and analog optimisation strategies. The second problem is that the desired analog process optimisation strategies may run counter to traditional CMOS scaling. State-of-the-art 90-nm foundry RF CMOS technology with various integrated passive elements and user-friendly design kit for RFIC design have been reported by Chang *et al.* [2].

Current SiGe technology provides easy access to different integrated active and passive components and for high-frequency applications, most important being the SiGe HBT itself and the passive inductor, capacitor and transmission line elements that are key to RF design [3]. Integrated inductors find application in many facets of radio frequency integrated circuits design including impedance matching, filtering, biasing and in oscillator circuits. Inductors are important, performance limiting components in monolithic RF circuits, such as voltage-controlled oscillators (VCOs), LNAs and passive-element filters. High-Q tank circuits in voltage-controlled oscillators reduce the phase noise. High-Q inductors in low noise amplifiers and power amplifiers (PAs) not only reduce noise but also improve gain, efficiency and input/output matching. Bandpass filters are crucial to many multi-function communication systems that are presently made of off-chip components based on SAW or ceramic technology, which are expensive, bulky, complex in assembly and problematic in reliability. In analog/mixed-signal design, performance is ultimately limited by the accuracy of the passive components in the technology [4–7].

Previously developed on-chip active inductors, such as gyrators, were implemented via impedance transformation. However, owing to difficult issues in power, noise, linearity and reproducibility, gyrators have not been popular in RF circuit design. To alleviate these problems, on-chip high-Q inductors using active magnetic coupling have also been investigated. The quality factor of the inductors is limited by the resistive losses in the spiral coil and by the substrate losses. With the emergence of RF and microwave applications for silicon integrated circuits, integration of spiral inductors with reasonable characteristics has become an urgent need. In particular, quality factors of integrated inductors in silicon ICs are typically low, and the understanding and optimisation of them have received intense attention.

This chapter gives a broad overview of on-chip spiral inductors. The design concept and modelling approach of the typical square-shaped spiral inductor are first addressed. This is followed by the discussions of advanced structures for the enhancement of inductor performance. Research results reported in the literature are reviewed to explain the recent development of passive components. The design of one- and two-port bipolar active inductors using SiGe/SiGeC BiCMOS technologies has been reported [8]. The prime motivation of this design is not to generate very high-Q values (in thousands), which involves very critical tuning of some sensitive design parameters; instead, the goal is to have a design that will provide moderately high-Q values (of the order of 100) that satisfy most of the requirements in Si-based RF applications. Various types of varactors, MIM capacitors and resistor structures currently in use are then introduced and the key characteristics of each type of these devices are discussed.

11.1 Inductor

Inductors are fundamental elements for the implementation of RF circuits and a critical component in analog/mixed-signal design. Small-valued, precise, high-Q inductors are employed in circuits such as RF transceivers. The most critical factor in inductor design is the optimisation of the inductor Q at the design frequency. Inductors are typically implemented by using circular or rectangular spirals. The conventional layouts of an on-chip passive inductor are shown in Figure 11.1. The shape could be square, hexagon, octagon or circular. The choice of layouts is always a trade-off between performance and simulation/optimisation efforts. Even though most electromagnetic (EM) simulators can handle a planar structure, the circular inductor requires much more mesh cells on the peripherals, which is extremely time consuming.

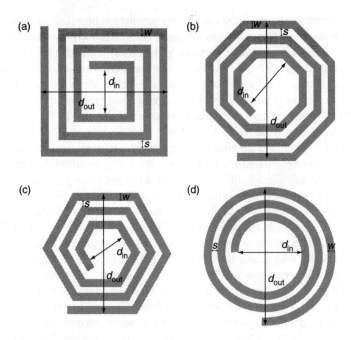

*Figure 11.1   Layouts of conventional on-chip passive inductors. (a) square,
(b) hexagonal, (c) octagon, and (d) circular*

From a performance point of view, however, the most optimum pattern is a circular spiral, because it suffers less resistive and capacitive losses. But the circular inductor is not widely used because only a few commercial layout tools support such a pattern. Hexagonal and octagonal structures are good alternatives, as they resemble closely to the circular structure, are easier to construct and are supported by most computer-aided design tools. It has been reported that the series resistance of the octagonal- and circular-shaped inductors is 10 per cent smaller than that of a square-shaped spiral inductor with the same inductance value.

Spiral inductors can be fabricated within a conventional MOS process with little modifications to the design rules. A minimum of two metal layers is required, one to form the spiral and one to form the underpass. To minimise parasitic capacitance to the substrate, the top metal layer is the usual choice for the main spiral. Structural parameters such as the outer dimension, number of turns, the distance between the centres of lines (or pitch) and substrate property are all important factors. On-chip inductors consume large silicon area, are sensitive to temperature variations, have limited Q-factor, and their inductance value is not precise, even if the technology is well characterised. In fact, determining the geometry and area required to deliver an optimised Q at a design frequency is not a straightforward process [9, 10].

Apart from the spiral inductor, two other kinds of on-chip inductors are in use. A gyrator, or active inductor, utilizes active components (i.e., transistors) to transform the impedance of a capacitor to inductance. Active inductors have been extensively used for low-frequency applications and in the design of mixers. For high-frequency applications, an active inductor with a high quality factor is necessary. In addition, other important specifications, such as noise figure and harmonic distortion components, can be maintained at low levels. On the basis of an active inductor, low-noise preamplifiers have also been developed. Larger, lower-Q devices have functions such as impedance matching and gain control. Significant research has been done on monolithic integration of inductors and there has been increasing use of inductors in state-of-the-art CMOS processes [11–14].

The active devices and capacitor required in the gyrator can be easily fabricated and occupy minimum space, but they consume a relatively large amount of power and introduce additional noise. The third on-chip inductor type is constructed with a bond wire [15]. It can offer a very high quality factor (30–60), but such an approach is likely to cause unwanted coupling to other devices on the substrate and may not be sufficiently robust for some RF applications.

However, the most difficult factor in inductor design is the minimisation of the impact of parasitic elements. Real inductors have parasitic resistance and capacitance. The parasitic resistance dissipates energy through ohmic loss, while the parasitic capacitance stores the unwanted energy. At high frequencies, the skin effect causes a non-uniform current distribution in the metal segments, which introduces (among other things) a frequency-dependent contribution to the parasitic resistance. Finally, electromagnetic effects caused by the Faraday effect introduce parasitic currents (eddy currents) in the silicon as well, adding an additional frequency-dependent term in the resistance [12, 16].

In addition to the layout, the substrate material affects the inductor performances. Minimising the substrate loss is a more complex task. As the frequency increases, to where the skin depth is of the order of the substrate thickness, eddy currents in the substrate become a major loss mechanism. To reduce this loss, one possibility is to build the inductors on a less lossy substrate. GaAs, silicon-on-insulator (SOI), silicon-on-sapphire (SOS), high-resistive silicon (HRS) or glass have all been used to reduce the substrate loss. Another approach to reduce the substrate effect is to remove it totally. This requires additional process to etch away the material beneath the inductor (suspended) and is fabricated by micro mechanical system (MEMS) technology. Suspended inductors have a higher Q factor and higher self-resonance. The crosstalk between the inductor and the other circuit elements can also be minimised. However, since the inductor is floating over a cavity, and only partially supported, the durability and robustness are questionable.

A patterned ground plane can be placed under the inductor. This approach partially isolates the eddy current loop degradation formed in the substrate and reduces crosstalk [17, 18]. However, the Q factor is only slightly increased, but the self-resonance is much lower owing to the higher capacitance to the ground, which makes the approach suitable only for the lower frequencies. Mitigating eddy current loss can be quite difficult. Electromagnetic simulation of a 1.5 (left) and a 4.5 (right) turn spiral inductors at 2 GHz demonstrates non-uniform current distribution in Figure 11.2. There are a number of potential techniques available for the optimisation

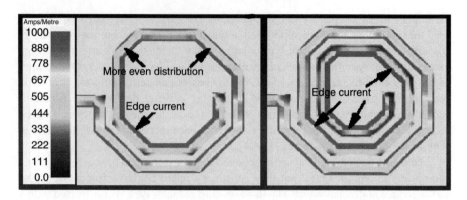

*Figure 11.2   Current crowding is more pronounced in multi-turn coils. Electromag-
netic simulation of a 1.5 (left) and 4.5 (right) turn spiral inductors
at 2 GHz. Simulations demonstrate non-uniform current distribution
[After D. Harame et al., Proc. 4th IEEE Intl. Caracas Conf. Devices,
Circuits and Systems, 2002 (D052-1–D052-17). ©2002 IEEE]*

of inductor design, including solid [12], patterned ground shields [4] and multi-level
metallisations to build vertical solenoids [19], as well as minimising doping levels
under the inductor.

11.1.1   Inductor model description

Quality factor characterises inductor loss at low frequency. The effective inductance shows the magnitude of the reactive part of the one-port input impedance with
frequency. The self-resonant frequency is the transition frequency where the input
impedance changes from inductive to capacitive.

Nguyen and Meyer [20] developed a simple π-model to describe the inductor's
behaviour, which can be considered as a section of the ladder model for interconnects.
An improved model, shown in Figure 11.3, was later developed by Ashby *et al.* [11].
This compact SPICE model accounts for more physical mechanisms. In this circuit,
R_s and L_s represent the respective series resistance and inductance of the metal line.
L_s comprises self inductance and mutual inductance. C_s is the capacitance between
metal lines. However, a practical inductor includes other components associated with
Si substrate, such as, R_{Si} and C_{Si}, the coupling resistance and capacitance. C_{ox} is the
capacitance of oxide layer underneath the spiral.

Except for the series inductance, all components in the model are parasitics
of the inductor and need to be minimised. Deficiencies of this simple model
include assumption of frequency-independence; parameters need to be extracted
from s-parameters at the operating frequency and parasitics associated with the overlaps and underpasses are not considered. Yue and Wong [21] reported an inductor
model similar to that in Figure 11.3(a), but with models parameters more relevant
to inductor geometry and processing. However, the model parameters need to be

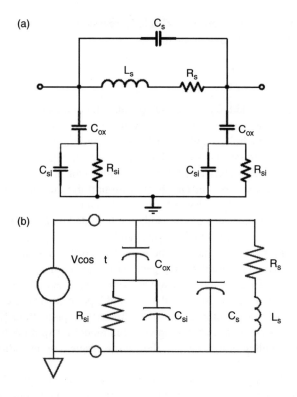

Figure 11.3 Lumped model equivalent circuit of a spiral inductor. For details, (see text)

extracted from empirical curve fitting rather than by association with underlying physics.

The inductance represents the magnetic energy stored in the device, although parasitic components may store energy as well. Numerical simulators computing the electromagnetic field distribution can therefore be used to calculate inductance by solving Maxwell's equations using a more computationally intensive three-dimensional finite element simulator. Another technique due to Greenhouse [22] divides the inductor into straight-line segments and sums the self-inductance of individual segments and mutual inductance between any two parallel segments. While the Greenhouse technique offers sufficient accuracy and adequate speed, it cannot provide an inductor design directly from specifications and is cumbersome for initial design. However, the analytical technique is less complicated and provides more physical insights.

Mohan *et al.* [23] have presented several new simple expressions for the d.c. inductance of square, hexagonal, octagonal and circular spirals (see Figure 11.1). The authors have also evaluated the accuracy of the expressions, by comparison with both three-dimensional field solver predictions and also with measurements. The first expression, called the modified Wheeler expression, is obtained by modifying a relevant expression for discrete inductors:

$$L_{mw} = K_1 \mu_o \frac{n^2 d_{avg}}{1 + K_2 \gamma} \tag{11.1}$$

where γ is the fill ratio and d_{avg} is the average diameter of the inductor. K_1 and K_2 are layout-dependent coefficients. The fill ratio γ represents how hollow the inductor is; for small γ, one gets a hollow inductor ($d_{out} \approx d_{in}$) and for a large γ, a full inductor ($d_{out} \gg d_{in}$). This expression is simple but gives very good accuracy.

The second expression is derived from electromagnetic principles by approximating the sides of the spiral by current sheets with uniform current distribution. This expression is intuitive and similar in form to inductance expressions for more conventional elements such as coaxial transmission lines and parallel wire transmission lines. The third expression is obtained by data-fitting techniques. Although it lacks the physically intuitive derivation of the other two approximations, it is very well suited for optimisation of circuits using geometric programming.

All three expressions match field solver simulations well, with typical errors of 1–2 per cent, and most errors smaller than around 3 per cent. These expressions are also good candidates for circuit design and optimisation applications [23]. They can be included in a physical, scalable lumped circuit model for spiral inductors, where, in addition to providing design insight, they allow efficient optimisation schemes to be employed.

Series resistance R_s, a key issue for inductor modelling (see Figure 11.3(a)), arises from the metal resistivity in the inductor and is closely related to the quality factor. When the inductor operates at high frequencies, the metal line suffers from the skin and proximity effects, and R_s becomes a function of frequency [24]. The series feed-forward capacitance, C_s, is approximated as the parallel-plate capacitance between the spiral and the centre-tap underpass. C_{ox} represents the capacitance between the spiral and the substrate. The silicon substrate is modelled by C_{Si} and R_{Si}. C_{ox}, C_{Si} and R_{Si} are proportional to the area covered by the spiral. C_{sub} and G_{sub} are the properties of the silicon substrate and are extracted from measured data. As a first-order approximation, the current density decays exponentially away from the metal-SiO$_2$ interface, and R_s can be expressed as [21]

$$R_s = \frac{\rho \cdot l_T}{w \cdot t_{eff}} \tag{11.2}$$

where ρ is the resistivity of the wire, l_T is total length and t_{eff} is given by

$$t_{eff} = \delta \cdot (1 - e^{t/\delta}) \tag{11.3}$$

where t is the physical thickness of the wire, and δ is the skin depth that is a function of the frequency

$$\delta = \sqrt{\frac{\rho}{\pi \mu f}} \tag{11.4}$$

where μ is the permeability in H/m and f is the frequency in Hz. R_s being a frequency-dependent component, it cannot be directly implemented in a time domain simulator, such as Cadence SPECTRE.

Yue *et al.* [10] have presented a physical model for planar spiral inductors on silicon with inductor geometry, allowing designers to predict and optimise the quality factor (see Figure 11.3(b)). Physical model parameters for an inductor on silicon are expressed as

$$C_s = n \cdot w^2 \cdot \frac{\epsilon_{ox}}{t_{oxM1-M2}} \tag{11.5}$$

$$C_{ox} = \frac{1}{2} \cdot l \cdot w \cdot \frac{\epsilon_{ox}}{t_{ox}} \tag{11.6}$$

$$C_{Si} = \frac{1}{2} \cdot l \cdot w \cdot C_{sub} \tag{11.7}$$

$$R_{Si} = \frac{2}{l \cdot w \cdot G_{sub}} \tag{11.8}$$

where ρ is d.c. metal resistivity, l is overall length of spiral, w is line width, δ is metal skin depth, t is metal thickness, N is number of turns, n is number of crossovers between spiral and centre-tap ($= N - 1$), $t_{oxM1-M2}$ is oxide thickness between spiral and centre-tap, t_{ox} is oxide thickness between spiral and substrate, C_{sub} is substrate capacitance per unit area and G_{sub} is substrate conductance per unit area.

11.1.2 Quality factor

For determining Q there are three possible approaches. The first approach calculates the ratio of the power stored to the power dissipated in the inductor. The second approach is equivalent to deriving the ratio of the imaginary part to the real part of the one-port input impedance, when the other port is grounded. The third approach arranges the inductor with a capacitor to find the ratio of the 3dB bandwidth to its centre resonant frequency. The energy storage and loss mechanisms in an inductor on silicon can be described by an equivalent energy model, where L_s, R_s, R_p and C_o represent the overall inductance, conductor loss, substrate loss and overall capacitance, respectively. Note that R_p and C_p represent the combined effects of C_{ox}, C_{Si} and R_{Si}, and hence are frequency dependent. Combining the energy terms according to the fundamental definition of Q yields

$$Q = \frac{\omega L_s}{R_s} \cdot \frac{R_p}{R_p + [(\omega L_s/R_s)^2 + 1]R_s} \cdot \left(1 - \frac{R_s^2 C_o}{L_s} - \omega^2 L_s C_o\right) \tag{11.9}$$

where $\omega L_s/R_s$ accounts for the magnetic energy stored and the ohmic loss in the spiral conductor. The second term is the substrate loss factor representing the energy dissipated in the semiconducting silicon substrate. The last term is the self-resonance factor describing the reduction in Q due to the increase in the peak electric energy with frequency and the vanishing of Q at the self-resonant frequency.

At relatively low frequency, Q is low because the substrate capacitance behaves like an open circuit. As the frequency goes higher, Q increases, and after peaking it decreases until the frequency reaches the self-resonant frequency. The effective inductance is close to the d.c. inductance value L_s, at low frequency, but increases with frequency owing to the parasitic capacitance to the ground. After reaching a maximum, it reduces and eventually becomes negative, as the component changes from inductive to capacitive. This frequency-dependent inductance applies in such a way that at low frequency, the ohmic loss of the coil dominates the real part of the equivalent input impedance. As the frequency goes higher, substrate effects start to take over. At the self-resonance frequency, the substrate resistor mainly determines the real part of the equivalent circuit. In order to maximise Q, a few observations can be made:

1. Using thicker metal or using a metal with greater conductivity (e.g., copper or gold) improves inductor Q by reducing R_s.
2. Placing the inductor as far from the substrate as possible minimises C_p and enhances the self-resonance factor.
3. Doping the substrate very lightly or very heavily results in a large R_p. This decreases the electrically induced losses in the substrate. Other high-frequency effects, however, induce loss in heavily doped substrates.
4. Operating at higher frequencies can improve Q, until other high-frequency effects become significant.

11.2   Parameter extraction

For parameter extraction, spiral inductors with different structural and process parameters have their two-port s-parameters measured using an HP8720B Network Analyzer and Cascade Microtech with substrate grounded. After de-embedding the parasitics of the probe pads the lumped elements in the inductor model (see Figure 11.3) are extracted from the complex propagation constant and characteristic impedance. A typical parameter extraction procedure for the lumped elements in the inductor model in shown in Figure 11.4.

To demonstrate the scalability of the model discussed above, Yue *et al.* [10] have tested inductors with various structural parameters. The measured and modelled values of the individual elements are plotted in Figure 11.5 for two 8-nH inductors. L_s and R_s are subject to eddy current effect in the conductor. At high frequencies, the penetration of magnetic field into the conductor is attenuated, which causes reduction in the magnetic flux internal to the conductor. However, L_s does not decrease significantly with increasing frequency because it is predominantly determined by the magnetic flux external to the conductor. Thus, it is valid to model L_s as a constant. The skin effect on R_s is more pronounced because R_s is inversely proportional to the effective cross-sectional area. C_s is independent of frequency since it represents the metal-to-metal cross-over capacitance between the spiral and the centre-tap; its extracted and modelled values are 26 and 28 fF, respectively. The frequency behaviours of R_p and C_p are governed by C_{ox}, C_{Si} and R_{Si}. At low frequencies, the electric field terminates

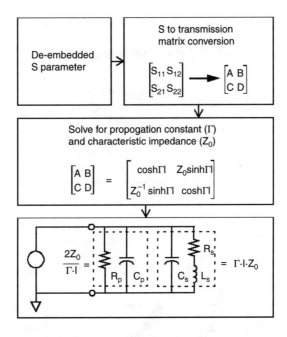

Figure 11.4   Testing and parameter extraction procedure for the lumped elements in the inductor model [After C. P. Yue et al., IEEE IEDM Tech. Dig., 1996 (155–158). ©1996 IEEE]

at the oxide-Si interface and C_p is primarily determined by C_{ox}. Since almost all energy is transmitted within the oxide layer along the spiral, little conduction current flows in the silicon substrate and thus R_p is large.

As frequency increases, the electric field starts to penetrate into the silicon substrate, which reduces C_p because of the series connection of oxide and silicon substrate capacitances. The roll-off in R_p signifies increasing energy transmission and hence more dissipation in the silicon substrate. At high frequencies, energy is transmitted mainly within the silicon substrate causing C_p and R_p to approach C_{Si} and R_{Si} respectively; C_{ox} is effectively short-circuited. Figure 11.6(a) illustrates that at low frequencies, Q is well described by $\omega L_s/R_s$. The rapid degradation of Q at high frequencies is a combined effect of the substrate loss and the self-resonance (see Figure 11.6(b)).

Figure 11.7(a) shows the effect of different metal schemes on Q. At 1 GHz, the effective thicknesses of 1 μm and 3 μm Al are 0.84 μm and 1.83 μm, respectively. After including the substrate factors, the ratio between their Q values at 1 GHz is 2. The single-level 3 μm and triple-level 1 μm cases have the same Q since the former suffers from strong self-induced eddy current and the latter from the combination of self and mutual eddy currents. Figure 11.7(b) illustrates that increasing oxide thickness improves the Q since the substrate loss and self-resonance effects are suppressed. But as frequency increases, the effect of C_{ox} vanishes and the Q merge together. Lowering

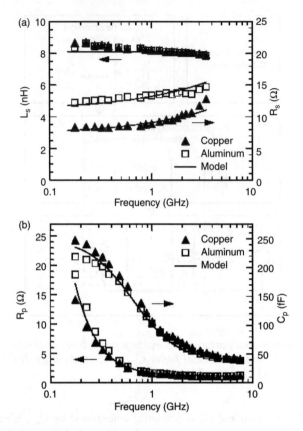

Figure 11.5   *The measured and modelled values of (a) L_s and R_s and (b) R_p and C_p for $1\,\mu m$ thick spirals in copper ($r_{Cu} = 2\,\mu\Omega\text{-}cm$) and aluminium ($r_{Al} = 3\,\mu\Omega\text{-}cm$). ($N = 7$, $w = 13\,\mu m$, $l = 4.4\,mm$, $t_{ox} = 4.5\,\mu m$, $t_{oxM1-M2} = 1.3\,\mu m$, $C_{sub} = 1.6 \times 10^{-3}$ $fF/\mu m^2$, $G_{sub} = 4 \times 10^{-8}\,S/\mu m^2$, line spacing ($s$) $= 7\,mm$, and outer dimension (OD) $= 300\,\mu\,m$) [After C. P. Yue et al., IEEE IEDM Tech. Dig., 1996 (155–158). ©1996 IEEE]*

the silicon substrate resistivity decreases R_{Si} and increases C_{Si}, causing the Q roll-off to occur at a lower frequency and a reduction of the self-resonant frequency (see Figure 11.8(a)). Figure 11.8(b) illustrates the effect of layout area on Q for the same inductance. The larger area inductors have higher Q at low frequencies because of lower series resistance. As frequency increases, the substrate effects begin to dominate and the larger area inductors actually have lower Q. In Figure 11.9, Q contour plots generated using the model described above along with some experimental data are presented. Q is plotted as a function of inductance and outer dimension of square spiral inductors. These contour plots serve as a design tool for achieving a specific inductance with the highest Q possible for a given technology at the frequency of interest.

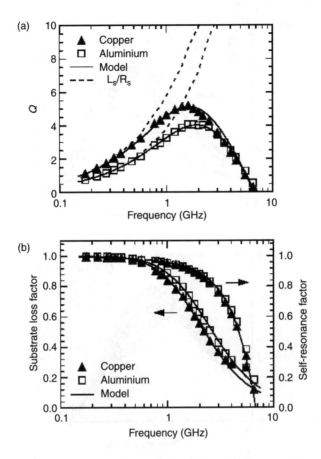

Figure 11.6   The measured and modelled values of (a) Q and (b) substrate loss factor and self-resonance factor for the inductors in Figure 11.5 [After C. P. Yue et al., IEEE IEDM Tech. Dig., 1996 (155–158). ©1996 IEEE]

11.3   Inductor simulation

Although the inductance value can be computed using analytical expressions discussed above or other similar expressions found in the literature, it is difficult to accurately predict analytically the losses associated with a spiral inductor. This necessitates the use of electromagnetic field solvers. While inductors are simple in physical structure, their design and modelling encounter the following challenges:

- electromagnetic field effect;
- non-uniform current distribution in the metal lines;
- components being bias as well as frequency dependent;
- significant substrate coupling, particularly for coils placed on a lossy substrate such as silicon;

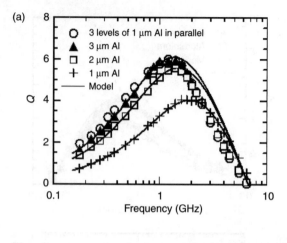

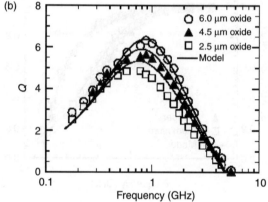

Figure 11.7   (a) The effect of metal scheme on Q. ($L_s = 8\,nH$, $t_{ox} = 4.5\,mm$, $\rho_{si} = 10\,\Omega\text{-}cm$, $N = 7$, $w = 13\,\mu m$, $s = 7\,\mu m$, $OD = 300\,\mu m$). (b) The effect of oxide thickness on Q. ($L_s = 8\,nH$, $t_{Al} = 3\,\mu m$, $\rho_{si} = 10\,\Omega\text{-}cm$, $N = 6$, $w = 24\,\mu m$, $s = 7\,\mu m$, $OD = 400\,\mu m$) [After C. P. Yue et al., IEEE IEDM Tech. Dig., 1996 (155–158). ©1996 IEEE]

- a wide range of dimensions and patterns;
- mutual inductance and capacitance due to adjacent metal lines;
- different topologies (i.e., coils placed on top, inside or bottom of the wafer).

   Inductors may be simulated using the freeware program ASITIC [25], but it was found that this program underestimated the magnetic substrate eddy current effects. By accurately calculating the fields in the substrate and the dielectric, Sonnet EM [26], a planar full-wave EM solver (method-of-moments) package, can accurately calculate inductance of a planar structure such as a spiral inductor. In Sonnet EM simulation, the conductive layer is assumed to be infinitesimally thin with a finite sheet resistivity.

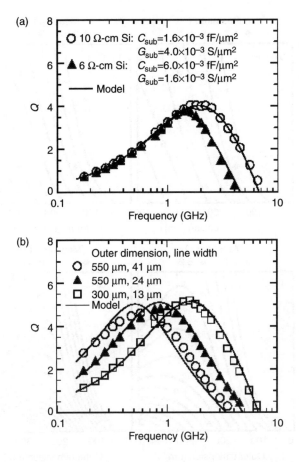

Figure 11.8   *(a) The effect of substrate resistivity on Q. ($L_s = 8\,nH$, $t_{ox} = 4.5\mu m$, $t_{Al} = 1\,\mu m$, $N = 7$, $w = 13\,\mu m$, $s = 7\,\mu m$, $OD = 300\,\mu m$). (b) The effect of layout area on Q. ($L_s = 8\,nH$, $t_{ox} = 4.5\mu m$, $t_{Cu} = 1\,\mu m$, $\rho_{si} = 10\Omega$-cm, $s = 7\,\mu m$) [After C. P. Yue et al., IEEE IEDM Tech. Dig., 1996 (155–158). ©1996 IEEE]*

11.3.1   ASITIC

Analysis and Simulation of Spiral Inductors and Transformers for ICs (ASITIC) is an interactive CAD tool that aids RF/microwave and high-speed digital engineers to analyse, model and optimise passive and interconnect metal structures residing on a lossy conductive substrate. This includes inductors, transformers, capacitors, transmission lines, interconnect and substrate coupling analysis. ASITIC allows the circuit designer to plan and optimise the layout of a chip in the presence of magnetic and electrical interaction and coupling through the substrate and oxide layers of the IC. Process engineers can also employ ASITIC to assess the effect of process changes to the quality of

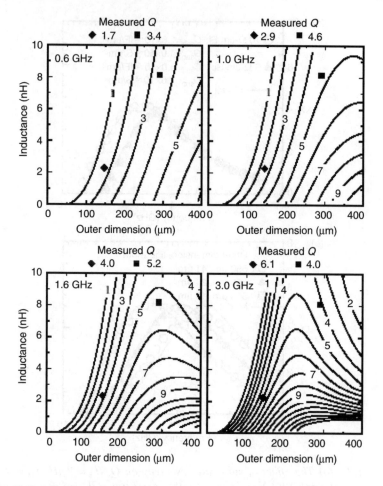

Figure 11.9 Contour plots of Q as a function of inductance and outer dimension
of square spiral inductors. ($t_{ox} = 4.5 \, \mu m$, $t_{Cu} = 1 \, \mu m$, $r_{Cu} = 2 \, \mu\Omega\text{-}cm$,
$r_{si} = 10 \, \Omega\text{-}cm$, $s = 7 \, \mu m$) [After C. P. Yue et al., IEEE IEDM Tech. Dig.,
1996 (155–158). ©1996 IEEE]

passive devices and isolation. In high-frequency circuits, the most important effects
tend to be the most difficult to simulate. Passive device modelling, substrate coupling
and the package parasitics fall under this category. Hence designers are forced to be
conservative, often leading to sub-optimal performance. ASITIC calculations include
the electrically induced losses and coupling as well as the magnetically induced eddy
current losses. Skin effect and proximity effects, or eddy currents in the metallisation,
are also included.

ASITIC input file describes the substrate and metal layers of an IC process that
utilizes three to four substrate layers (bulk, epi, oxide and passivation) and three to
six metal layers. ASITIC can output s-parameters for inductors, transformers, bond

pads and capacitors and generate parameters for SPICE circuit simulation. The final design can be imported into layout tools.

11.4 Active inductors

Active inductors [27–29] solve many of the problems associated with the on-chip inductor design, for example, area requirement is drastically reduced, standard low-cost technology can be used with confidence and high value of quality factor (more than 100) can be easily obtained by controlling the associated losses via negative resistance feedback technique [30,31]. By impedance transformation, the capacitive load can be transformed into an inductive reactance to achieve high Q. However, it is imperative to generate a high Q inductance within other stringent RF requirements such as stability, reproducibility, noise and linearity. Since field-effect transistors utilize space more effectively they find application in active inductor design. However, the significant internal ohmic losses can degrade their performance [8]. Several approaches to solve this problem have been introduced, including some promising ones that achieve high-Q values by using negative resistance.

Active inductors using GaAs-based FET or bipolar processes have been investigated [32–37] and Q values of the order of several thousand have been reported [38–40]. In spite of the advantage of the semi-insulating substrate of GaAs to reduce losses, an economic choice for the VLSI industry is the low-cost Si-based processes. Therefore, realisation of high-Q active inductor using low resistivity Si-substrate is in high demand. Efforts have also been made to design active inductors in Si-based CMOS processes, with Q values more than several hundred [41, 42] and even an enhanced value of more than one thousand [43]. However, as the application circuits are intended to be used at ultra-high frequency, SiGe/SiGeC HBTs are presently an alternative to CMOS because of their higher cut-off frequencies. Although various aspects of bipolar active inductor design are investigated [44, 45], for Si-based processes, reported Q value is 47 with 20 nH inductance at 1.8 GHz [46]. Indeed, the potential of Si-heterostructure-based design of bipolar active inductors in advanced BiCMOS process, which is the technology of choice in the present day, has not yet been explored in detail [8].

Compared to lateral CMOS transistors, a vertical bipolar transistor is expected to suffer less substrate loss, because comparatively less active area is exposed to the substrate and highly doped buried layers act as shields for the substrate currents. The schematics of the circuits, shown in Figures 11.10–11.12, are different topologies of one-port bipolar active inductors. The circuit shown in Figure 11.10(a) is the well-known cascade configuration with the collector of T1 fed back to base of T2. Base terminal of T1 and the emitter terminal of T2 are grounded. This connection realises a popular gyrator C-type active inductor [35].

The Smith chart plot of Figure 11.13(a) shows s_{11} or the input reflection coefficient data simulated from the one-port type-I bipolar active inductor (see Figure 11.11(a)), with $R_0 = 35\ \Omega$, designed using SiGe transistors. Most of the curve lies around the periphery of upper half circle in the impedance Smith chart signifying high-Q

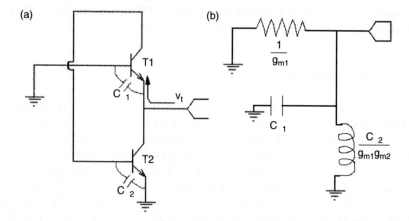

Figure 11.10   One-port bipolar active inductor in gyrator-C configuration: (a) Circuit diagram (b) its equivalent circuit [After A. Chakravorty, Compact Modeling of Si-Heterostructure Bipolar Transistors and Active Inductor Design for RF Applications, PhD Thesis, IIT Kharagpur 2005]

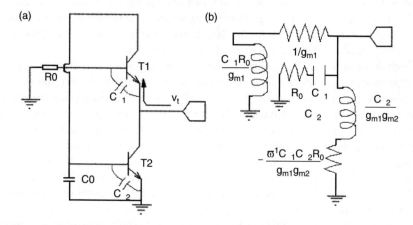

Figure 11.11   One-port high-Q active inductor incorporating negative resistance (type-I): (a) Circuit diagram (b) its equivalent circuit [After A. Chakravorty, Compact Modeling of Si-Heterostructure Bipolar Transistors and Active Inductor Design for RF Applications, PhD Thesis, IIT Kharagpur 2005]

inductive behaviour. The operating current is 1.9 mA and the transistors are biased to operate in the linear region. The operating frequency range under investigation lies is between 0.5 and 15 GHz, out of which 4–10 GHz range is of importance due to enhanced Q values. The cut-off frequency f_T at the specified operating current lies above 24 GHz.

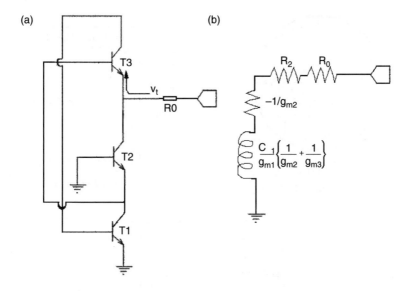

Figure 11.12   *One-port high-Q active inductor incorporating constant negative resistance (type-II): (a) Circuit diagrams (b) its simple equivalent circuit [After A. Chakravorty, Compact Modeling of Si-Heterostructure Bipolar Transistors and Active Inductor Design for RF Applications, PhD Thesis, IIT Kharagpur 2005]*

Figures 11.13(b)–(d) show the equivalent inductance (L), resistance (R) and the quality factor (Q) values of the designed inductor for a wide range of frequencies. The tuning element R_0 is varied from 30 to 40 Ω and there is still room for critical tuning. For $C_0 = 60$ fF, the peak Q varies from 14 to 57 at about 7.5 GHz as R_0 sweeps as shown in Figure 11.13(d). Here reasonable high-Q values are achieved without critical tuning. It is observed that a suitable design of type-I inductor can provide Q values more than 35 between $f = 6.5$ and 8.5 GHz. From Figure 11.13(a), it is also clear that the input reflection coefficient curve, in spite of being on the periphery of the Smith chart, remains near the low reactance values. This signifies that this circuit produces low inductance value. Figure 11.13(b) shows that the inductance value (L) of about 0.2 nH is achieved in the high-Q frequency band. Though this range of inductance value is low, it may find its utility in some specific applications.

11.5   Capacitors

The metal-insulator-metal (MIM) capacitors have attracted much attention in silicon integrated circuit applications because of their high-conductive electrode and low parasitic capacitance. Analog/mixed-signal processes use three major types of capacitors: polySi-insulator-polySi (PIP), MIM and MOS-style (depletion or accumulation).

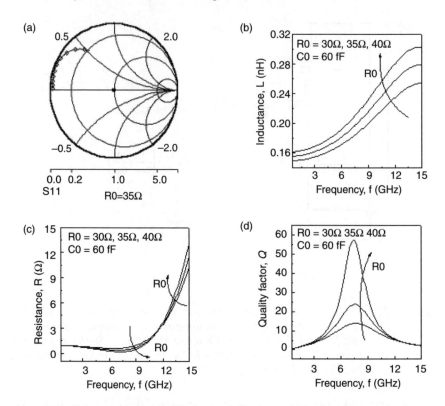

Figure 11.13   *Simulation results of one-port bipolar active inductor using SiGe transistors: (a) s$_{11}$ plot in Smith chart, (b) inductance (L), (c) resistance (R), and (d) quality factor (Q) over frequencies for different R$_o$ values [After A. Chakravorty, Compact Modeling of Si-Heterostructure Bipolar Transistors and Active Inductor Design for RF Applications, PhD Thesis, IIT Kharagpur 2005]*

By far, the most popular analog/mixed-signal capacitor is the MIM. MIM capacitors have the inherent advantage that they are metal and, if implemented at the last metal layer, have the entire inter-layer dielectric (ILD) stack between them and the substrate. MIM capacitors are generally implemented in commercial CMOS Cu-damascence processes [47, 48] and the capacitors show excellent linearity with voltage and temperature that are the requirement for analog applications. However, the selection of the inner-layer dielectric is crucial. SiN is a popular choice due to its availability in the back-end process although the low-temperature-deposited SiN is known to show higher relaxation recovery voltages than oxide [49]. PECVD SiN displays significant sensitivity to operation frequency, bias voltage and temperature when compared to oxide [50]. SiN also displays frequency-dependent shifts that are consistent with bulk nitride traps located within a tunnelling distance of the nitride metal interface.

The main problem of MIM capacitor is its non-scalability. One of the restrictions with MIM devices is that process technologies do not scale the vertical spacing in the back-end nearly as fast as the lateral spacing. The reason is that digital circuit designs cannot tolerate large increases in the wiring capacitance from generation to generation. Lateral flux (finger) capacitors solve this problem by using the lateral capacitance (between the metal lines) rather than the vertical capacitance (between the different ILD layers). As a consequence, the capacitance is under design control and scales more effectively with the technology [4]. For a comprehensive review on MIM capacitor for RF circuit applications, the reader may refer to a paper by Ng *et al.* [51] in which the authors provide an overview of MIM capacitor integration issues.

Silicon oxide and silicon nitride are dielectrics that are commonly used in conventional MIM capacitors. Though they can provide good voltage linearity properties and low temperature coefficients, their capacitance density will be limited owing to low dielectric constants ($\epsilon_r \sim 3.9$ for SiO_2, ~ 7 for Si_3N_4). A 30 nm thick SiO_2 MIM capacitor has shown a voltage linearity of 20 ppm/V.

A high capacitance density is required for a MIM capacitor in order to minimise chip area. Adoption of high-k material is a very efficient way to increase the capacitance density. HfO_2 is now being researched as a very promising candidate for gate dielectrics in MOSFET applications. Furthermore, a very high performance HfO_2 MIM capacitor, with a capacitance density of 3.0 fF/μm, has been demonstrated [52]. Further increase in capacitance density can be implemented by scaling down the dielectric thickness, which, however, may result in higher voltage coefficients of capacitance [53].

Another limitation of the MIM device is the thickness of the insulator region. In contrast, MOS devices can take advantage of thin gate oxide processes to achieve high capacitance per unit area. However, since one of the contacts is formed in silicon, the series resistance of a MOS capacitor is quite large. In addition, the very high gate leakage currents of modern scaled oxides at 180 nm node and beyond make gate-oxide-based MOS devices excessively leaky for conventional applications.

11.6 Varactor

The varactor is one of the most important components in RFIC applications such as VCOs. The type of varactor device available depends on the technology. They include standard p-n junction varactors quasi-hyperabrupt varactors or MOS varactors. Some technologies offer multiple types of varactors. Key characteristics of varactors include low losses (high Q), sufficient capacitive tuning range, low parasitic capacitance, high unit capacitance and high linearity at RF frequencies.

The most common type of varactor is a p-n junction, constructed using a typical bipolar transistor collector–base structure. To enhance junction varactor performance, an optional new quasi-hyperabrupt (QHA) varactor, formed from the intrinsic npn collector–base junction with a specially tailored implant has been reported [54]. The QHA varactor presents a larger capacitive tuning ratio, a good quality factor (Q)

performance and excellent linearity. The scalable quasi-hyperabrupt junction varactor comprises the intrinsic npn collector–base junction with a specially tailored implant to enhance its performance. The QHA varactor provides an increase in the capacitance tuning range and excellent linearity while maintaining high Q.

MOS varactor is a thin-oxide MOS varactor device with n^+ source and drain shorted together. However, instead of n^+ source and drain diffusions being in a p-well, as with n-FETs, the MOS varactor is built in an n-well. The MOS varactor includes n^+ subcollector (NS) and deep trench (DT) levels. The variable capacitance is achieved by controlling the gate-to-diffusion/n-well potential (V_{gd}), which takes the silicon surface under the gate from depletion to electron accumulation. Generally speaking, the MOS varactor is modelled as two capacitors in series: a gate-oxide capacitor and a depletion capacitor. With a positive voltage V_{gd}, the MOS varactor behaves like a parallel-plate capacitor with an accumulation layer underneath the silicon surface. The device capacitance equals the gate-oxide capacitance. If the device is reverse-biased, a depletion layer is formed in the n-well directly under the gate. The device capacitance decreases because of the depletion capacitance in series with the gate-oxide capacitance. The MOS-varactor model supports variable channel lengths and widths and includes gate resistance, channel resistance, gate-oxide capacitance, depletion capacitance fringe capacitance and parasitic substrate capacitance.

11.7   Resistor

Precision polySi and metal thin film resistors are key passive elements in analog circuits. The simultaneous presence of both poly- and metal resistors can add value in a process, because the metal resistors are at the top of the stack and the poly resistors at the bottom. Two widely separated locations allow designers to choose a resistor that minimises parasitics for their particular circuit. Also, the presence of a front-end resistor may enable in-line or early learning electrical evaluation on key circuit elements.

The key figures of merit for resistors are tolerance, parasitic capacitance, temperature coefficients and voltage coefficients. Traditionally, resistors are single crystal silicon or polySi made from implants and layers already existing in the process (e.g. gate poly). This simplifies the process but makes it more difficult to optimise separate parts of the process. Metal thin film resistors can be built at any of the traditional metal layers. In addition, a metal thin film resistor can be built as a by-product of the MIM capacitor process. TaN is frequently used as well, owing to its ready availability in a Cu-damascene process as a Cu-diffusion barrier.

PolySi resistors exist in both silicided and unsilicided versions. Since the resistance of polySi-silicided (polycide) resistors tends to be quite low (5–15 ohms/sq) and the voltage coefficient tends to be quite high (100–600 ppm/V) there is a strong tendency to use the unsilicided (or silicide-blocked) resistors [55]. The silicide-blocking layer is usually an oxide or nitride and is frequently chosen to leverage a pre-existing layer elsewhere in the process. Existence of a silicide-blocking layer

Table 11.2   *Passives in a 0.18 μm SiGe BiCMOS process [After D. Harame et al.,*
Proc. 4th IEEE Intl. Caracas Conf. Devices, Circuits and Systems, 2002
(D052-1–D052-17)]

Resistors	R_s (Ω/Sq.)	TCR (ppm/C)
Subcollector	8.1	1430
n^+ Diffusion	72	1910
p^+ Diffusion	105	1430
p^+-PolySi	270	50
p-PolySi	1600	−1178
TaN	142	−750
Capacitors	C_p (fF/μm^2)	V_{cc} (+5/−5 ppm/V)
MIM	1	<45
MOS	2.6	−7500/−1500
Inductor	L (nH)	max Q at 5 GHz
Al spiral inductor	≥0.7	18
Varactor	Tuning range	Q @ 0.5
c-b junction	1.64:1	90
MOS accumulation	3.1:1	300

also enables devices such as silicide-blocked diffusion resistors and silicide-blocked
MOS devices [56]. The sheet resistance, as well as the thermal and voltage coefficients
of silicide-blocked polySi resistors, are very process dependent. Implant conditions,
grain boundary size, thermal activation and end-cap silicide quality can all impact
the key polySi resistor parameters. As a consequence, reported values for the major
resistor parameters vary widely. These key passive elements in the 0.18 μm SiGe BiC-
MOS technology are a good example of advanced passives development and they are
listed in Table 11.2. Passive component development summary is given in Table 11.3.

11.8   Interconnects

The increasing complexity of ICs demands a greater number of integrated transistors
and gates, which require more wiring for interconnectivity. In addressing interconnect
performance limits, it is important to understand what is meant by interconnect performance. It is used synonymously with the operating frequency or speed, assuming
IC architecture and design remain constant. The international technology roadmap for
semiconductors (ITRS) projects performance improvement of advanced ULSI circuits
likely to saturate beyond the 100-nm technology node, owing to the rapidly dominating interconnect performance limitations, unless a paradigm shift from present
IC architecture is introduced [57]. Considerable efforts are being made to push the
interconnect performance limit.

Table 11.3 *Passive component development summary [After D. Harame et al., Proc. 4th IEEE Intl. Caracas Conf. Devices, Circuits and Systems, 2002 (D052-1–D052-17).]*

Device	Type	Issues	Solutions
Inductors	Spiral	Q	Thick dielectric module/thick metal 10–20 Ω-cm substrates
Capacitors	Poly-Insulator-Si Poly-Insulator-Poly Metal–Insulator–Metal	Density Reliability Voltage rating V_{cc} Q	Optimise processes Offer variety
Resistors	polySi BEOL thin film Single-crystal	Tolerance TCR Parasitic C Current rating	Process control Offer variety Layout options for low C
Varactor	Junction MOS (accumulation)	tuning range Q linearity	Offer variety Layout in model
Interconnect	Transmission line Local interconnect	CMOS compatibility Electromigration loss	Low levels CMOS upper levels – thick Copper

Interconnect optimisation has two conflicting requirements: it must remain fully compatible with the base CMOS process for large digital circuit design as well as have thick low resistance metals spaced far from the substrate by thick dielectrics as shown in the previous section on inductors. The trend in VLSI CMOS metallisation is to reduce the metal and dielectric thickness of each generation by almost a factor of 2. The RF/analog solution to this apparent conflict is to have both CMOS compatibility and improved passives by creating thick metal/dielectric add-on modules that are added on top of the existing base CMOS metal layers. The evolutionary development of add-on thick dielectric/metal modules in a 0.5 μm SiGe BiCMOS process is shown in Figure 11.14.

The trend is towards reduced line widths together with larger die size, greater number of interconnect layers and GHz clock frequencies. As signal frequencies continue to rise, the effect of package parasitics also become important. As a consequence, the electrical characteristics of the interconnections are becoming important

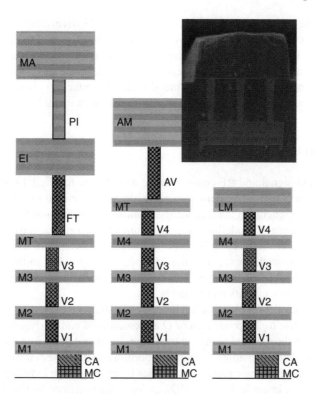

Figure 11.14   *Cross-sectional view of metallisation development in a 0.5 μm SiGe BiCMOS process. The inset shows an SEM micrograph of the top (two) thick metal layers The difference in these three technologies is the thick metallisation for passives improvement and analog scaling. Cross-section is not to scale [After D. Harame et al., Proc. 4th IEEE Intl. Caracas Conf. Devices, Circuits and Systems, 2002 (D052-1–D052-17). ©2002 IEEE]*

factors in determining the behaviour of integrated circuits. In fact, future submicron integrated circuits will behave more and more like giant microwave circuits.

As VLSI technology shrinks to deep sub-micron (DSM) geometries (below 90 nm), the parasitic due to interconnects is becoming a limiting factor in determining circuit performance. An accurate modelling of interconnect parasitic resistance (R), capacitance (C) and inductance (L) is thus essential in determining various chip interconnect-related issues such as delay, crosstalk, IR drop, power dissipation and so on. Cu has become the interconnect metal of choice because of its lower resistivity as compared to Al, again to reduce resistance. Much work is also being pursued in low dielectric constant (low-k) materials to replace deposited SiO_2, as an inter-layer (ILD) and inter-metal dielectric (IMD) to reduce interconnect capacitance.

The first step is to add one layer of thick dielectric/metal (3.0 μm oxide and 4.0 μm aluminium-AM layer). As technology evolved an improved thick dielectric/metal add-on module with two thick metal layers was created (3.0 μm oxide, 3.0 μm Cu-E1 layer, 4.0 μm oxide and 4.0 μm Al-MA layer). The second layer of thick metal allows more complex inductor structures including stacked or transformer inductors. Stacked inductor designs can significantly increase both the inductance and Q as well as reduce the area. For example, with two thick metal layers the inductor Q is increased 40 per cent over inductors of the same inductance with one thick metal layer.

11.9   Summary

Analog, mixed-signal and RF continue to be a challenge for digital CMOS designers and manufacturers. The performance is ultimately limited by the accuracy of the passive components in the technology. Conflicting scaling methodologies, complex measurement and modelling support requirements, a multiplicity of interacting features and increasingly complex process integration issues are the challenges to overcome to support the next generation of product designs. The ability to accurately construct and model passives with Qs > 15–20 at frequencies >10 GHz represents a key enabler for new circuits and products.

A physical model for planar spiral inductors on silicon is presented. Physical phenomena important to the prediction of Q are considered and analysed. The scalable inductor model shows good agreement with measured and published data. However, as the eddy current loss is approximately proportional to the cube of the inductor diameter, strategies to minimise resistive parasitics by making large inductors (as is common in GaAs) are less effective in CMOS owing to the more conductive Si substrates.

Although the use of Cu or multi-tiered interconnect architecture yields lower RC delay, there are still other performance limitations. Power consumption and dissipation, for instance, is rapidly becoming unmanageable as the clock frequencies continue to rise. Any increase in interconnect loading significantly increases the power consumption in high-performance chips. Interconnect scaling has significant implications for computer-aided design (CAD) methodologies and tools that are causing the design cycles to increase.

References

1 D. L. Harame, K. M. Newton, R. Singh, S. L. Sweeney, S. E. Strang, J. B. Johnson, *et al.*, 'Design automation methodology and rf/analog modeling for rf CMOS and SiGe BiCMOS technologies', *IBM J. Res. Dev.*, vol. 47, pp. 139–175, 2003.

2 C.-S. Chang, C.-P. Chao, J. G. J. Chern, and J. Y.-C. Sun, 'Advanced CMOS technology portfolio for RF IC applications', *IEEE Trans. Electron Devices*, vol. 52, pp. 1271–1285, 2005.

3 D. Harame, A. Joseph, D. Coolbaugh, G. Freeman, K. Newton, S. M. Parker, *et al.*, 'The emerging role of SiGe BiCMOS technology in wired and wireless communications', in *Proc. IEEE Int. Caracas Conf. Devices, Circuits and Systems*, pp. D052-1–D052-17, 2002.

4 T. H. Lee, and S. S. Wong, 'CMOS RF integrated circuits at 5 GHz and beyond', *Proc. IEEE*, vol. 88, pp. 1560–1571, 2000.

5 D. J. Allstot, and W. C. Black, Jr., 'Technology design considerations for monolithic MOS switched-capacitor filtering systems', *Proc. IEEE*, vol. 71, pp. 967–986, 1983.

6 J. J. Wikner, and N. Tan, 'Influence of circuit imperfections on the performance of DACs', *Analog Integrated Circuits Signal Proc.*, vol. 18, pp. 7–20, 1999.

7 J. L. McCreary, 'Matching properties and voltage and temperature dependence of MOS capacitors', *IEEE J. Solid-State Circuits*, vol. SC-16, pp. 608–616, 1981.

8 A. Chakravorty, *Compact Modeling of Si-Heterostructure Bipolar Transistors and Active Inductor Design for RF Applications*. PhD thesis, Indian Institute of Technology, Kharagpur, 2005.

9 J. Long, and M. Copeland, 'Modeling, characterization and design of monolithic inductors for Silicon RF ICs', in *IEEE CICC Proc.*, pp. 185–188, 1996.

10 C. P. Yue, C. Ryu, J. Lau, T. H. Lee, and S. S. Wong, 'A physical model for planar spiral inductors on silicon', in *IEEE IEDM Tech. Dig.*, pp. 155–158, 1996.

11 K. B. Ashby, W. C. Finley, J. J. Bastek, and S. Moinian, 'High Q inductors for wireless applications in a complementary silicon bipolar process', in *IEEE BCTM Proc.*, pp. 179–182, 1994.

12 J. N. Burghartz, M. Soyuer, K. A. Jenkins, M. Kies, M. Dolan, K. Stein, *et al.*, 'Integrated RF components in a SiGe bipolar technology', *IEEE J. Solid-State Circuits*, vol. 32, pp. 1440–1445, 1997.

13 J. N. Burghartz, D. C. Edelstein, K. A. Jenkins, C. Jahnes, C. Uzoh, E. J. O'Sullivan, *et al.*, 'Monolithic spiral inductors fabricated using a VLSI Cu-Damascene interconnect technology and low-loss substrates', in *IEEE IEDM Tech. Dig.*, pp. 99–102, 1996.

14 J. N. Burghartz, M. Soyuer, and K. A. Jenkins, 'Microwave inductors and capacitors in standard multilevel interconnect silicon technology', *IEEE Trans. Microwave Theory Tech.*, vol. 44, pp. 100–104, 1996.

15 S.-M. Oh, C.-W. Kim, and S.-G. Lee, 'A 74%, 1.56-2.71 GHz, wide-tunable LC-tuned VCO in 0.35 μm CMOS technology', *Microwave and Optical Technol. Lett.*, vol. 37, pp. 98–100, 2003.

16 J. N. Burghartz, A. E. Ruehli, K. A. Jenkins, M. Soyuer, and D. Nguyen-Ngoc, 'Novel substrate contact structure for high-Q silicon-integrated spiral inductors', in *IEEE IEDM Tech. Dig.*, pp. 55–58, 1997.

17 S. Stefanou, J. S. Hamel, P. Baine, M. Bain, B. M. Armstrong, H. S. Gamble, *et al.*, 'Cross-talk suppression Faraday cage structure in silicon-on-insulator', in *IEEE Intl. SOI Conf. Proc.*, pp. 181–182, 2002.

18 S. Stefanou S, J. S. Hamel, P. Baine, M. Bain, B. M. Armstrong, H. S. Gamble, *et al.*, 'Ultralow silicon substrate noise crosstalk using metal Faraday cages in an SOI technology', *IEEE Trans. Electron Devices*, vol. 51, pp. 486–491, 2004.

19 J. N. Burghartz, M. Soyuer, and K. A. Jenkins, 'Integrated RF and microwave components in BiCMOS technology', *IEEE Trans. Electron Devices*, vol. 43, pp. 1559–1570, 1996.

20 N. M. Nguyen and R. G. Meyer, 'Si IC-compatible inductor and LC passive filters', *IEEE J. Solid-State Circuits*, vol. 25, pp. 1028–1031, 1990.

21 C. P. Yue, and S. S. Wong, 'Physical modeling of spiral inductors on silicon', *IEEE Trans. Electron Devices*, vol. 47, pp. 560–568, 2000.

22 H. M. Greenhouse, 'Design of planar rectangular microelectronic inductors', *IEEE Trans. Parts, Hybrids, and Packaging*, vol. 10, pp. 101–109, 1974.

23 S. S. Mohan, M. M. Hershenson, S. P. Boyd, and T. H. Lee, 'Simple accurate expressions for planar spiral inductances', *IEEE J. Solid–State Circuits*, vol. 34, pp. 1419–1424, 1999.

24 Y. Cao, R. A. Groves, N. D. Zamdmer, J. O. Plouchart, R. A. Wachnik, X. Huang, *et al.*, 'Frequency-independent equivalent circuit model for on-chip spiral inductors', in *IEEE CICC Proc.*, pp. 217–220, 2002.

25 A. M. Niknejad, 'ASITIC: Analysis and Simulation of Spiral Inductors and Transformers for ICs, Univ. of California, Berkeley', 2001.

26 Sonnet USA, Inc., 'SONNET Manual', 1999.

27 S. Hara, T. Tokumitsu, T. Tanaka, and M. Aikawa, 'Broad-band monolithic microwave active inductor and its application to miniaturized wide-band amplifiers', *IEEE Trans. Microwave Theory and Tech.*, vol. MTT-36, pp. 1920–1924, 1988.

28 S. Hara, T. Tokumitsu, and M. Aikawa, 'Lossless broad-band monolithic microwave active inductors', *IEEE Trans. Microwave Theory and Tech.*, vol. MTT-37, pp. 1979–1984, 1989.

29 F. E. van Vliet, F. L. M. van den Bogaart, J. L. Tauritz, and R. G. F. Baets, 'Systematic analysis, synthesis and realization of monolithic microwave, active inductors', in *IEEE MTT-S Dig.*, pp. 1659–1662, 1996.

30 B. Hopf, and I. Wolff, 'Narrow band MMIC active filter using negative resistance circuits in coplanar line technique', in *Proc. EuMC*, pp. 1110–1112, 1995.

31 W. Li, B. L. Ooi, Q. J. Xu, and P. S. Kooi, 'High Q active inductor with loss compensation by feedback network', *Electron. Lett.*, vol. 35, pp. 1328–1329, 1999.

32 C. Leifso, J. W. Haslett, and J. G. McRory, 'Monolithic tunable active inductor with independent Q control', *IEEE Trans. Microwave Theory and Tech.*, vol. 48, pp. 1024–1029, 2000.

33 C. Leifso, and J. W. Haslett, 'A fully integrated active inductor with independent voltage tunable inductance and series-loss resistance', *IEEE Trans. Microwave Theory and Tech.*, vol. 49, pp. 671–676, 2001.

34 K. W. Kobayashi, and A. K. Oki, 'A novel heterojunction bipolar transistor VCO using an active tunable inductance', *IEEE Microwave and Guided Wave Lett.*, vol. 4, pp. 235–237, 1994.

35 R. Kaunisto, P. Alinikula, and K. Stadius, 'Active inductors for GaAs and bipolar technologies', *Analog Integrated Circuits and Signal Proc.*, vol. 7, pp. 35–48, 1995.

36 R. Kaunisto, P. Alinikula, K. Stadius, and V. Porra, 'A low-power HBT MMIC filter based on tunable active inductors', *IEEE Microwave and Guided Wave Lett.*, vol. 7, pp. 209–211, 1997.

37 G. F. Zhang, C. S. Ripoll, and M. L. Villegas, 'GaAs monolithic microwave floating active inductor', *Electron. Lett.*, vol. 27, pp. 1860–1862, 1991.

38 P. Alinikula, R. Kaunisto, and K. Stadius, 'Monolithic active resonators for wireless applications', in *IEEE MTT-S Dig.*, pp. 1151–1154, 1994.

39 H. Hayashi, and M. Muraguchi, 'A novel loss compensation technique for high-Q broad-band active inductors', in *IEEE Microwave and Millimeter Wave Circuit Symp.*, pp. 103–106, 1996.

40 S. Lucyszyn, and I. D. Robertson, 'Monolithic narrow-band filter using ultrahigh-Q tunable active inductors', *IEEE Trans. Microwave Theory and Tech.*, vol. 42, pp. 2617–2622, 1994.

41 D. R. Akbari, A. Payne, and C. Toumazou, 'A high-Q RF CMOS differential active inductor', in *IEEE Intl Conf. Electronics, Circuits and Systems Dig.*, pp. 157–160, 1998.

42 M. Grozing, A. Pascht, and M. Berroth, 'A 2.5 V CMOS differential active inductor with tunable L and Q for frequencies up to 5 GHz', in *IEEE MTT-S Symp. Dig.*, pp. 575–578, 2001.

43 K. V. Chiang, K. Y. Lam, W. W. Choi, K. W. Tam, and R. Martins, 'A modular approach for high Q microwave CMOS active inductor design', *Proc. IEEE*, pp. 41–44, 2000.

44 C. F. Campbell, and R. J. Weber, 'Design of a broadband microwave BJT active inductor circuit', *Proc. IEEE*, pp. 407–409, 1992.

45 D. P. Anderson, R. J. Weber, and S. F. Russell, 'Bipolar active inductor realizability limits, distortion, and bias considerations', *Proc. IEEE*, pp. 241–244, 1997.

46 G. D'Angelo, L. Fanucci, A. Monorchio, A. Monterastelli, and B. Neri, 'High-quality active inductors', *Electron. Lett.*, vol. 35, pp. 1727–1728, 1999.

47 M. Armacost, A. Augustin, P. Felsner, Y. Feng, G. Friese, J. Heidenreich, *et al.*, 'A high reliability metal insulator metal capacitor for 0.18 μm copper technology', in *IEEE IEDM Tech. Dig.*, pp. 157–160, 2000.

48 P. Zurcher, P. Alluri, P. Chu, A. Duvallet, C. Happ, R. Henderson, *et al.*, 'Integration of thin film MIM capacitors and resistors into copper metallization based RF-CMOS and Bi-CMOS technologies', in *IEEE IEDM Tech. Dig.*, pp. 153–156, 2000.

49 J. W. Fattaruso, M. D. Wit, G. Warwar, K. Tan, and R. K. Hester, 'The effect of dielectric relaxation on charge-redistribution A/D converters', *IEEE J. Solid-State Circuits*, vol. 25, pp. 1550–1561, 1990.

50 J. Babcock, S. Balster, A. Pinto, C. Dirnecker, P. Steinmann, R. Jumpertz, *et al.*, 'Analog characteristics of metal-insulator-metal capacitors using PECVD nitride dielectrics', *IEEE Electron Devices Lett.*, vol. 22, pp. 230–232, 2001.

51 C. H. Ng, C.-S. Ho, S.-F. Chu, and S.-C. Sun, 'MIM capacitor integration for mixed-signal/RF applications', *IEEE Trans. Electron Dev.*, vol. 52, pp. 1399–1409, 2005.

52 H. Hu, C. Zhu, Y. F. Lu, M. F. Li, B. J. Cho, and W. K. Choi, 'A high performance MIM capacitor using HfO$_2$ dielectrics', *IEEE Electron Dev. Lett.*, vol. 23, pp. 514–516, 2002.

53 A. Kar-Roy, C. Hu, M. Racaneli, C. A. Compton, P. Kempf, G. Jolly, *et al.*, 'High density metal insulator metal capacitors using PECVD nitride for mixed signal and RF circuits', in *Proc. IEEE Int. Interconnect Technology Conf.*, pp. 245–247, 1999.

54 W. Ni, K. Carrig, and X. Yuan, 'The variety of varactor devices available in IBM SiGe BiCMOS technologies', *MicroNews*, vol. 2nd Quarter, pp. 5–8, 2002.

55 K. J. Kuhn, S. Ahmed, P. Vandervoorn, A. Murthy, B. Obradovic, K. Raol, *et al.*, 'Integration of mixed-signal elements into a high-performance digital CMOS process', *Intel Technol. J.*, vol. 6, pp. 31–41, 2002.

56 A. Salman, R. Gauthier, W. Stadler, K. Esmark, M. Muhammad, C. Putnam, *et al.*, 'Characterization and investigation of the interaction between hot electron and electrostatic discharge stresses using NMOS devices in 0.13 μm CMOS technology', in *IEEE Rel. Symp. Proc.*, pp. 219–225, 2001.

57 Semiconductor Industry Association, 'International Technology Roadmap for Semiconductors.' San Jose, Calif., 2005.

Index